Physics and Mathematical Tools

METHODS AND EXAMPLES

Physics and Mathematical Tools

METHODS AND EXAMPLES

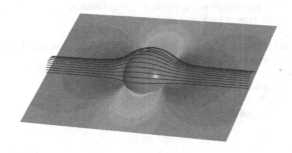

Angel Alastuey
CNRS, France & ENS de Lyon, France

Maxime Clusel
CNRS, France & Université de Montpellier, France

Marc Magro
ENS de Lyon, France

Pierre Pujol
Université Paul Sabatier, France

 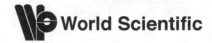

Published by

World Scientific Publishing Co. Pte. Ltd.

5 Toh Tuck Link, Singapore 596224

USA office: 27 Warren Street, Suite 401-402, Hackensack, NJ 07601

UK office: 57 Shelton Street, Covent Garden, London WC2H 9HE

Library of Congress Cataloging-in-Publication Data

Names: Alastuey, Angel.

Title: Physics and mathematical tools : methods and examples / by Angel Alastuey
  (CNRS, France & ENS de Lyon, France) [and three others].

Other titles: Physique et outils mathematiques. English.

Description: New Jersey : World Scientific, 2016. | Includes bibliographical references and index.

Identifiers: LCCN 2015036789 | ISBN 9789814713238 (hardcover : alk. paper) |
  ISBN 9789814713245 (softcover : alk. paper)

Subjects: LCSH: Mathematical physics. | Mathematical instruments.

Classification: LCC QC20 .P4913 2016 | DDC 530.15--dc23

LC record available at http://lccn.loc.gov/2015036789

British Library Cataloguing-in-Publication Data

A catalogue record for this book is available from the British Library.

Originally published in French as "Physique et outils mathématiques: méthodes et exemples" by Angel Alastuey,
Marc Magro et Pierre Pujol. © EDP Sciences 2009. A co-publication with EDP Sciences, 17, av. du Hoggar F-91944
Les Ulis, France

Printed in Singapore

Foreword

As this book was completed and we were checking the proofs, our co-author Maxime CLUSEL suddenly and tragically died. We were particularly shocked and saddened by this terrible event. We would like to emphasize both the prominent role of Maxime in the writing of the manuscript and his rich and charming personality. His hard, patient and inspired work during more than a year was a decisive element in the success of the project. Throughout our collaboration, we also appreciated his infinite kindness and his broad open-mindedness. Although he was quite reserved, we know he displayed indisputable artistic skills, in particular for photography. Everybody who had the chance to meet him will remember his deep and true friendship. We hope that this book testifies a little of his everlasting passion for physics and science.

A. Alastuey, M. Magro, P. Pujol

Contents

List of exercises

Chapter 1 : pages 47-52

(1) Response functions associated with linear operators
(2) Responce function for a RLC circuit
(3) Charged Brownian particle
(4) Absorption line
(5) Application of Kramers-Kronig relations in astrophysics
(6) Sum rules
(7) Response to noise
(8) Kramers-Kronig relations for a metal
(9) Signal propagation in dielectric media

Chapter 2 : pages 111-124

(1) Green's function G_∞ for the Laplacian in 3d
(2) Green's function G_∞ for the Laplacian in dimension $d \geq 3$
(3) Green's functions for the Laplacian in 1d and 2d
(4) Symmetry of Laplacian Green's functions with homogeneous Dirichlet BC
(5) Special Neumann Green's function of the Laplacian
(6) Sum rules and resolvent
(7) Conductive plane
(8) Green's functions for Laplace operator in spherical coordinates
(9) Point charge in a conducting sphere
(10) Point charge and dielectric sphere
(11) Green's function G_∞ for Laplace operator in cylindrical coordinates
(12) Oseen's tensor
(13) Green's function and elasticity theory
(14) Discrete Laplacian and resistors network
(15) Method of image charges for a bidimensional problem
(16) Semi-cylindrical warehouse exposed to wind
(17) Dirac operator
(18) Mercury perihelion precession
(19) Harmonic oscillator in the presence of an impurity

Chapter 3 : pages 206-212

(1) Uniqueness of solutions for diffusion and wave equations
(2) Reciprocal relations
(3) Equation for long cables
(4) Neumann conditions in diffraction theory
(5) Green's function for d'Alembert operator in $2 + 1$ dimensions
(6) Green's function for d'Alembert operator in $1 + 1$ dimensions
(7) Laplacian Green's function G_∞ in dimension $d \geq 3$
(8) Heat diffusion in a ball
(9) From Dirichlet to Robin boundary conditions
(10) Robin conditions for heat equation
(11) Cattaneo's equation in 3D
(12) Klein-Gordon equation

Chapter 4 : pages 256-261

(1) Asymptotic behaviour of Bessel function J_0
(2) Binomial coefficients
(3) Asymptotic behaviour of Helmholtz Green's function
(4) Isothermal-isobaric ensemble
(5) Evolution of a wave packet and group velocity
(6) From Cattaneo to diffusion Green's function
(7) Ising model with long-range interactions
(8) Bernoulli random walk
(9) Harmonic oscillator and number theory

Introduction

The purpose of this book is to present several general methods with a wide range of applications in physics. Chosen methods make use of analytical properties of susceptibilities in linear response theory, Green's functions to solve partial differential equations, and saddle point method to estimate integrals of all kinds.

Various examples are presented in order to illustrate how these methods can be successfully applied to many problems in electromagnetism, classical or quantum mechanics, statistical physics or quantum field theory, etc. An inventory of applications would actually amount to enumerate almost all areas of physics. This broad range of possible applications has inspired a transverse presentation of these methods in a general framework which is not specific to a particular branch or domain. This unifying point of view explains the structure of each chapter. The first part is devoted to the presentation of general properties highlighting the universality of some mechanisms. A few examples are then presented in the second part of each chapter. They consolidate the understanding of general mechanisms by making connections between different problems. They also have their own interests and motivations. Finally a third part complements each chapter by exercises with indications of solutions. The presentation adopted here emphasizes on discussions and physical examples without hiding mathematical subtleties.

Beyond their interest for the resolution of practical problems, these methods present outstanding features discussed in detail in each chapter. We highlight here some of them. Analytic properties of susceptibilities do not depend on the system under consideration and its more or less complex intrinsic dynamics. A pole in the admittance of an RLC circuit is for example somehow similar to a pole in the dielectric constant of a material medium. This common analytical framework paves the way for simple modelling.

The dynamical eigenmodes of a system can be associated with the singularities arising in its susceptibility in response to a weak monochromatic forcing. Relaxation is controlled by dissipative mechanisms. The presence of dissipation requires the forcing to provide energy in order to maintain oscillations. In some systems dissipation is explicitly introduced in the equations of motion, such as in hydrodynamics by means of a viscous force. Dissipation is induced by subtle mechanisms in conservative systems at thermodynamic equilibrium. Analytic signatures of dissipation are however identical in all cases! Note that energy absorption is also possible in the absence of dissipation by resonant effects.

Static and dynamic Green's functions are the fundamental tools for solving linear partial differential equations. They are the building blocks providing the general solutions for arbitrary sources by using the superposition principle. Causal Green's functions also exhibit a dual character: On the one hand, they can be interpreted as response functions, but on the other hand, they describe diffusion/propagation of physical quantities.

Imposing boundary conditions is essential for defining univocally the solution of a partial differential equation. One should keep in mind that akin boundary conditions could actually lead to very different solutions. These boundary conditions must be adjusted to the physical characteristics of the considered situation. For instance the evolution of particle density through diffusion inside a box crucially depends on the nature of the enclosing walls. Reflecting or absorbing walls are taken into account via boundary conditions which ultimately lead to different behaviours.

Symmetry properties play an important role when solving partial differential equations. These properties must be analysed by examining both the involved operator, the shape of boundaries and the nature of boundary conditions.

Green's functions associated with resolvents are extremely useful. They allow one to analyse spectral properties of operators, and they appear as natural ingredients for perturbative expansions. They are moreover simply related to dynamical Green's functions, which can then be computed more easily.

Finally the saddle point method is of great practical importance due to its countless applications. It embraces a wide range of physical interpretations depending on the field under consideration. It for example leads to Landau mean-field theory of paramagnetic-ferromagnetic transition, or to semi-classical approximation in quantum mechanics!

Acknowledgements

We would like to thank François Delduc and François Gieres for their comments and suggestions. We are grateful to Emmanuel Lévêque for providing us with the picture on the cover.

Chapter 1

Linear response and analyticity

Regardless of the field – classical mechanics, quantum mechanics, electromagnetism, hydrodynamics, thermodynamics, statistical physics, etc.– a physicist often has to determine the response of a system to a small external perturbation. Examples include the current induced in a circuit by applying an AC voltage, the polarisation of an atom submitted to a time-dependent electric field, or even the flow of a fluid induced by a pressure gradient in a capillary. Although these systems pertain to different fields of physics, their behaviour must follow simple common and fundamental principles. In all these cases for instance, the response function must satisfy the causality principle imposing that a perturbation applied at a given time only acts on the system state at later times. Moreover, if the perturbation is sufficiently weak, one can generally assume that the system response is a linear function of the perturbation.

In this chapter we show how a few simple principles can be used to establish general results on the structure and properties of the linear response. This approach presents two major advantages. First of all it highlights the generic character of the obtained features, regardless of the precise nature and the complexity of the system under consideration. This universal aspect is illustrated by detailed studies of several systems submitted to excitations oscillating in time at a given frequency. The second major advantage of this approach may seem surprising a priori: Even if the principles underlying the general properties of the linear response are simple, their consequences are far from being irrelevant and could even be dramatic in some cases! It might be useful at this stage to explain the origin of the title of this chapter, by illustrating these remarks in an example. One of the essential properties of the linear response is related to the analyticity of a key physical quantity, namely the susceptibility as a function of the excitation frequency. Without going into details, let us simply mention here that the susceptibility satisfies several fundamental identities, called Kramers-Kronig relations, arising from its analytical properties. These relationships have an unexpected scope. They for instance lead to a standard result of electromagnetism, namely that the propagation speed of electromagnetic waves in any dispersive medium is lower than the speed of light in vacuum.

In the first part of this chapter, the general properties of response to a small excitation are derived from the triptych "linearity-causality-stationarity". These general properties are then made explicit through different examples. The specific interest of each example stems from the use of at least one specific property, both for understanding involved mechanisms and for their phenomenological description. These examples are moreover selected from different areas of physics (electronics, quantum mechanics, electromagnetism, fluid mechanics, statistical physics and finally plasma physics) to highlight the strength of the methods. Their unifying character is also the source of numerous and fruitful analogies. Each example naturally presents an interest on its own.

The examples, resulting from more and more complex system dynamics, are presented in order of nearly increasing difficulty. The first example is thus a basic linear system with few degrees of freedom, i.e. an RLC circuit: its dynamics is identical to that of a damped harmonic mechanical oscillator. We then move on to study continuous systems, hence presenting an infinite number of degrees of freedom. We start with the example of a dielectric material to illustrate the application of analyticity properties to phenomenological modelisations. The next example is an oscillatory flow in a capillary forced by a pressure gradient. We eventually study the linear response of systems with many degrees of freedom. Note that this problem of equilibrium statistical mechanics constitutes a field by itself! A first example is the study of a plasma subject to a charge density wave within a mean-field approach *à la* Vlasov. We present various derivations of classical and quantum Kubo formulas for the electrical conductivity. Let us conclude by quoting a last example, presented in Chapter 3 and devoted to the hydrogen atom submitted to a time-dependent electric field: it also provides an application of perturbative expansions in terms of Green's functions.

1.1   General properties

1.1.1   *Problem statement*

Let us consider a generic system with an arbitrary number of degrees of freedom. This system is initially in a stationary state $\mathcal{E}_0$. The physical quantities related to this state are then time independent. We then submit this system to a small homogeneous and time-dependent perturbation, $F(t)$, with the following initial conditions:

$$\left\{ \begin{array}{c} \text{State}(t \to -\infty) = \mathcal{E}_0 \\ F(t \to -\infty) = 0 \end{array} \right\}.$$

We want to study the evolution of a physical quantity or observable A under the previous forcing. Our goal is to determine the form of the system response to a perturbation, that is to say the relation between $A(t)$ and $F(t')$. $\mathcal{E}_0$ could be for instance a mechanical oscillator at rest while F is a force and A, the displacement of this oscillator with respect to its equilibrium position.

◊ **General form of the response function** ◊

As it is always possible to subtract from the physical quantity A its equilibrium value $A(\mathcal{E}_0)$, we will set $A(\mathcal{E}_0) = 0$ from now on without loss of generality. The system response to the perturbation must obey simple physical principles: this is the reason why it has generic properties common to many situations. Let us successively study these properties.

Linearity Even for systems driven by non-linear equations of motion, it is natural to assume that the response $A(t)$ is linear in $F(t')$ if the perturbation is weak enough. One can then write:

$$A(t) = \int_{-\infty}^{+\infty} dt' \, K_0(t, t') \, F(t'). \tag{1.1}$$

This amounts to keeping only the first term in the Taylor expansion of functional $A_{[F]}(t)$ in powers of F,

$$A_{[F]}(t) = A_0 + \int_{-\infty}^{+\infty} dt' \, \frac{\delta A_{[F]}(t)}{\delta F(t')}\bigg|_{F=0} F(t') + \cdots ,$$

where $A_0 = A_{[F=0]} = A(\mathcal{E}_0) = 0$ by assumption, while $K_0(t, t') = \frac{\delta A_{[F]}(t)}{\delta F(t')}\big|_{F=0}$. K_0 is called the functional derivative[1] of $A_{[F]}$ evaluated at $F = 0$.

Note actually that two fundamental assumptions must be satisfied to ensure the validity of the previous expansion. First the state $\mathcal{E}_0$ has to be stable: the system must remain in the neighbourhood of $\mathcal{E}_0$ in its subsequent evolution when slightly moved away from this state. It of course excludes unstable stationary states, such as an inverted pendulum with its centre of gravity above its attachment point. Moreover the differentiability of $A_{[F]}$ implies a sufficient regularity in the dependence of the functional A with respect to F. Conversely, non-differentiability is the signature of singular behaviours such as those appearing at the critical point of a phase transition in thermodynamics✠.

It is also important to note that the system response is not local in time. In other words, the value of observable A at time t in principle depends on values of the perturbation F at other times t'.

> ✠ **Comment:** For example, the susceptibility diverges at the critical temperature for a magnetic system exhibiting a transition between ferromagnetic and paramagnetic phases. Magnetisation is then no longer proportional to the applied external magnetic field.

The linearity assumption therefore requires that K_0 be an intrinsic quantity specific to the system and the state $\mathcal{E}_0$. It also depends on the observable A under

---

[1]The interested reader can refer to Appendix H where we recall definition and properties of the functional derivative. This concept is however not used in this chapter.

consideration. It does not depend on the precise time evolution of F but only on the way the system is forced. The properties of this quantity will be studied throughout this section.

Causality   This is a fundamental principle of physics that goes beyond the linearity assumption. Using the example of a mechanical oscillator, it simply states that the force applied at a given time t_0 only acts on the displacement of the oscillator for subsequent times. In other words, the response of the system at a given time generally depends only on perturbation applied at earlier times, namely:

$$K_0(t, t') = 0 \text{ for } t < t'. \tag{1.2}$$

We will show below that this simple property has fundamental consequences. For the time being we insert condition (1.2) into expression (1.1), to obtain

$$A(t) = \int_{-\infty}^{t} dt' \, K_0(t, t') \, F(t'). \tag{1.3}$$

Stationarity   Since $\mathcal{E}_0$ is a stationary state, K_0 is invariant under time translation, i.e.

$$K_0(t, t') = K_0(t - t').$$

The dependence of K_0 on the sole difference $t - t'$ indeed ensures the invariance of K_0 if t and t' are simultaneously shifted by Δt. It means that this time shift Δt of perturbation $F(t)$ leads to the same time shift for the response $A(t)$. More explicitly, the system response to perturbation $F_2(t) = F_1(t - \Delta t)$ is given by

$$A_2(t) = \int_{-\infty}^{t} dt' K_0(t, t') F_2(t') = \int_{-\infty}^{t} dt' K_0(t, t') F_1(t' - \Delta t).$$

A simple change of variable in the previous integral shows that if $K_0(t, t') = K_0(t - t')$, then $A_2(t) = A_1(t - \Delta t)$.

Using the above properties one can write the system response as

$$A(t) = \int_{-\infty}^{t} dt' \, K_0(t - t') \, F(t'),$$

or

$$A(t) = \int_{0}^{+\infty} d\tau \, K_0(\tau) \, F(t - \tau). \tag{1.4}$$

◊ **Properties of the response function and interpretation** ◊

The function $K_0(\tau)$ plays the role of a response function with memory since it determines the contribution of the perturbation at time $t - \tau$ to the quantity A at a later time t. In order to clarify the meaning of K_0, we consider the case where the excitation is a pulse, i.e. $F(t) = F_0\delta(t - t_0)$ where $\delta(t - t_0)$ is the Dirac distribution centred at t_0. The pulse action is to weakly remove the system from its stationary state $\mathcal{E}_0$ at time t_0, while later evolution of the system is controlled by its own internal dynamics. Following equation (1.4), the response to this type of excitation is then given by

$$A_{\text{pulse}}(t) = F_0 K_0(t - t_0) \text{ for } t > t_0. \tag{1.5}$$

The corresponding evolution of observable A is simply proportional to the function $K_0(\tau)$. It confirms that the response function K_0 is indeed specific to the observable A under consideration. Moreover we can immediately infer the qualitative behaviour of $K_0(\tau)$ when $\tau \to \infty$ and when $\tau \to 0$, from simple arguments about the intrinsic evolution of the system without external forcing.

Long times Since $\mathcal{E}_0$ is stable, $A_{\text{pulse}}(t)$ remains small (of order F_0) at all time, and $K_0(\tau)$ is therefore bounded for $\tau \geq 0$. If state $\mathcal{E}_0$ is weakly stable[2], the system oscillates endlessly around $\mathcal{E}_0$ and $K_0(\tau)$ therefore oscillates without decreasing as $\tau \to \infty$. This situation is typical of conservative mechanical systems such as a harmonic oscillator.

If the stability is strong, the system returns to its initial state $\mathcal{E}_0$ in the course of its evolution. The response $A_{\text{pulse}}(t)$ then tends to zero as $t \to \infty$, leading to the decay of $K_0(\tau)$ for large τ, $\lim_{\tau \to \infty} K_0(\tau) = 0$. This behaviour reflects a progressive memory loss in the system. It results from dissipative mechanisms which directly control the decay rate of $K_0(\tau)$ at large times. A stronger dissipation leads to a faster damping of perturbations, namely a fast decay of $K_0(\tau)$ when $\tau \to \infty$. In the case of a particle immersed in a highly viscous fluid for instance, the velocity response to an exciting force decreases very rapidly. The characteristic time scale of this decay is inversely proportional to the viscous friction. Conduction electrons in a metal constitutes another example implying more complex physics. The dissipation processes are here due to energy transfer from electrons to the lattice by collisions✠.

---

[2]Concepts of stability in weak and strong sense characterising a stable fixed point of a dynamical system are presented in the excellent book by V. Arnold on differential equations [5].

Increasing the collision frequency makes $K_0(\tau)$ decay faster. The expected behaviours for these two examples are therefore very similar. This analogy is at the origin of Drude model for conduction, where the effect of electron-lattice collisions is precisely described by fluid friction at a phenomenological level.

✠ **Comment:** These collisions generate phonons pumping energy from the electrons. This transfer mechanism is dominant at room temperature for many metals. At very low temperatures (few Kelvins), collisions with impurities become dominant. They tend to induce localisation of backscattered electrons by quantum interference. This quantum effect related to disorder reduces the conductivity and can be described by an additional friction term in the phenomenological equation of motion for a conduction electron (see the book [45]).

Short times   Under the action of a pulse, $A(0^+)$ takes a non-zero value whose amplitude is fixed by system inertia (in a broad sense). The response of a soccer ball to a kick for instance is controlled at short time by the mass of the ball. Thus given the form of the response (1.5), $K_0(0^+)$ must roughly speaking be inversely proportional to the system inertia. The shape of $K_0(\tau)$ at short time is given by free evolution of the elements composing the system. Their mutual interactions can generally be treated using a perturbative expansion. Note that this form is specific to the observable A under consideration while relaxation behaviours at long times are controlled by collective effects inherent to the system internal dynamics, independently of the particular observable.

The dynamics of A is in many cases governed by a linear differential equation: the response function K_0 is then a particular Green's function for this differential equation. More specifically $K_0(t - t')$ is the causal Green's function associated with this differential equation. We refer the reader to Chapters 2 and 3 devoted to the study of Green's functions for further details. In addition Appendix C presents a systematic and useful method to compute the response function K_0 in these situations.

1.1.2   *Definition of the susceptibility*

Consider now a perturbation of the form

$$F(t) = \mathrm{Re}\,(F_z\ e^{-izt}),$$

where Re denotes the real part, with $z = \omega + i\epsilon$, ω real, $\epsilon > 0$ and $F_z \in \mathbb{C}$. The perturbation takes the form of a damped sinusoid for $t < 0$ since $\epsilon > 0$, so that $F(-\infty) = 0$. Complex frequency z is thus taken in the upper half complex plane. The response to a monochromatic excitation at purely real frequency $z = \omega$ is defined as the limit $\epsilon \to 0^+$, which physically corresponds to an adiabatic introduction of the perturbation at $t \to -\infty$. Expression (1.4) of the response can then be written as:

$$A(t) = \int_0^{+\infty} d\tau \; K_0(\tau) \; \mathrm{Re}\left(F_z \; e^{-iz(t-\tau)}\right), \tag{1.6}$$

$$= \mathrm{Re}\left(\chi(z) \; F_z \; e^{-izt}\right), \tag{1.7}$$

where $\chi(z)$ is the susceptibility defined as

$$\boxed{\chi(z) = \int_0^{+\infty} d\tau \; K_0(\tau) \; e^{iz\tau}.} \tag{1.8}$$

We used in (1.7) the fact that response function K_0 is real. By virtue of the linearity of the response, $A(t)$ is monochromatic as $F(t)$ with the same complex frequency z. We can rewrite $A(t)$ given by expression (1.7) as $A(t) = \mathrm{Re}\left(A_z \; e^{-izt}\right)$ with a complex amplitude

$$A_z = \chi(z)F_z. \tag{1.9}$$

From now on we may sometimes omit the real parts in some linear manipulations.

Note that susceptibility $\chi(z)$ is a complex quantity even for $z = \omega$ real. Its modulus measures the ratio between the amplitudes of A and F, while its argument gives the phase shift between F and A. One can identify in expression (1.8) of $\chi(z)$ the Laplace transform of K_0 for the argument $s = -iz$. A reminder of Laplace transform properties is given in Appendix B.

It often happens in practice that the forcing is monochromatic, so that the introduction of susceptibility $\chi(z)$ is quite natural. If the forcing introduced adiabatically at $t = -\infty$ is now an arbitrary function of time, Fourier analysis allows to decompose $F(t)$ as a sum of monochromatic excitations. Linear response (1.4) then obviously reduces to the corresponding sum of responses to the different monochromatic components of $F(t)$. Each monochromatic response is controlled by the susceptibility evaluated at the corresponding frequency. The knowledge of $\chi(z)$ thus allows to obtain the response to any type of excitation.

1.1.3 *Analyticity of the susceptibility*

In this section we show how fundamental properties such as causality, combined with the behaviours of $K_0(\tau)$, confer analyticity properties to the susceptibility. The consequences of these analyticity properties are numerous and commensurate with the breadth of the concept of analytic functions.

◊ **Analyticity in** $\mathbb{C}^+$ ◊

The analytical properties of the susceptibility clearly depend on the behaviour of $K_0(\tau)$ when $\tau \to +\infty$. The discussion in the previous paragraph however indicates that K_0 is bounded for any stable state $\mathcal{E}_0$. Then for $z = \omega + i\epsilon$ with $\epsilon > 0$, $\chi(z)$ defined by the formula (1.8) is analytic by Leibniz integral rule/differentiation under the integral sign theorem. This lack of singularity on $\mathbb{C}^+$ implies in particular that the susceptibility is finite. This property may simply be interpreted in terms of energy: in fact, when $\epsilon > 0$, the energy injected in the system by the perturbation $F(t) = F_z\, e^{-izt}$ is finite and so are the variations of the system physical quantities.

◊ **Real axis** ◊

For $\epsilon \to 0^+$, that is to say for z real, only a strong decrease of $K_0(\tau)$ for $\tau \to +\infty$ can ensure the analyticity of $\chi(z)$ up to the real axis. From a physical point of view, and according to the previous discussion on the nature of the stability of $\mathcal{E}_0$, we must distinguish two cases according to the presence or not of dissipation in the system.

In the presence of dissipation   As justified above, the dissipation manifests itself by a sufficiently fast decay of $K_0(\tau)$, typically an exponential decay

$$K_0(\tau) \propto e^{-\lambda \tau} \text{ when } \tau \to \infty. \tag{1.10}$$

Positive constant λ is the inverse of a relaxation time characteristic of the dissipative mechanism at play. For a particle immersed in a fluid for instance, the response function on the velocity satisfies property (1.10) where λ is proportional to the viscous friction coefficient. Behaviour (1.10) then generally implies that $\chi(z)$ is analytic on the real axis.

Zero or weak dissipation   In the total absence of dissipation, the response function generally oscillates without decaying. If ω coincides with one of the characteristic frequencies of these oscillations, $\chi(\omega)$ diverges: this is the resonance phenomenon, with a divergence of the response associated with an infinite injection of energy in the system ✠.

This situation occurs for example in the case of an undamped harmonic oscillator. In some cases weak dissipation is at play in the system, in the sense that $K_0(\tau)$ indeed decreases for large time but slower than an exponential, i.e. typically as a power law,

$$K_0(\tau) \sim \frac{\mathrm{Cst}}{\tau^p} \text{ when } \tau \to \infty.$$

> ✠ **Comment:** If the quantity A keeps increasing, one ends up leaving the framework of linear response. The non-linear terms in the equations of forced motion can no longer be neglected, and they are the ones that may lead to the saturation of A at an amplitude no longer being proportional to the forcing intensity F_0. In practice, the breakdown threshold of the system may be exceeded before the saturation has been reached, not without dramatic consequences sometimes ...

Then $\chi(z)$ diverges in $z = 0$ for $p \leq 1$. If $p > 1$, $\chi(z)$ is finite at $z = 0$ but it has singular derivatives. Point $z = 0$ is then typically the branch point of a cut.

In summary, $\chi(z)$ is always analytic on $\mathbb{C}^+$ and singularities are possible on $\mathbb{R}$ if dissipation is absent or weak enough.

◇ Analytic continuation ◇

In $\mathbb{C}^-$, the integral representation of χ in terms of K_0 (1.8) is no longer necessarily defined. However χ can generally be extended by analytic continuation. Several possibilities can appear:

Absence of singularities Suppose that χ is analytic on $\mathbb{C}^-$. The definition (1.8) of χ shows that $\chi(z) \to 0$ when $z \to \infty$ with $\mathrm{Im}\, z > 0$ (see §1.1.5 for a detailed study of the asymptotic behaviour of $\chi(z)$ at infinity). We conclude that $\chi(z)$ must diverge when $z \to \infty$ with $\mathrm{Im}\, z < 0$. Indeed $\chi(z)$ would otherwise be a bounded entire function and therefore constant by virtue of Liouville's theorem on analytic functions. Since it also vanishes at infinity on $\mathbb{C}^+$, it would necessarily be zero everywhere.

Presence of singularities in $\mathbb{C}^-$ Singularities of χ in $\mathbb{C}^-$ can be of different natures, such as branch cuts or poles. These poles typically correspond to specific modes of the system in the state $\mathcal{E}_0$ as we shall see later on several examples.

This study of the analyticity of $\chi(z)$ provides a framework allowing to address other general properties of the susceptibility.

1.1.4 *Parity properties and dissipation*

Let us consider the limit where the frequency z is real, i.e. $z = \omega + i\epsilon$ with $\epsilon \to 0^+$. Susceptibility $\chi(\omega)$ thus defined from $\chi(z)$ has a real part $\chi'(\omega)$ as well as an imaginary part $\chi''(\omega)$. Let us recall that the presence of an imaginary part shows that the response is generally out of phase with the forcing. If $K_0(\tau)$ decreases at large time, then we can take the limit $\epsilon \to 0^+$ under the sum sign in integral representation (1.8) of $\chi(z)$. This procedure gives

$$
\begin{aligned}
\chi'(\omega) &= \int_0^\infty d\tau \, \cos(\omega\tau) K_0(\tau), \\
\\
\chi''(\omega) &= \int_0^\infty d\tau \, \sin(\omega\tau) K_0(\tau).
\end{aligned}
\tag{1.11}
$$

Obviously χ' and χ'' are respectively even and odd functions of ω. If $K_0(\tau)$ does not tend to zero as $\tau \to \infty$, then the above integral representations for $\chi'(\omega)$ and $\chi''(\omega)$ are no longer valid. These quantities are however finite, except possibly at singular points of $\chi(z)$ located on the real axis, and they satisfy the same parity properties as those given by expressions (1.11). It is quite common that the power provided on average over a period $T = 2\pi/\omega$ by forcing is proportional to $\chi'(\omega)$ or to $\omega\chi''(\omega)$, as discussed for each example presented later. This power is generally positive, which means that the external operator provides energy to the system by imposing a forcing. This is obviously the case in the presence of dissipation in its internal dynamics. It should be noted that other mechanisms of energy absorption induced in particular by resonances are also possible, a situation typical of conservative systems: Away from these resonances, the average power provided by the forcing then vanishes✠. All these considerations imply that $\chi'(\omega) \geq 0$ for all ω, or $\chi''(\omega) \geq 0$ for $\omega > 0$.

✠ **Comment:**   Naturally the forcing must provide a finite amount of energy, from its introduction at $t = -\infty$ to the time considered, to remove the system from its initial state $\mathcal{E}_0$ and induce oscillations of finite amplitude for the observable A. Once these oscillations are switched on however, their maintenance requires no energy input on average over a period $T = 2\pi/\omega$.

1.1.5   *Low and high frequency behaviours*

Various mathematical properties of susceptibility $\chi(z)$ do not only have a purely technical interest but also –and specially– provide an interpretation to better understand and characterise the system considered. Thus the behaviours of the susceptibility at low and high frequencies have specific physical interpretations.

◊ **Low frequencies** ◊

Let us first consider the low frequency limit $z \to 0$ of $\chi(z)$. If $K_0(\tau)$ decreases sufficiently fast at large time to be integrable, then $\chi(0)$ is finite and reduces to

$$\chi(0) = \int_0^\infty \mathrm{d}\tau \, K_0(\tau). \tag{1.12}$$

This is nothing but the static susceptibility controlling the response to a constant forcing F_0. Note that this quantity can often be determined from a completely static approach. The integral representation (1.12) then provides a constraint on response function $K_0(\tau)$. This turns out to be particularly useful in cases where this function is not exactly known, and for which a phenomenological model is introduced.

If $\chi(z)$ is analytic at $z = 0$, it can be expanded as a power series in z and the corresponding coefficients are the moments of $K_0(\tau)$ (which then decreases sufficiently rapidly as previously discussed in §1.1.3). If $z = 0$ is a singular point, the expansion of $\chi(z)$ for small z contains singular terms determined by the nature of the decay of $K_0(\tau)$ (see §1.1.3). Again these properties can be used for phenomenological purposes.

◊ **High frequencies** ◊

The behaviour of the susceptibility is related to the structure of $K_0(\tau)$ at short time. Assume that $K_0(\tau)$ is infinitely differentiable with respect to τ, and also that all corresponding derivatives admit a Laplace transform for z in the upper half complex plane. Returning back to integral representation (1.8) of the susceptibility, we obtain by integration by parts:

$$\chi(z) = \frac{iK_0(0)}{z} - \frac{1}{iz} \int_0^{+\infty} \mathrm{d}\tau \, K_0'(\tau) \, e^{iz\tau}. \tag{1.13}$$

It is clear that an iteration of this operation provides an expansion of susceptibility as a Laurent series around infinity, that is to say in inverse powers of z. We thus find the asymptotic behaviour of $\chi(z)$ for large z, i.e.

$$\boxed{\chi(z) = \frac{iK_0(0)}{z} - \frac{K_0'(0)}{z^2} + O(1/z^3).} \tag{1.14}$$

Expansion (1.14), *a priori* valid for $\operatorname{Im} z > 0$, can be extended onto the real axis in general, provided that $\chi(\omega)$ does not present any singularity at infinity on the real axis. Since K_0 is real, we get the asymptotic expansions of the real and imaginary parts[✠] of $\chi(\omega)$.

$$\chi'(\omega) = -\frac{K_0'(0)}{\omega^2} + O(1/\omega^4), \qquad (1.15)$$

and

$$\chi''(\omega) = \frac{K_0(0)}{\omega} + O(1/\omega^3), \qquad (1.16)$$

which contain respectively only inverse even and odd powers.

✠ Comment: This is just a physicist's trick! On a mathematical level, it is not possible to state a set of necessary and sufficient conditions on $K_0(\tau)$ ensuring the validity of (1.14) for $z = \omega$ real. A possible sufficient condition is that the integrals of $K_0(\tau)$ and its derivatives are absolutely convergent. But even when $K_0(\tau)$ does not decrease at infinity for example, the recipe stated here almost always works...

The decay of $\chi(\omega)$ at high frequencies is a consequence of the system inertia, preventing it to follow the corresponding excitation. The previous asymptotic expansions involve the short time behaviour of $K_0(\tau)$, whose physical nature has been discussed above. Note that amplitude $K_0(0)$ of the dominant term in expression (1.16) is inversely proportional to the system inertia. In addition, amplitude $K_0'(0)$ of the dominant term in expansion (1.15) is in general fixed by the free evolution without any action of the internal interactions. These amplitudes are therefore easily computable, so that high frequencies expansions are particularly useful for the *ad hoc* construction of $\chi(\omega)$.

If the assumptions set out on $K_0(\tau)$ are not met, the expansion of $\chi(z)$ as a Laurent series at large z may contain singular terms. For example only the p first terms in expansion (1.14) are in $1/z^k$, with $1 \leq k \leq p$, if the derivative of $K_0(\tau)$ of order p is not well defined at $\tau = 0$. The next term in the asymptotic expansion which tends to zero faster than $1/z^p$ is not equivalent to an inverse integer power z.

1.1.6 *Kramers-Kronig relations*

This paragraph is devoted to the so-called Kramers-Kronig relations, satisfied by the susceptibility. We will first establish these relations before discussing their interests and applications.

Suppose first that $\chi(z)$ is an analytic function on the upper half complex plane $\mathbb{C}^+$ and on $\mathbb{R}$, as it is the case in the presence of sufficiently strong dissipation in the system. Consider now a closed contour $\mathcal{C}$ on the complex plane as shown in Figure 1.1: it is made of segment $[-R, R]$ on the real axis complemented by the semicircle of radius R and centre 0 in the upper half complex plane. For $z \in \mathbb{C}^+$, it is clear

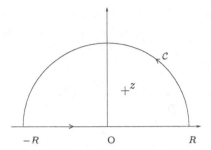

Fig. 1.1   Contour $\mathcal{C}$ around the point z.

that the function $f(z')$ defined by

$$f(z') = \frac{\chi(z')}{(z' - z)}$$

is analytic in the entire upper half complex plane except at $z' = z$, where it has a simple pole with residue $\chi(z)$. Applying Cauchy theorem then leads to

$$\chi(z) = \frac{1}{2\pi i} \oint_{\mathcal{C}} dz' \, \frac{\chi(z')}{(z' - z)}.$$

Since $\chi(z)$ goes to 0 when $z \to \infty$ with $\operatorname{Im} z > 0$, the integral over the semicircle tends to 0 when $R \to \infty$ by virtue of Jordan's lemma. We then have:

$$\chi(z) = \frac{1}{2\pi i} \int_{-\infty}^{\infty} d\omega' \, \frac{\chi(\omega')}{(\omega' - z)}. \tag{1.17}$$

Let us now take the limit $\epsilon = \operatorname{Im} z \to 0^+$ in the above equation with $\omega = \operatorname{Re} z$. We obtain for the left-hand side $\chi(\omega)$ by continuity of $\chi(z)$ at $z = \omega$. To take the limit on the right-hand side, we proceed as follows. Let us fix initially ϵ. As χ is analytic on the real axis by assumption, there exists $\nu \in \mathbb{R}$ small enough such that χ is analytic on the disk of centre ω and radius ν. It is then useful to deform the path $\mathbb{R}$ of integral (1.17) into $\mathcal{D}_\nu \bigcup \mathcal{C}_\nu$, where $\mathcal{D}_\nu$ consists of the half-lines $]-\infty, \omega - \nu]$ and $[\omega + \nu, \infty[$, while $\mathcal{C}_\nu$ is the half-circle of centre ω and radius ν in the lower half-plane (see Figure 1.2). Since $f'(s)$ is analytic inside the domain between $\mathbb{R}$ and $\mathcal{D}_\nu \bigcup \mathcal{C}_\nu$ we can use Cauchy theorem to obtain:

$$\int_{-\infty}^{\infty} d\omega' \, \frac{\chi(\omega')}{(\omega' - z)} = \int_{-\infty}^{\omega - \nu} d\omega' \, \frac{\chi(\omega')}{(\omega' - z)} + \int_{\omega + \nu}^{\infty} d\omega' \, \frac{\chi(\omega')}{(\omega' - z)} + \int_{\mathcal{C}_\nu} dz' \, \frac{\chi(z')}{(z' - z)}.$$

It is then easy to take the limit $\epsilon \to 0^+$ in the right-hand side to get:

$$\lim_{\epsilon \to 0^+} \int_{-\infty}^{\infty} d\omega' \, \frac{\chi(\omega')}{(\omega' - z)} = \int_{-\infty}^{\omega - \nu} d\omega' \, \frac{\chi(\omega')}{(\omega' - \omega)} + \int_{\omega + \nu}^{\infty} d\omega' \, \frac{\chi(\omega')}{(\omega' - \omega)}$$
$$+ \int_{\mathcal{C}_\nu} dz' \, \frac{\chi(z')}{(z' - \omega)}. \tag{1.18}$$

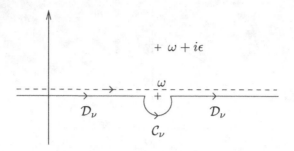

Fig. 1.2   Contour $\mathcal{D}_\nu \bigcup \mathcal{C}_\nu$.

Equation (1.18) is valid for any sufficiently small ν and does not depend on its precise value. The RHS can be evaluated taking the limit $\nu \to 0^+$. For the integral over $\mathcal{C}_\nu$, it is legitimate to replace $\chi(z')$ by its Taylor expansion in the neighbourhood of $z' = \omega$ since $\chi(z')$ is analytic at $z' = \omega$, leading to:

$$\chi(z') = \chi(\omega) + \frac{\mathrm{d}\chi}{\mathrm{d}z}(\omega)(z' - \omega) + \cdots .$$

Setting $z' = \omega + \nu e^{i\theta}$ on $\mathcal{C}_\nu$ with $\theta \in [-\pi, 0]$, one can then easily show that

$$\lim_{\nu \to 0^+} \int_{\mathcal{C}_\nu} \mathrm{d}z' \, \frac{\chi(z')}{(z' - \omega)} = i\pi\chi(\omega). \tag{1.19}$$

For the integral over $\mathcal{D}_\nu$, the limit $\nu \to 0^+$ defines the principal part (PP), i.e.

$$\lim_{\nu \to 0^+} \left[\int_{-\infty}^{\omega - \nu} \mathrm{d}\omega' \, \frac{\chi(\omega')}{(\omega' - \omega)} + \int_{\omega + \nu}^{\infty} \mathrm{d}\omega' \, \frac{\chi(\omega')}{(\omega' - \omega)} \right]$$
$$= \mathrm{PP} \int_{-\infty}^{\infty} \mathrm{d}\omega' \, \frac{\chi(\omega')}{(\omega' - \omega)}. \tag{1.20}$$

We then find, plugging the results (1.19) and (1.20) in identity (1.18):

$$\lim_{\epsilon \to 0^+} \int_{-\infty}^{\infty} \mathrm{d}\omega' \, \frac{\chi(\omega')}{(\omega' - \omega - i\epsilon)} = \mathrm{PP} \int_{-\infty}^{\infty} \mathrm{d}\omega' \, \frac{\chi(\omega')}{(\omega' - \omega)} + i\pi\chi(\omega). \tag{1.21}$$

Note that each integral on the left-hand side of definition (1.20) of the principal part diverges if $\chi(\omega) \neq 0$. Yet these divergences, of the form $\pm\chi(\omega)\ln\nu$, compensate each other exactly so that the sum of the two integrals remains finite when $\nu \to 0^+$. Note also that result (1.21) is actually a consequence of the well-known identity in the sense of distributions:

$$\boxed{\lim_{\epsilon \to 0^+} \frac{1}{(x - i\epsilon)} = \mathrm{PP}\, \frac{1}{x} + i\pi\delta(x).} \tag{1.22}$$

Taking into account equation (1.21), the limit $\epsilon \to 0^+$ for each term of identity (1.17) leads to

$$\chi(\omega) = \frac{1}{2\pi i} \mathrm{PP} \int_{-\infty}^{\infty} \mathrm{d}\omega' \, \frac{\chi(\omega')}{(\omega' - \omega)} + \frac{1}{2}\chi(\omega).$$

Decomposing $\chi(\omega')$ inside the integral in terms of its real and imaginary parts, $\chi(\omega') = \chi'(\omega') + i\chi''(\omega')$, we finally obtain the **Kramers-Kronig relations** (KK):

$$
\chi'(\omega) = \frac{1}{\pi} \, \mathrm{PP} \int_{-\infty}^{\infty} d\omega' \, \frac{\chi''(\omega')}{(\omega' - \omega)},
$$
$$
\chi''(\omega) = -\frac{1}{\pi} \, \mathrm{PP} \int_{-\infty}^{\infty} d\omega' \, \frac{\chi'(\omega')}{(\omega' - \omega)}.
$$
(1.23)

The Kramers-Kronig relations are of crucial importance. As direct consequences of causality and dissipation, they are very general and independent of the dynamical complexity of each system. These exact formulas are also very useful from a phenomenological viewpoint. In some specific situations it is indeed possible to experimentally obtain the dissipative part of the susceptibility, say its imaginary part to fix ideas. KK relations then allow to automatically obtain the corresponding real part. This method is well illustrated on the example of §1.2.2.

Note that Kramers-Kronig relations can easily be generalised to the case where $\chi(z)$ has simple poles on the real axis. Let us imagine that $\chi(z)$ has simple poles $\omega_1, ..., \omega_n$ with residue $R_1, ..., R_n$ respectively. It is then enough to repeat the preceding argument, this time with an integration path $\mathcal{C}$ in the complex plane bypassing these singularities from above, as shown in Figure 1.3. Then $\chi(z)$ admits a Laurent

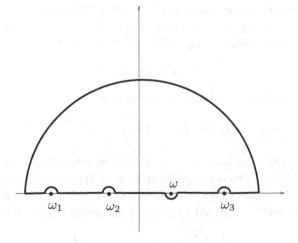

Fig. 1.3 Contour used in the case where χ has three poles ω_1, ω_2 and ω_3 on the real axis.

series expansion in the vicinity of each pole ω_k:

$$
\chi(z') = \frac{R_k}{(z' - \omega_k)} + \text{regular part.}
$$

Just as in the previous case, it is possible to perform the integral over the semicircle bypassing the considered singularity by replacing $\chi(\omega')$ by the previous Laurent series expansion. We then find, sending the radius of each semicircle to zero:

$$\chi(\omega) = \frac{1}{2\pi i} \, \mathrm{PP} \int_{-\infty}^{\infty} d\omega' \, \frac{\chi(\omega')}{(\omega' - \omega)} + \frac{1}{2} \left(\chi(\omega) - \sum_{k=1}^{n} \frac{R_k}{(\omega_k - \omega)} \right)$$

(the $-$ sign in front of the residues comes from the anti-clockwise direction of integration on the semicircles bypassing the corresponding poles ω_k). By decomposing each residue in its real and imaginary parts, $R_k = R_k' + i R_k''$, we finally obtain the generalisation of Kramers-Kronig relations:

$$\chi'(\omega) + \sum_{k=1}^{n} \frac{R_k'}{(\omega_k - \omega)} = \frac{1}{\pi} \, \mathrm{PP} \int_{-\infty}^{\infty} d\omega' \, \frac{\chi''(\omega')}{(\omega' - \omega)},$$

$$\chi''(\omega) + \sum_{k=1}^{n} \frac{R_k''}{(\omega_k - \omega)} = -\frac{1}{\pi} \, \mathrm{PP} \int_{-\infty}^{\infty} d\omega' \, \frac{\chi'(\omega')}{(\omega' - \omega)}.$$

$$(1.24)$$

Note that the principal parts in these relations take into account the n singularities present in the corresponding integrands at $\omega' = \omega_k$ in addition to that at $\omega' = \omega$.

1.1.7 *Sum rules*

Suppose that $K_0(\tau)$ satisfies the assumptions described in §1.1.5 ensuring the validity of the asymptotic expansions of $\chi'(\omega)$ and $\chi''(\omega)$ when $\omega \to \infty$. Suppose in addition that $K_0(\tau)$ decays fast enough to ensure that $\chi(z)$ is indeed analytic onto the real axis. We then deduce from expansion (1.16) of $\chi''(\omega)$ that

$$\lim_{\omega \to \infty} \omega \chi''(\omega) = K_0(0). \tag{1.25}$$

The second Kramers-Kronig relation moreover gives

$$\lim_{\omega \to \infty} \omega \chi''(\omega) = \lim_{\omega \to \infty} \frac{1}{\pi} \, \mathrm{PP} \int_{-\infty}^{\infty} d\omega' \, \frac{\omega \chi'(\omega')}{(\omega - \omega')}. \tag{1.26}$$

It is legitimate to take the limit $\omega \to \infty$ under the integral sign in the right-hand side of equation (1.26). The corresponding limit of the integrand $\omega \chi'(\omega')/(\omega - \omega')$ for ω' fixed indeed reduces to $\chi'(\omega')$, which is integrable on ω'. The left-hand side of equation (1.26) can be replaced by $K_0(0)$ from result (1.25). We then obtain

$$\frac{1}{\pi} \int_{-\infty}^{\infty} d\omega \, \chi'(\omega) = K_0(0). \tag{1.27}$$

This identity is a **sum rule**. It constrains the zeroth order moment of $\chi'(\omega)$ in terms of a quantity more easily accessible and therefore it has a similar interest as the asymptotic expansions for large ω established on p. 12.

It is possible to obtain in the same way other sum rules for χ' and χ''. In particular the reader may show by similar arguments (see Exercise 1.6 on p. 50) sum rules for the first moment of χ'',

$$\boxed{\frac{1}{\pi} \int_{-\infty}^{\infty} d\omega\, \omega \left[\chi''(\omega) - \frac{K_0(0)}{\omega}\right] = K_0'(0),}$$ (1.28)

and for the second moment of χ',

$$\boxed{-\frac{1}{\pi} \int_{-\infty}^{\infty} d\omega\, \omega^2 \left[\chi'(\omega) + \frac{K_0'(0)}{\omega^2}\right] = K_0''(0).}$$ (1.29)

Note that by virtue of asymptotic behaviours (1.16) and (1.15) respectively, integrands $\omega(\chi''(\omega) - K_0/\omega)$ and $\omega^2(\chi'(\omega) + K_0'(0)/\omega^2)$ decay at least as $1/\omega^2$ for large ω: this indeed ensures the convergence of integrals in the left-hand side of expressions (1.28) and (1.29).

Similar manipulations allow us to obtain analogous sum rules for higher moments of $\chi(\omega)$. As indicated on p. 12, if a derivative of order p of $K_0(\tau)$ is singular, only the first p terms of asymptotic expansion (1.14) are well defined. The previous sum rules then remain valid for the first p moments of $\chi(\omega)$.

1.1.8 *Inhomogeneous perturbations*

So far we have not considered the spatial dependence for perturbation $F(t)$. However, the properties previously established simply generalise to the case of an inhomogeneous perturbation.

Let us consider an infinitely large system. In practice it is a finite system large enough for the boundary effects on its bulk properties to be neglected. These properties can be intrinsically characterised by a procedure similar to the thermodynamic limit where system boundaries are sent to infinity. Suppose again that the system is in a stationary state $\mathcal{E}_0$ for $t = -\infty$. Now imagine that forcing $F(\mathbf{r}, t)$ has a spatial dependence. In order to study the behaviour of a physical quantity $A(\mathbf{r}, t)$ now depending not only on time but also on the position, response function $K_0(\mathbf{r}, \mathbf{r}', t, t')$ is naturally introduced such that

$$\boxed{A(\mathbf{r}, t) = \int_{-\infty}^{t} dt' \int d\mathbf{r}'\, K_0(\mathbf{r}, \mathbf{r}', t, t')\, F(\mathbf{r}', t').}$$ (1.30)

The properties of K_0 are again intrinsic to the system in initial state $\mathcal{E}_0$. Thus if $\mathcal{E}_0$ is stationary and invariant under spatial translation, response function K_0 is a function of the difference of arguments, i.e.

$$K_0(\mathbf{r}, \mathbf{r}', t, t') = K_0(\mathbf{r} - \mathbf{r}', t - t').$$

Now suppose that the perturbation is a monochromatic plane wave of the form

$$F(\mathbf{r}', t') = \mathrm{Re}\left(F_{\mathbf{k},z}\, e^{i\mathbf{k}\cdot\mathbf{r}' - izt'}\right).$$

In an analogous manner to the homogeneous case, observable A can then also be written as

$$A(\mathbf{r}, t) = \text{Re} \left(\chi(\mathbf{k}, z) \quad F_{\mathbf{k}, z} e^{i\mathbf{k}.\mathbf{r} - izt} \right),$$

with

$$\chi(\mathbf{k}, z) = \int_0^\infty \mathrm{d}\tau \int \mathrm{d}\mathbf{r}' \; K_0(\mathbf{r}', \tau) \; e^{-i\mathbf{k}.\mathbf{r}' + iz\tau}. \qquad (1.31)$$

The response in A then also has the spatio-temporal structure of a monochromatic plane wave similar to the forcing, with the same wave number $\mathbf{k}$ and the same complex frequency z. It is clear that all the arguments used in the homogeneous case remain valid for $\mathbf{k}$ fixed. The equivalent of $K_0(\tau)$ is nothing but the Fourier transform $\tilde{K}_0(\mathbf{k}, \tau)$ of $K_0(\mathbf{r}, \tau)$ given by

$$\tilde{K}_0(\mathbf{k}, \tau) = \int \mathrm{d}\mathbf{r}' \; K_0(\mathbf{r}', \tau) \; e^{-i\mathbf{k}.\mathbf{r}'}.$$

In particular Kramers-Kronig relations are now written as

$$\chi'(\mathbf{k}, \omega) = \frac{1}{\pi} \text{PP} \int_{-\infty}^{\infty} \mathrm{d}\omega' \; \frac{\chi''(\mathbf{k}, \omega')}{(\omega' - \omega)},$$

$$\chi''(\mathbf{k}, \omega) = -\frac{1}{\pi} \text{PP} \int_{-\infty}^{\infty} \mathrm{d}\omega' \; \frac{\chi'(\mathbf{k}, \omega')}{(\omega' - \omega)},$$

and there are sum rules similar to those of the homogeneous case.

In the case where the forcing has an arbitrary dependence on time and position, it is wise to break it down into a sum of monochromatic plane waves, the sum covering both frequencies and wave numbers. By linearity of the response, each monochromatic component of A, characterised by the wave number $\mathbf{k}$ and the frequency ω, is then simply proportional to the corresponding component of F with a proportionality factor which is nothing but $\chi(\mathbf{k}, \omega)$. In an analogous manner to $\chi(\omega)$ in the homogeneous case, susceptibility $\chi(\mathbf{k}, \omega)$ therefore plays a central role in the response to an arbitrary inhomogeneous forcing.

1.2 Applications and examples

1.2.1 *Admittance of a RLC circuit*

◇ Presentation ◇

The simplest examples of linear response come from classical systems with a small number of degrees of freedom whose internal dynamic is linear. A typical example of such systems is an RLC circuit, whose states are described by the charge of the capacitor Q and the intensity in the circuit I. These variables are analogous to the position and the velocity of a particle in one dimension. The steady-state $\mathcal{E}_0$

is characterised by zero charge and zero current. Let us immediately note that the current response to a voltage falls within the general framework introduced in the first part of this chapter. For a current to appear in the circuit, it is indeed necessary to apply an external electric field to the charge carriers. In practice, this field can be induced by a voltage generator connected across the circuit. The external forcing F is the applied voltage U and adiabatically switched on. We will study the current flowing through the circuit in response to the application of this external voltage. The role of the observable A is therefore held here by intensity I.

$$\diamond \textbf{ Study and resolution } \diamond$$

Let us recall the differential equation governing the evolution of the circuit shown in Figure 1.4. Let U be the voltage across the circuit, I the current intensity and

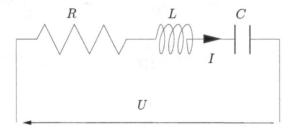

Fig. 1.4 *RLC* circuit.

Q, the capacitor charge. The voltages across the resistance, the inductive coil and the capacitor are respectively RI, $L\,dI/dt$ and Q/C. Since by charge conservation $I = dQ/dt$, Kirchhoff's current law leads to

$$\boxed{\frac{\mathrm{d}^2Q}{\mathrm{d}t^2} + \frac{R}{L}\frac{\mathrm{d}Q}{\mathrm{d}t} + \frac{1}{LC}Q = \frac{U}{L}.} \tag{1.32}$$

In the situation considered here, the initial conditions are sent to $t \to -\infty$ and written as $Q(-\infty) = 0$ and $I(-\infty) = 0$. The applied voltage $U(t)$ also vanishes at $t = -\infty$. When U is a monochromatic excitation of the form $U = U_z\,e^{-izt}$, with $z = \omega + i\epsilon$ and $\epsilon > 0$, the susceptibility $\chi(z)$ controlling the current response is directly accessible without using the response function $K_0(\tau)$. We show that this susceptibility indeed exhibits general analyticity properties. We then give the form of $K_0(\tau)$ which can be obtained in three different ways.

Susceptibility Due to the linearity of the electronic circuit, the intensity $I(t)$ oscillates at the same frequency z as the voltage, i.e. $I = I_z\,e^{-izt}$. We then obtain, inserting this monochromatic form in the derivatives involved in differential equation (1.32):

$$I_z = \frac{U_z}{Z},$$

where $Z = R - i(Lz - 1/(Cz))$ is the circuit complex impedance. The susceptibility is then simply equal to the admittance, so that

$$\chi(z) = \frac{iCz}{LCz^2 + iRCz - 1}. \tag{1.33}$$

The susceptibility $\chi(z)$ is here a rational fraction. Introducing

$$\delta = -R^2 C^2 + 4LC,$$

its poles are located at (see Figure 1.5)

$$z_1 = \frac{-iRC - \sqrt{\delta}}{2LC} \quad ; \quad z_2 = \frac{-iRC + \sqrt{\delta}}{2LC} \quad \text{if} \quad \delta > 0 \quad \text{(underdamping)},$$

or

$$\widetilde{z}_1 = -i\frac{RC - \sqrt{|\delta|}}{2LC} \quad ; \quad \widetilde{z}_2 = -i\frac{RC + \sqrt{|\delta|}}{2LC} \quad \text{if} \quad \delta < 0 \quad \text{(overdamping)}.$$

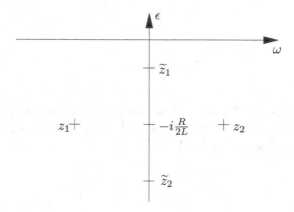

Fig. 1.5   The poles (z_1, z_2) or $(\widetilde{z}_1, \widetilde{z}_2)$ of the admittance $\chi(z)$ of a RLC circuit are located in the lower half complex plan.

Since $RC > \sqrt{|\delta|}$ when $\delta < 0$, the two poles of $\chi(z)$ are always in the lower half complex plane. Therefore, χ is analytic in $\mathbb{C}^+$. Moreover the poles of $\chi(z)$ clearly correspond to eigenmodes of the RLC circuit.

Since the rational fraction (1.33) is analytic in $\mathbb{C}^+$, including the real axis, Kramers-Kronig relations are necessarily satisfied. It is however interesting from a technical standpoint to explicitly check them on this simple example. We indicate some steps of the corresponding calculations. For $z = \omega$ real, the real and imaginary parts of $\chi(z)$ are

$$\chi'(\omega) = \frac{R}{[R^2 + (L\omega - \frac{1}{C\omega})^2]} \quad \text{and} \quad \chi''(\omega) = \frac{(L\omega - \frac{1}{C\omega})}{[R^2 + (L\omega - \frac{1}{C\omega})^2]}. \quad (1.34)$$

It is convenient to use the partial fraction decomposition of $\chi'(\omega)$ and $\chi''(\omega)$ to perform the integrations involved in the KK relations.

We then rewrite the principal parts as integrals over $]-L, \omega - \epsilon]$ and $[\omega + \epsilon, L[$ and then take the limits $L \to +\infty$ and $\epsilon \to 0^+$. The verification is completed using the identity ✠

$$\lim_{L \to +\infty} \int_{-L}^{L} d\omega' \frac{1}{(\omega' - z_0)}$$
$$= i\pi \, \text{sign} \, [\text{Im}(z_0)] \quad (1.35)$$

where $\text{sign}(x)$ is the sign of x.

✠ **Comment:** It is enough, in order to establish the identity (1.35), to absorb the real part of z_0 by translating the variable ω', and then compute the integral of the even part in ω' using the residue theorem. The limit of the integral, then understood as the principal part, vanishes when z_0 is real.

Response function The first method to determine $K_0(\tau)$ is to compute the current $I(t)$ induced by an arbitrary voltage $U(t)$. The basic differential equation (1.32) for $Q(t)$ is then solved by the method of variation of constants[3]. The response function is identified in the integral expression obtained for $I(t) = dQ/dt$, takes the form (1.4), p. 4, leading to

$$K_0(\tau) = \frac{e^{-\frac{R}{2L}\tau}}{L\sqrt{\delta}} \left[\sqrt{\delta} \, \cos\left(\frac{\sqrt{\delta}}{2LC}\tau\right) - RC \, \sin\left(\frac{\sqrt{\delta}}{2LC}\tau\right) \right] \quad \text{for } \delta > 0 \quad (1.36)$$

and

$$K_0(\tau) = \frac{e^{-\frac{R}{2L}\tau}}{L\sqrt{|\delta|}} \left[\sqrt{|\delta|} \, \text{ch}\left(\frac{\sqrt{|\delta|}}{2LC}\tau\right) - RC \, \text{sh}\left(\frac{\sqrt{|\delta|}}{2LC}\tau\right) \right] \quad \text{for } \delta < 0. \quad (1.37)$$

These expressions reduce to $K_0(0) = 1/L$ for $\tau = 0$, in accordance with the physical interpretation of $K_0(0)$ and of the coil L as describing electrical inertia in the usual electromechanical analogy. The high frequencies behaviour of $\chi(z) \sim i/(Lz)$ obtained from the formula (1.33) is thus in agreement with the general asymptotic behaviour (1.14), p. 11.

The reader is encouraged to recover the expressions (1.36) and (1.37) by considering a forcing pulse $U_{\text{pulse}}(t) = F_0 \delta(t - t_0)$. It is then necessary to study the free relaxation of the circuit without forcing, namely the current intensity $I_{\text{pulse}}(t)$ from the initial conditions induced by the pulse, $Q(t_0^+) = 0$ and $I(t_0^+) = (F_0/L)$. The response function is then obtained by writing $K_0(\tau) = F_0^{-1} I_{\text{pulse}}(t_0 + \tau)$.

There is finally a third method to find the expressions (1.36) and (1.37) for $K_0(\tau)$. Indeed, as the susceptibility χ has already been calculated directly, one just

---

[3]This method is recalled in Appendix C, p. 272.

has to take the inverse Laplace transform of $\chi(z)$ as proposed in Exercise 1.2, p. 48.

◇ Interpretation ◇

Although simple, this example already highlights the rich physical content of response functions. In particular it illustrates the general relation, discussed in paragraph 1.1.3 between dissipation and analyticity properties of the susceptibility. Indeed the response function $K_0(\tau)$ decreases exponentially fast at large time when $R \neq 0$. Presence of dissipation for $R \neq 0$, also manifests itself by the need to constantly provide energy to the circuit to prevent current oscillations from damping. Thus the power supplied by the forcing and averaged over a period $T = 2\pi/\omega$,

$$\overline{\mathcal{P}} = \frac{1}{T} \int_0^T dt\ U(t)\ I(t),$$

is equal to

$$\overline{\mathcal{P}} = \frac{\chi'(\omega)}{2}\ |U_\omega|^2. \tag{1.38}$$

This power is positive as expected, since $\chi'(\omega)$ is positive according to formula (1.34).

When $R = 0$, that is to say in the absence of dissipation, we have $\delta = 4LC > 0$ so that expressions (1.36) and (1.33) respectively reduce to

$$K_0(\tau) = \frac{1}{L}\ \cos(\tau/\sqrt{LC}) \qquad \text{and} \qquad \chi(z) = \frac{iCz}{LCz^2 - 1}.$$

So the response function no longer decays to zero at infinity anymore, while the susceptibility is no longer analytic on $\mathbb{R}$ since there are singularities at $\omega = \pm\omega_0$ where $\omega_0 = 1/\sqrt{LC}$ is the LC circuit eigenfrequency. For $\omega \neq \omega_0$, since the real part of the susceptibility $\chi'(\omega)$ vanishes, the power provided $\overline{\mathcal{P}}$ vanishes too, in agreement with the absence of dissipation. The limit behaviour when $\omega \to \pm\omega_0$ must be carefully considered by taking

$$z = \pm\omega_0 + i\epsilon \quad \text{with} \quad \epsilon \to 0^+.$$

We then find in this case

$$\chi(\pm\omega_0 + i\epsilon) \sim \frac{1}{2L\epsilon}.$$

The susceptibility thus diverges while remaining real, meaning that the power $\overline{\mathcal{P}}$ also diverges at frequency ω_0. This corresponds of course to the resonance of the forced LC oscillator.

Let us finally note that the results obtained on this particular example are generic for any dynamical system at a stable fixed point seen as a steady state. When such a system is subjected to a weak external forcing, it remains in the vicinity of the fixed point, and the corresponding evolution equations take a linear form similar to the differential equation (1.32). The susceptibility $\chi(z)$ is then a rational fraction whose poles are located in the lower half complex plane or on the real axis, and controlled by intrinsic relaxation eigenmodes of the fixed point considered.

1.2.2   *Absorption and dispersion in a dielectric*

$\Diamond$ **Presentation** $\Diamond$

A dielectric medium is immersed in an external electric field
$$\mathbf{E}_{\text{ext}}(t) = \text{Re}\big(\mathbf{E}_z e^{-izt}\big),$$
oscillating at the complex frequency z, and whose wavelength is very large compared with the sample size. This condition allows one to neglect all propagation phenomenon. Under the usual assumptions of local equilibrium, the polarisation of microscopic origin $\mathbf{P}(\mathbf{r}, t)$ is simply proportional to the total electric field $\mathbf{E}(\mathbf{r}, t)$, which is the sum of $\mathbf{E}_{\text{ext}}(t)$ and the electric field created by polarisation charges[4]. The polarisation and the local electric field oscillate at the same complex frequency since Maxwell equations are linear. The total field reads $\mathbf{E}(\mathbf{r}, t) = \text{Re}\big(\mathbf{E}_z(\mathbf{r})e^{-izt}\big)$ so that the constitutive relation becomes
$$\mathbf{P}(\mathbf{r}, t) = \epsilon_0 \, \text{Re}\Big([\varepsilon(z) - 1]\, \mathbf{E}_z(\mathbf{r})e^{-izt}\Big), \tag{1.39}$$
where the complex dielectric permittivity $\varepsilon(z)$ is an intrinsic property of the medium. In addition to the frequency z, it depends only on the density, assumed to be homogeneous, and the temperature. In equation (1.39), for a given neighbourhood of an arbitrary point $\mathbf{r}$, the field $\text{Re}(\mathbf{E}_z(\mathbf{r})e^{-izt})$ acts as an external forcing, uniform at the mesoscopic scale of this neighbourhood. Therefore the quantity
$$\chi(z) = \varepsilon(z) - 1$$
is a susceptibility in the general sense.

Let us now consider $z = \omega$ real and denote respectively $\varepsilon'(\omega) - 1$ and $\varepsilon''(\omega)$ the real and imaginary parts of $\varepsilon(\omega) - 1$. From the previous remark, these quantities have the general properties of susceptibilities. Under certain experimental conditions, it may be possible to gain access to $\varepsilon''(\omega)$ without having a direct way of measuring $\varepsilon'(\omega)$. Indeed $\varepsilon''(\omega)$ is related to the dissipation, and therefore to the absorption in the material which can be measured experimentally.

We shall show how the analyticity properties of the susceptibility can in principle allow to determine the real part $\varepsilon'(\omega)$ and then to modelise the full permittivity $\varepsilon(z)$.

$\Diamond$ **Analysis and solution** $\Diamond$

A sufficiently precise experimental determination of $\varepsilon''(\omega)$ gives access in principle to $\varepsilon'(\omega)$ using Kramers-Kronig relation[5]
$$\varepsilon'(\omega) = 1 + \frac{1}{\pi} \, \text{PP} \int_{-\infty}^{\infty} d\omega' \, \frac{\varepsilon''(\omega')}{\omega' - \omega}.$$
Here we will instead model the shape of this function in the neighbourhood of one of these peaks using a simple parametrisation. We detail the principle and then provide some ideas for a more comprehensive modelisation of the whole spectrum.

---

[4]As in the static case the polarisation charges are localised on the sample boundaries if the medium is homogeneous.

[5]This of course assumes that $\varepsilon''(\omega)$ is known on a broad enough range of frequencies.

Modelling of an absorption peak   Let us consider a given peak of $\varepsilon''(\omega)$ in the neighbourhood of a given frequency ω_0 whose typical shape is shown in Figure 1.6.

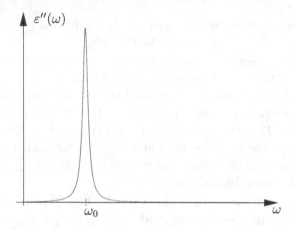

Fig. 1.6   Typical shape of $\varepsilon''(\omega)$ in the neighbourhood of a peak centred in ω_0.

The peak width is small compared to ω_0, and its amplitude is very high compared to other values of $\varepsilon''(\omega)$. It is reasonable to imagine that this peak is the signature on the real axis $z = \omega$ of a singularity of $\chi(z)$ in the lower complex half plane at a point close to ω_0. Suppose that this singularity is a simple pole at $z_p = \omega_p - i\nu_p$ with $\nu_p > 0$, with a residue $-C$. Based on the parity properties of $\chi'(\omega)$ and $\chi''(\omega)$ presented in section 1.1.4, p. 9, there is another simple pole at $\widetilde{z}_p = -\omega_p - i\nu_p$ (see Figure 1.7) with a real and positive residue C. It is then legitimate to write, in

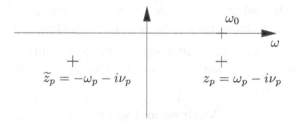

Fig. 1.7   The susceptibility (1.40) has simple poles in z_p and $\widetilde{z}_p$. The fact that the pole z_p is close to ω_0 is the origin of the peak of $\varepsilon''(\omega)$ presented on Figure 1.6.

close neighbourhoods of $\pm\omega_0$:

$$\chi(z) = \frac{C}{z + \omega_p + i\nu_p} - \frac{C}{z - \omega_p + i\nu_p} = \frac{-2C\omega_p}{z^2 + 2i\nu_p z - (\omega_p^2 + \nu_p^2)}. \tag{1.40}$$

Setting $\varepsilon''(\omega) = \operatorname{Im}\chi(\omega)$ we then get

$$\varepsilon''(\omega) = \frac{4C\omega_p\nu_p\omega}{\left[(\omega_p^2 + \nu_p^2 - \omega^2)^2 + 4\nu_p^2\omega^2\right]}. \tag{1.41}$$

The closed form (1.41) depends on three parameters C, ω_p and ν_p. Then suppose that these parameters are well fitted to the peak position, its width and its amplitude measured on the experimental curve. We then find that ν_p is the peak half-width at half height, and then

$$\omega_p = \sqrt{\omega_0^2 - \nu_p^2} \approx \omega_0, \qquad (1.42)$$

and C is set by the peak height. Obviously, the above procedure also provides a simple analytical parametrisation for $\varepsilon'(\omega) = 1 + \operatorname{Re}\chi(\omega)$ in the neighbourhood of the peak under consideration. Plots of $\varepsilon''(\omega)$ and $\varepsilon'(\omega) - 1$ are shown in Figure 1.8.

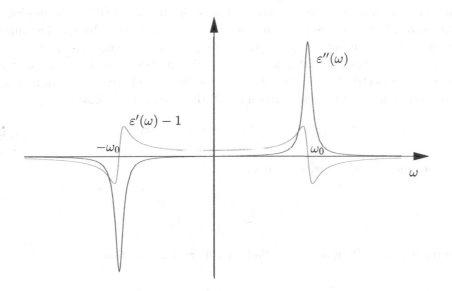

Fig. 1.8 Real and imaginary parts of $\chi(z)$ corresponding to expression (1.40).

Global constraints The overall structure of $\varepsilon''(\omega)$ is in general very rich, showing a large number of peaks. Following the previous strategy, it is then natural to sum the contributions (1.40) corresponding to the various peaks to get $\chi(z)$. It is wise however to add a simple *ad hoc* form, centred on $\omega = 0$, allowing to take into account other experimental measurements or some sum rules described in section 1.1.7. Static value $\varepsilon'(0)$ for instance is nothing but the usual dielectric constant that can be measured by other techniques. We determine the fitting parameters by imposing the corresponding constraints to the empirical form chosen. It is eventually possible to obtain a very reliable representation of the permittivity over a wide frequency range.

$\Diamond$ **Interpretation** $\Diamond$

It turns out that the previous model can be interpreted by a simple phenomenology at the microscopic level. We first consider an absorption line corresponding to an electronic excitation mode in atoms or molecules. We then briefly mention other situations before concluding.

Thomson's model for atomic electrons   Let us consider a dielectric medium composed of atoms, with a number density n, and an absorption peak corresponding to an electronic transition at frequency ω_0. We then assume that the polarisation is the sum of each atom contributions, which is better justified at low density. Consider moreover that the involved transition is characterised by the motion of a single so-called excitable electron, whose position is denoted by $\mathbf{r}_e$. The dipole carried by the atom is then $\mathbf{p} = q\mathbf{r}_e$ where q is the electron charge. The total polarisation can then be written as $\mathbf{P} = nq\mathbf{r}_e$. When $\mathbf{P}$ oscillates at frequency z, so does $\mathbf{r}_e$. The electric field is almost uniform at the atomic scale. Relation (1.39), expression (1.40) for $\chi(z)$ and relation (1.42) then allow us to show that

$$\left(-z^2 - 2i\nu_p z + \omega_0^2\right)\mathbf{r}_{ez}e^{-izt} = \frac{q}{m_{\mathrm{exc}}}\mathbf{E}_z e^{-izt}, \tag{1.43}$$

where we introduced the effective mass

$$m_{\mathrm{exc}} = \frac{nq^2}{2\epsilon_0 C\sqrt{\omega_0^2 - \nu_p^2}}.$$

Returning to $\mathbf{r}_e = \mathrm{Re}[\mathbf{r}_{ez}e^{-izt}]$ we find the differential equation

$$\frac{d^2\mathbf{r}_e}{dt^2} + 2\nu_p \frac{d\mathbf{r}_e}{dt} + \omega_0^2 \mathbf{r}_e = \frac{q}{m_{\mathrm{exc}}}\mathbf{E}(t). \tag{1.44}$$

In other words, everything happens as if the excitable electron were a classical particle of charge q and mass m_{exc} in the electric field $\mathbf{E}(t)$, also submitted to the restoring force of a spring of stiffness $m_{\mathrm{exc}}\omega_0^2$ and a viscous force $-2m_{\mathrm{exc}}\nu_p d\mathbf{r}_e/dt$:   this is the well-known Thomson's model for an atomic electron.

✠ **Comment:**   Although this differential equation is the same as the one presented on p. 19, we must note that the corresponding susceptibilities (1.40) and (1.33) have different behaviours, in particular in the limit $z \to \infty$. This is explained by the fact that the admittance (1.33) is analogous to the susceptibility associated with speed $(d\mathbf{r}/dt)$ while we consider here the susceptibility associated with the position. Inertia implies then that the admittance is proportional to $1/z$ while $\chi(z)$ varies as $1/z^2$ at high frequency.

Extensions At low density, typically in the gas phase, it is justified to consider the atoms as independent and the previous phenomenology gives access to the energy and the life time of the excited electronic level. This is illustrated by the quantum study of the polarisability of the hydrogen atom in section 3.2.6. At higher density and at the level of Clausius-Mossotti approximation, the experimental determination of ω_0 and ν_p provides estimates for the displacement and the broadening of atomic transition by collective effects for instance[6].

When the medium is composed of molecules, absorption peaks also appear in $\varepsilon''(\omega)$ corresponding to the excitation of rotational and vibrational modes. We can then proceed with a similar phenomenology, providing interesting information about the characteristics of these modes.

In conclusion this example illustrates the usefulness and the consequences of analyticity properties of the susceptibility. It provides simple and accurate parametrisations, while paving the way for a reliable phenomenology of complex microscopic effects.

1.2.3   *Oscillating flow in a capillary*

◇ **Presentation** ◇

Consider a capillary containing an incompressible viscous fluid, and excited at one of its ends by a vibrating membrane. We do not take into account any cavitation phenomenon, which is justified at low enough frequency. We also assume that the induced oscillating flow is laminar. This is valid if the Reynolds number is not too large, ensuring the absence of turbulence [28]. The capillary is a cylinder of radius R and length L. The fluid is at rest in the absence of external excitation and characterised by a constant pressure P_0, a constant density ρ, and a zero velocity $\mathbf{v} = \mathbf{0}$: this defines the steady state $\mathcal{E}_0$. The vibrating membrane then imposes a pressure $P_0 + P = \delta P_z e^{-izt}$ at one end, taken at $x = 0$, while the pressure at the other end $x = L$ is constant and equal to P_0 as shown in Figure 1.9. Here it is the pressure $\delta P_z e^{-izt}$ which plays the role of the forcing $F(t)$, and we will study the response of the fluid flow rate $Q(t)$, taken as dynamical observable $A(t)$.

◇ **Analysis and solution** ◇

In order to calculate the flow rate we first determine the flow velocity and briefly describe its characteristics. We then deduce the susceptibility of interest, and review its analytic properties. In order to simplify notation we do not explicitly denote the real parts that must be taken to obtain the physical quantities of interest.

---

[6]The Clausius-Mossotti relation relates the atomic susceptibility to the polarisability for a dense liquid phase as discussed in the book [36].

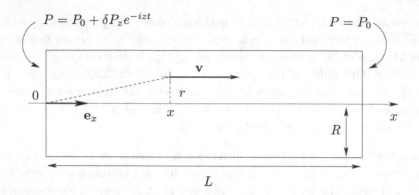

Fig. 1.9 A capillary of length L is subjected to a vibrating membrane which imposes an oscillating pressure at the end $x = 0$.

Determination of the velocity field Taking into account the cylindrical symmetry, the speed depends only on t, x and $r = \sqrt{y^2 + z^2}$. Furthermore, it is assumed to be everywhere directed along the Ox axis, i.e. $\mathbf{v} = v_x(x, r, t)\mathbf{e}_x$ with the boundary condition $\mathbf{v} = 0$ at $r = R$, as a consequence of the viscous nature of the fluid. Since the latter is incompressible, the equation of conservation of matter gives

$$\frac{\partial v_x}{\partial x} = 0,$$

leading to $v_x = v_x(r, t)$. Navier-Stokes equation,

$$\partial_t \mathbf{v} + (\mathbf{v} \cdot \boldsymbol{\nabla})\mathbf{v} = -\frac{1}{\rho}\boldsymbol{\nabla}P + \nu \Delta \mathbf{v},$$

where ν is the kinematic viscosity, then leads for each component to

$$\frac{\partial v_x}{\partial t} = -\frac{1}{\rho}\frac{\partial P}{\partial x} + \nu \left(\frac{\partial^2 v_x}{\partial y^2} + \frac{\partial^2 v_x}{\partial z^2} \right), \tag{1.45}$$

$$0 = -\frac{1}{\rho}\frac{\partial P}{\partial y} \quad \text{and} \quad 0 = -\frac{1}{\rho}\frac{\partial P}{\partial z}. \tag{1.46}$$

The last two equations (1.46) show that the pressure P depends only on x and t, $P = P(x, t)$. Since v_x depends only on r and t, equation (1.45) then implies that $\partial P/\partial x$ depends only on time t, so that taking into account the boundary conditions on pressure one gets

$$\boxed{P(x, t) = P_0 + \delta P_z e^{-izt}\,\frac{(L - x)}{L}.}$$

The velocity necessarily oscillates at the same frequency z as the overpressure, since the above equations are linear. Writing $v_x(r, t) = \phi(r)e^{-izt}$, equation (1.45) becomes

$$\frac{\mathrm{d}^2\phi}{\mathrm{d}r^2} + \frac{1}{r}\frac{\mathrm{d}\phi}{\mathrm{d}r} + \frac{iz}{\nu}\phi = -\frac{\delta P_z}{\eta L}, \tag{1.47}$$

where $\eta = \nu\rho$ is the dynamic viscosity.

The ordinary differential equation (1.47) for ϕ, with the boundary condition $\phi(R) = 0$ can easily be solved as follows. The constant function

$$\frac{i\delta P_z}{z\rho L} \tag{1.48}$$

is clearly a particular solution. The homogeneous equation without right-hand side is written as

$$\phi'' + \frac{1}{\xi}\phi' + \phi = 0, \tag{1.49}$$

where $\phi' = \mathrm{d}\phi/\mathrm{d}\xi$ and ξ is the dimensionless quantity

$$\xi = \left(\frac{iz}{\nu}\right)^{\frac{1}{2}} r.$$

The analytic function $(iz/\nu)^{\frac{1}{2}}$ is defined by the choice of determination

$$(iz/\nu)^{\frac{1}{2}} = \sqrt{|z|/\nu} \; e^{i(\pi/4 + \arg(z)/2)},$$

with a branch cut on the negative imaginary half-axis starting at the branch point $z = 0$. The introduction of this branch cut is a necessary calculus trick which has no effect on the final result, as we shall later check. The homogeneous equation (1.49) is of second order and therefore admits two independent solutions. One of them is the Bessel function of order zero of the first kind J_0,

$$\phi_1(\xi) = J_0(\xi) = \frac{1}{2\pi}\int_0^{2\pi} \mathrm{d}\theta\, e^{i\xi\cos\theta}.$$

The reader can directly check this result by differentiation under the integral sign and integration by parts. To determine another independent solution ϕ_2, we use the method of the Wronskian $W = \phi_1'\phi_2 - \phi_2'\phi_1$. The latter is a solution of the differential equation

$$W' + \frac{1}{\xi}W = 0,$$

giving $W(\xi) = 1/\xi$ up to a multiplicative constant. Thus, ϕ_2 is a solution of the first order differential equation

$$J_0'\phi_2 - J_0\phi_2' = \frac{1}{\xi}, \tag{1.50}$$

which can be solved[7] by the usual method of variation of constants. It turns out that $\phi_2(\xi)$ presents a logarithmic singularity at $\xi = 0$, as shown by simple inspection of equation (1.50) in the limit $\xi \to 0$, with $J_0(\xi) \to 1$ and $J_0'(\xi) \to 0$. Since the fluid cannot have an infinite velocity at the centre of the capillary, the second solution of the homogeneous equation must be disregarded. The physical solution

---

[7]In fact, ϕ_2 is proportional to the so-called Neumann function N_0 or Bessel function of the second kind of order zero. We do not need here the properties of this function, but the interested reader can refer to the books listed in the bibliography of this chapter.

of the problem therefore reduces to the sum of the particular solution (1.48) and a constant times $J_0(\xi)$. This constant is determined by imposing that ϕ vanishes on the capillary boundary, finally giving

$$\mathbf{v}(r,t) = \phi(r)e^{-izt}\mathbf{e}_x = \frac{i\delta P_z}{z\rho L}\left(1 - \frac{J_0\left[\left(\frac{iz}{\nu}\right)^{1/2}r\right]}{J_0\left[\left(\frac{iz}{\nu}\right)^{1/2}R\right]}\right)e^{-izt}\mathbf{e}_x. \tag{1.51}$$

Velocity profiles   Before determining the flow, let us study the velocity profiles corresponding to the result (1.51). It is natural to define the cutoff frequency $\omega_c = \nu/R^2$ setting the limit between the viscous regime ($\omega \ll \omega_c$) and the inertial regime ($\omega \gg \omega_c$).

At high frequency $\omega \gg \omega_c$, we have

$$J_0(e^{i\frac{\pi}{4}}u) \simeq \sqrt{\frac{2}{\pi u}}\exp\left(\frac{i\pi}{8} - \frac{i}{\sqrt{2}}u + \frac{u}{\sqrt{2}}\right) \qquad \text{when } u \to +\infty,$$

so that the velocity reads

$$v_x(r,t) \simeq \frac{\delta P_z}{\omega\rho L}\left[\sin(\omega t) - \sqrt{\frac{r}{R}}\exp\left(-\frac{R-r}{\delta}\right)\sin\left(\omega t - \frac{R-r}{\delta}\right)\right]. \tag{1.52}$$

We have introduced the charactric depth $\delta = R\sqrt{2\omega_c/\omega} = \sqrt{2\nu/\omega}$ of the boundary layer. Note that $\delta \ll R$ since the frequency is high. Let us fix a time t and start from the edge of capillary. At $r = R$ the velocity vanishes as imposed by the viscosity. It increases for $R - r \ll \delta$ away from the boundary. The velocity is essentially constant beyond the distance δ. The velocity profile exhibits fingers localised near the boundaries as shown in Figure 1.10. Note that highly oscillatory velocity profiles can appear at certain times. Indeed, for $\omega t \equiv 0\,[\pi]$, the velocity is proportional to $\sin[(R-r)/\delta]$ which varies very rapidly in space. Such a profile is plotted in Figure 1.11.

At low frequency, $\omega \ll \omega_c$, the velocity profile can easily be obtained from the series expansion of the Bessel function

$$J_0(\xi) = 1 + \sum_{p=1}^{\infty}\frac{(-1)^p}{(p!)^2}\left(\frac{\xi}{2}\right)^{2p}. \tag{1.53}$$

Viscous effects then essentially determine the profile. In particular the velocity profile is parabolic at zero frequency,

$$v_x(r,t) = \frac{\delta P_z}{2\rho L\omega_c}\left[1 - (r/R)^2\right].$$

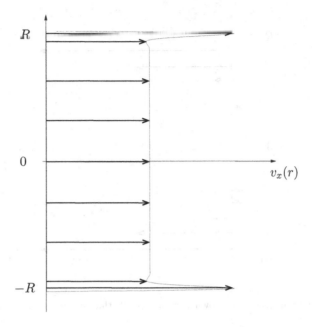

Fig. 1.10   Velocity profile (1.51) at high frequency $\omega \gg \omega_c$.

Flow rate and susceptibility   The flow rate induced through a section S of the capillary is

$$Q(t) = \rho \int_S d\Sigma \, \mathbf{n} \cdot \mathbf{v} = 2\pi\rho \int_0^R dr \, r \, \phi(r) \, e^{-izt}. \qquad (1.54)$$

Using expression (1.51) for $\phi(r)$, we find that the flow rate response is proportional to the forcing pressure,

$$Q(t) = \chi(z)\delta P e^{-izt},$$

via the susceptibility[8]

$$\boxed{\chi(z) = \frac{i\pi R^2}{Lz}\left(1 - \frac{2J_1\left[\left(\frac{iz}{\nu}\right)^{1/2}R\right]}{\left(\frac{iz}{\nu}\right)^{1/2}RJ_0\left[\left(\frac{iz}{\nu}\right)^{1/2}R\right]}\right)} \qquad (1.55)$$

where J_1 is the Bessel function of the first kind of order 1 defined by

$$J_1(\xi) = \frac{e^{-i\frac{\pi}{2}}}{2\pi}\int_0^{2\pi} d\theta \, e^{i\xi\cos\theta + i\theta}.$$

Properties of this susceptibility will now be studied in detail.

---

[8]The explicit calculation of the low rate is based on the identity $\xi J_0(\xi) = J_1(\xi) + \xi J_1'(\xi)$. The reader amateur of Bessel functions can demonstrate using the integral representations of J_0 and J_1.

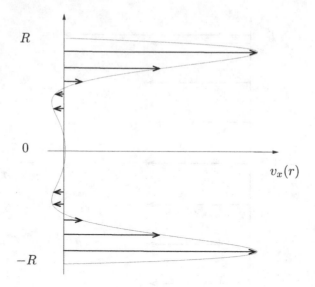

Fig. 1.11 At certain times, as shown here at $t = 0$, the velocity profile (1.51) can be highly oscillatory.

Analyticity of the susceptibility Since Bessel functions J_0 and J_1 are entire functions, i.e. analytic on $\mathbb{C}$, $\chi(z)$ is analytic in $\mathbb{C}$ by the composition theorem, except maybe:

(i) On the branch cut that defines $(iz/\nu)^{1/2}$. In fact, when one turns by 2π around $z = 0$ between $-(\pi/2)^+$ and $(3\pi/2)^-$, $(iz/\nu)^{1/2}$ changes its sign. Since J_1 is odd and J_0 even, $\chi(z)$ is continuous at the cut. The same is true for all derivatives of $\chi(z)$ which therefore remains analytic on the branch cut of auxilliary function $(iz/\nu)^{1/2}$.

(ii) At $z = 0$. Using the power series expansion (1.53) of $J_0(\xi)$ and that for $J_1(\xi) = -J_0'(\xi)$, it is then easy to show that we have at $z = 0$

$$\chi(z) = \frac{\pi R^4}{8\nu L} + \text{Power series in } z = \frac{\pi R^4}{8\nu L} + O(z), \qquad (1.56)$$

where this series admits a finite radius of convergence. This shows that $\chi(z)$ remains analytic at $z = 0$ as well.

(iii) At points where $J_0[(iz/\nu)^{1/2}R]$ vanishes. The zeros of $J_0(\xi)$, denoted ξ_n with $n \in Z^*$, are located on the real axis[9]. These zeros give rise to simple poles of $\chi(z)$ of the form

$$\boxed{z_n = -i\frac{\nu}{R^2}\xi_n^2 = -i\omega_c\xi_n^2.} \qquad (1.57)$$

---

[9]For large n it is possible to derive the asymptotic formula $\xi_n = n\pi - \pi/4 + O(1/n)$, from the result of Exercise 4.1, p. 256.

All poles of the susceptibility are on the negative imaginary axis.

Finally, $\chi(z)$ is analytic on $\mathbb{C}$, except at the points z_n of the lower half-plane which are simple poles.

For real frequencies $z = \omega$, $\chi'(\omega)$ and $\chi''(\omega)$ are obtained by taking the real and imaginary parts of expression (1.55). At low frequencies $\omega \ll \omega_c$ one gets

$$\frac{\chi'(\omega)}{\chi(0)} = 1 - \frac{11}{384}(\omega/\omega_c)^2 + O\big((\omega/\omega_c)^4\big)$$

$$\frac{\chi''(\omega)}{\chi(0)} = \frac{1}{6}(\omega/\omega_c) + O\big((\omega/\omega_c)^3\big)$$

with static susceptibility $\chi(0) = \pi R^4/(8\nu L)$, which gives back Poiseuille's formula $Q = \big(\pi R^4/(8\nu L)\big)\,\delta P_z$. Asymptotic behaviours at high frequencies $\omega \gg \omega_c$ are obtained using $J_1[(i\omega/\nu)^{1/2}R] \simeq iJ_0[(i\omega/\nu)^{1/2}R]$, i.e.

$$\frac{\chi'(\omega)}{\chi(0)} \simeq 8\sqrt{2}(\omega_c/\omega)^{\frac{3}{2}} \qquad \text{and} \qquad \frac{\chi''(\omega)}{\chi(0)} \simeq 8(\omega_c/\omega).$$

We plotted the variations of $\chi'(\omega)$ and $\chi''(\omega)$ in Figure 1.12.

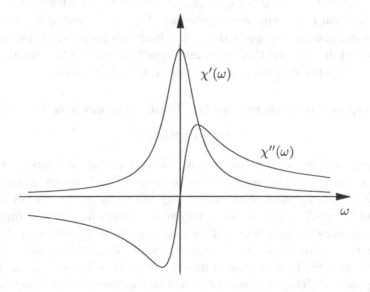

Fig. 1.12   Real and imaginary parts of the susceptibility $\chi(\omega)$ given by (1.55).

◊ **Interpretation** ◊

In this example from fluid mechanics and involving an infinite number of degrees of freedom, the susceptibility is no longer a rational fraction as in systems such as the RLC circuit seen in section 1.2.1. However the general properties of susceptibilities are still present, and find here their physical origin in the dissipation due to the viscosity. In particular, $\chi(z)$ is analytic in the complex upper half-plane including the real axis. Its singularities are infinitely many simple poles z_n located in the lower complex half-plane. Their positions are controlled by the viscosity, so that they accumulate at $z = 0$ when viscosity vanishes.

As for the RLC circuit, the presence of dissipation is also apparent by the need to provide energy to the fluid in order to maintain the oscillations. The average power $\overline{\mathcal{P}}$ thus provided by the membrane coincides with the power of the pressure force,

$$\overline{\mathcal{P}} = \frac{1}{T} \int_0^T \mathrm{d}t \int_S \mathrm{d}\Sigma \, P(t) \, \mathbf{n} \cdot \mathbf{v}(r, t) = \frac{1}{T} \int_0^T \mathrm{d}t \, P(t) \, Q(t),$$

with $T = 2\pi/\omega$. To compute this time integral we have to work with real parts of the complex expressions because the integrant is not a linear form on the quantities involved. An elementary calculation gives

$$\overline{\mathcal{P}} = \frac{\chi'(\omega)}{2\rho} |\delta P_z|^2,$$

since only the part $\chi'(\omega) P(t)$ of $Q(t)$ in phase with $P(t)$ contributes to the time average. Low and high frequency expansions of $\chi'(\omega)$ as well as its numeric representation in Figure 1.12 suggest that $\chi'(\omega)$ is always positive, in agreement with the presence of dissipation. However a mathematical proof of this positivity is not straightforward, since it involves inequalities on Bessel functions.

1.2.4   *Response of a plasma within Vlasov approximation*

◊ **Introduction** ◊

This example focuses on the study of a model of plasma consisting of a single species of mobile point charges, of charge q, mass m and density ρ, immersed in a rigid uniform background of charge density $-q\rho$ ensuring the overall neutrality. This model is called jellium or one component plasma in the literature. It can reasonably describe many physical situations, such as the conduction electrons in a metal or the carbon nuclei in the core of a white dwarf. Here we shall assume that the charges are classical charges interacting with each other by the two-body Coulomb potential. They are also submitted to the electrostatic potential created by the rigid bath. Suppose that the plasma is originally at thermal equilibrium. We then propose to determine the response of the internal charge density $Q(\mathbf{r}, t)$ to an external electric potential of the form $\phi_{\text{ext}}(\mathbf{r}, t) = \text{Re}[\phi_z \exp(i\mathbf{k} \cdot \mathbf{r} - izt)]$ neglecting any phenomenon of electromagnetic propagation, so this potential is created by an external charge distribution given by Poisson equation, i.e.

$$Q_{\text{ext}}(\mathbf{r}, t) = -\epsilon_0 \mathbf{k}^2 \, \text{Re}[\phi_z \exp(i\mathbf{k} \cdot \mathbf{r} - izt)].$$

Note that this distribution is present inside the system and corresponds to a charge density wave. Using results from the general analysis, the plasma response must take the form

$$Q(\mathbf{r}, t) = \text{Re}\left[\chi(\mathbf{k}, z) \, \phi_z e^{i\mathbf{k} \cdot \mathbf{r} - izt}\right], \tag{1.58}$$

with a susceptibility $\chi(\mathbf{k}, z)$ depending also on the wave number $\mathbf{k}$.

Determining the plasma response falls within the general problem of Hamiltonian systems at thermal equilibrium, excited by an external perturbation. For this class of systems, susceptibility $\chi(\mathbf{k}, z)$ can be expressed in terms of dynamical correlation functions at equilibrium[10]. Here we compute $\chi(\mathbf{k}, z)$ in the framework of a mean field theory. This approach captures several essential physical features. It allows in addition to avoid the complexity of the N-body problem. Indeed the plasma state is then fully described by the distribution function $f(\mathbf{r}, \mathbf{p}, t)$ in the phase space of a single particle of position $\mathbf{r}$ and impulsion $\mathbf{p}$. This function corresponds to a probability density in phase space. The charge density reads

$$Q(\mathbf{r}, t) = q\left[-\rho + \int d\mathbf{p} f(\mathbf{r}, \mathbf{p}, t)\right]. \tag{1.59}$$

The steady-state $\mathcal{E}_0$ at temperature T and density ρ is characterised by equilibrium distribution f_0 written as

$$f_0(\mathbf{r}, \mathbf{p}) = f_0(\mathbf{p}) = \rho\left(\frac{\beta}{2\pi m}\right)^{3/2} \exp(-\beta \frac{\mathbf{p}^2}{2m}), \tag{1.60}$$

with $\beta = 1/(k_B T)$. This distribution brings up the Maxwell-Boltzmann distribution of velocity $\mathbf{v} = \mathbf{p}/m$, which takes the well-known Gaussian shape. The charge density (1.59) is identically zero in the state $\mathcal{E}_0$ for $f_0(\mathbf{p})$. In the general case, it can be rewritten as

$$Q(\mathbf{r}, t) = q \int d\mathbf{p}\left[f(\mathbf{r}, \mathbf{p}, t) - f_0(\mathbf{r}, \mathbf{p})\right]. \tag{1.61}$$

We assume throughout this study that the plasma is infinitely extended so that there is no boundary effect on the quantities under consideration.

◊ Analysis and solution ◊

We first introduce the Vlasov equation governing the evolution of $f(\mathbf{r}, \mathbf{p}, t)$ under the effect of the Coulomb interaction and the force field derived from the potential $\phi_{\text{ext}}(\mathbf{r}, t)$. We then determine the difference $f(\mathbf{r}, \mathbf{p}, t) - f_0(\mathbf{p})$ to first order in ϕ_{ext} and deduce the susceptibility $\chi(\mathbf{k}, z)$.

---

[10]The corresponding formulas are established in the example presented in section 1.2.5 dealing with the conductivity of a material.

Vlasov equation Let us introduce the equation of evolution for the distribution $f(\mathbf{r}, \mathbf{p}, t)$ of a particle submitted to a force $\mathbf{F}(\mathbf{r}, t)$ depending only on position and time. This equation expresses the conservation of the total number of states for the particle in phase space. It is obtained from the usual conservation equation

$$\frac{\partial f}{\partial t} + \operatorname{div} \mathbf{J} = 0,$$

where

$$\mathbf{J} = (f \ (\mathrm{d}\mathbf{r}/\mathrm{d}t), f \ (\mathrm{d}\mathbf{p}/\mathrm{d}t)) = (f \ \mathbf{p}/m, f \ \mathbf{F}(\mathbf{r}, t))$$

is the current density in the 6-dimension space $(\mathbf{r}, \mathbf{p})$, div is the divergence operator in this space, and where we used the equations of motion for $\mathbf{r}$ and $\mathbf{p}$. This equation takes the form

$$\frac{\partial f}{\partial t} + \frac{\mathbf{p}}{m} \cdot \frac{\partial f}{\partial \mathbf{r}} + \mathbf{F}(\mathbf{r}, t) \cdot \frac{\partial f}{\partial \mathbf{p}} = 0. \tag{1.62}$$

Here the force $\mathbf{F}(\mathbf{r}, t)$ includes not only the external force $-q\partial\phi_{ext}/\partial\mathbf{r}$, but also the electrostatic force associated with the electric field created by the internal charge distribution $Q(\mathbf{r}, t)$. The evolution equation (1.62) then takes the so-called Vlasov form

$$
\begin{aligned}
\frac{\partial f}{\partial t} + \frac{\mathbf{p}}{m} \cdot \frac{\partial f}{\partial \mathbf{r}} &+ \left[\int \mathrm{d}\mathbf{r}' \frac{q}{4\pi\epsilon_0} \frac{(\mathbf{r} - \mathbf{r}')}{|\mathbf{r} - \mathbf{r}'|^3} Q(\mathbf{r}', t) \right] \cdot \frac{\partial f}{\partial \mathbf{p}} \\
&= \frac{\partial f}{\partial \mathbf{p}} \cdot \operatorname{Re}\left(iq\mathbf{k} \ \phi_z e^{i\mathbf{k}\cdot r - izt} \right).
\end{aligned}
\tag{1.63}
$$

Note that in the last term of the left side of this equation, $Q(\mathbf{r}', t)$ also depends on the distribution of f via (1.61). This is a mean-field type term because it corresponds to the approximate expression for the total force exerted by all other particles, obtained by neglecting their correlations with the particle under consideration✠.

✠ **Comment:** The Vlasov equation can be established within the framework of the N-body problem. One must then start from the Liouville equation in the total phase space of $6N$ dimensions. As explained in [31, 73], integration over all positions and momenta of $N - 1$ particles leads to the first equation of the Bogoliubov-Born-Green-Kirkwood-Yvon (BBGKY) hierarchy connecting the one and two particles distributions $f^{(1)}$ and $f^{(2)}$. The Vlasov equation is then obtained via the factorisation $f^{(2)} \to f^{(1)} f^{(1)}$.

First order induced distribution We introduce the difference

$$\delta f = f - f_0.$$

Once this difference is assumed to be small, the next step is to linearise the Vlasov equation (1.63), retaining only the first order terms in the perturbation. Obviously f_0 is a solution of the equation without second member. In addition, $Q(\mathbf{r}', t)$ is at least of first order since the plasma is locally neutral in state $\mathcal{E}_0$. Then we can replace $\partial f/\partial \mathbf{p}$ by $\partial f_0/\partial \mathbf{p} = -\beta \mathbf{p} f_0/m$ both in the mean field term and the external forcing. The linearised Vlasov equation is then written as

$$\frac{\partial}{\partial t}\,\delta f + \frac{\mathbf{p}}{m}\cdot\frac{\partial}{\partial \mathbf{r}}\,\delta f - \beta f_0\,\frac{\mathbf{p}}{m}\cdot\int d\mathbf{r}'\,\frac{q^2}{4\pi\epsilon_0}\frac{(\mathbf{r}-\mathbf{r}')}{|\mathbf{r}-\mathbf{r}'|^3}\int d\mathbf{p}'\,\delta f(\mathbf{r}',\mathbf{p}',t)$$
$$= -i\beta q\mathbf{k}\cdot\frac{\mathbf{p}}{m}f_0\phi_z e^{i\mathbf{k}\cdot\mathbf{r}-izt}. \quad (1.64)$$

From now on we omit to explicitly indicate real parts unless it is absolutely necessary.

Given the initial conditions at $t_0 = -\infty$ on the one hand and the linearity of evolution equation (1.64) on the other hand, δf oscillates in time at the same complex frequency z as ϕ_{ext}. In addition, in an infinite volume, translation invariance leads to seek δf of the form of a plane wave

$$\delta f(\mathbf{r}, \mathbf{p}, t) = \psi(\mathbf{p})\phi_z e^{i\mathbf{k}\cdot\mathbf{r}-izt},$$

similar to ϕ_{ext}. Moreover, the Fourier transform of $q^2(\mathbf{r}-\mathbf{r}')/(4\pi\epsilon_0|\mathbf{r}-\mathbf{r}'|^3)$ is equal to[11] $-iq^2\mathbf{k}/(\epsilon_0 k^2)$. One then obtains

$$(-iz + i\mathbf{k}\cdot\frac{\mathbf{p}}{m})\,\psi(\mathbf{p}) + i\beta q^2\frac{\mathbf{p}}{m}\cdot\frac{\mathbf{k}}{\epsilon_0 k^2}\,f_0(\mathbf{p})\int d\mathbf{p}'\,\psi(\mathbf{p}') = -i\beta q\mathbf{k}\cdot\frac{\mathbf{p}}{m}f_0(\mathbf{p}),$$

or in other words

$$\psi(\mathbf{p}) = \psi_{\text{id}}(\mathbf{p})\left[1 + \frac{q}{\epsilon_0 k^2}\int d\mathbf{p}'\,\psi(\mathbf{p}')\right], \quad (1.65)$$

where

$$\psi_{\text{id}}(\mathbf{p}) = \frac{\beta q\,\mathbf{k}\cdot\mathbf{p}}{(mz - \mathbf{k}\cdot\mathbf{p})}\,f_0(\mathbf{p}) \quad (1.66)$$

is the form of ψ for an ideal gas without Coulomb interactions between the particles. A term by term integration of the integral equation (1.65) for ψ readily gives the constant $\int d\mathbf{p}\psi(\mathbf{p})$ in terms of ψ_{id}. We finally get

$$\delta f(\mathbf{r}, \mathbf{p}, t) = \frac{\psi_{\text{id}}(\mathbf{p})}{1 - (q/\epsilon_0 k^2)\int d\mathbf{p}\psi_{\text{id}}(\mathbf{p})}\,\phi_z e^{i\mathbf{k}\cdot\mathbf{r}-izt}. \quad (1.67)$$

---

[11]This result amounts to calculate the Fourier transform of the gradient of the Coulomb potential, which can be readily obtained by multiplying the transform of the potential by $-i\mathbf{k}$. The reader can then see Chapter 2, p. 68 where the latter transform is computed.

Susceptibility and analyticity properties   One just has to insert expression (1.67) for δf in integral (1.61) for Q to obtain the following expression for the susceptibility

$$\boxed{\chi(\mathbf{k}, z) = \frac{\chi_{\text{id}}(\mathbf{k}, z)}{1 - \chi_{\text{id}}(\mathbf{k}, z)/(\epsilon_0 k^2)}.}\tag{1.68}$$

In this expression $\chi_{\text{id}}(\mathbf{k}, z) = q \int d\mathbf{p}\, \psi_{\text{id}}(\mathbf{p})$ is the susceptibility of an ideal gas,

$$\boxed{\chi_{\text{id}}(k, z) = \beta q^2 \rho (2\pi m k_B T)^{-\frac{1}{2}} \int_{-\infty}^{+\infty} dp_x e^{-\beta \frac{p_x^2}{2m}} \frac{k p_x}{mz - k p_x},}\tag{1.69}$$

where we used expression (1.60) for f_0 and rotation invariance.

For $\mathbf{k}$ fixed the ideal susceptibility χ_{id} is analytic with respect to z in the upper half complex plane, as shown by differentiating the integral representation (1.69) under the integral sign. We conclude that susceptibility χ is analytic in this half-plane by virtue of the theorem of composition on the one hand and because the denominator $1 - \chi_{\text{id}}(k, z)/(\epsilon_0 k^2)$ is always finite since χ_{id} has a non-zero imaginary part. This last property can be established by deriving the integral representation of imaginary part χ_{id}'' from formula (1.69): It does not vanish for $\text{Im}\, z > 0$.

In order to obtain $\chi(k, \omega)$ on the real axis one has to evaluate $\chi(k, z)$ with $z = \omega + i\epsilon$ in the limit $\epsilon \to 0^+$. To do so we can use the standard identity (1.22) seen on page p. 14, allowing to extract the real and imaginary parts of the integral representation (1.69) of $\chi_{\text{id}}(k, \omega)$, that is to say

$$\chi_{\text{id}}'(k, \omega) = -\beta q^2 \rho \left(\frac{\beta m}{2\pi}\right)^{1/2} \text{PP} \int_{-\infty}^{+\infty} dv \frac{v\, e^{-\frac{\beta m v^2}{2}}}{v - \omega/k}\tag{1.70}$$

and

$$\chi_{\text{id}}''(k, \omega) = -\beta q^2 \rho \left(\frac{\pi \beta m}{2}\right)^{1/2} \frac{\omega}{k}\, \exp\left[-\frac{\beta m \omega^2}{2k^2}\right].\tag{1.71}$$

The corresponding expression for $\chi(k, \omega)$ follows immediately by replacing $\chi_{\text{id}}(k, \omega)$ by $\chi_{\text{id}}'(k, \omega) + i\chi_{\text{id}}''(k, \omega)$ in expression (1.68). An important property satisfied by $\chi(k, \omega)$ is that $\chi(k, \omega)$ remains analytic in ω on the real axis for $k \neq 0$. Singularities may appear at $k = 0$ as discussed below.

High and low frequencies   The asymptotic behaviour of $\chi_{\text{id}}'(k, \omega)$ at high frequencies is readily obtained by replacing $1/(v - \omega/k)$ by its expansion $-(k/\omega)(1 + kv/\omega + ...)$ in the integrand of formula (1.70), leading to

$$\chi'(k, \omega) \simeq \frac{q^2 \rho k^2}{m\omega^2}.\tag{1.72}$$

Moreover $\chi_{\text{id}}''(k, \omega)$ decreases as fast as a Gaussian when $\omega \to \infty$. From formula (1.68), we deduce that $\chi(k, \omega)$ behaves as $\chi_{\text{id}}(k, \omega)$ at high frequency. Interactions become negligible compared to pure inertia effects in this limit, in agreement with the general discussion p. 12.

Note that the Gaussian decay of $\chi''(k, \omega)$ at high frequencies is faster than any inverse power of ω and comes from the behaviour of the response function $K_0(\tau)$ at small τ.

The corresponding development of $K_0(\tau)$ contains indeed ✠ only odd powers of τ as a result of time-reversal invariance of the Hamiltonian of the system under consideration. So according to the general analysis p. 11, all coefficients in the expansion of $\chi''(k, \omega)$ in inverse powers of ω vanish.

> ✠ **Comment:** By comparison there is a dissipative force in Thomson's model seen in section 1.2.2, so that the system is not invariant under transformation $t \to -t$. The expansion of $K_0(\tau)$ for small τ contains all the powers of τ, and the susceptibility eventually has an algebraic decay.

At low frequencies, $\chi''_{\mathrm{id}}(k, \omega)$ vanishes as ω while $\chi'_{\mathrm{id}}(k, \omega)$ tends to a finite value. We then obtain the static value

$$\chi(k, 0) = \frac{-\beta q^2 \rho}{1 + \beta q^2 \rho / (\varepsilon_0 k^2)}. \tag{1.73}$$

It differs from the ideal value $-\beta q^2 \rho$ by a factor due to interactions.

◊ **Interpretation** ◊

Analytic properties of $\chi(k, z)$ have several important physical consequences. We first study the screening and then interpret energy absorption in this conservative system, as well as the onset of plasmon modes.

Screening The total potential ϕ_{tot} is the sum of the external potential ϕ_{ext} and the induced potential ϕ_{ind} created by the charge distribution $Q(\mathbf{r}, t)$. The dielectric permittivity $\varepsilon(\mathbf{k}, z)$ is defined through the usual relation

$$\phi_{\mathrm{tot}} = \frac{\phi_{\mathrm{ext}}}{\varepsilon(\mathbf{k}, z)}.$$

Expressing ϕ_{ind} in terms of susceptibility χ and ϕ_{ext}, we obtain a simple relation between $\varepsilon(k, z)$ and $\chi(k, z)$. Given the expression (1.68) for $\chi(k, z)$ we eventually find, in the Vlasov approximation,

$$\varepsilon(k, z) = 1 - \frac{\chi_{\mathrm{id}}(k, z)}{\epsilon_0 k^2}. \tag{1.74}$$

In the static limit at zero frequency $\chi_{\mathrm{id}}(k, 0)$ reduces to $-\beta q^2 \rho$ so that

$$\varepsilon(k, 0) = 1 + \frac{\beta q^2 \rho}{\epsilon_0 k^2}. \tag{1.75}$$

When $k \to 0$, $\varepsilon(k,0)$ diverges, and ϕ_{tot} therefore vanishes. Induced charge $Q(\mathbf{r})$ then matches exactly with $-Q_{\text{ext}}(\mathbf{r})$ at any point. This screening effect is specific to conducting systems. In addition formula (1.75), obtained in the framework of Vlasov approximation, becomes exact in the large wavelength limit.

Energy absorption   The power provided to the system is proportional to $\varepsilon''(k,\omega)$ which using the formula (1.74), reduces to $-\chi''_{\text{id}}(k,\omega)/(\varepsilon_0 k^2)$. As the imaginary part $\chi''_{\text{id}}(k,\omega)$ is negative on the real axis, the power delivered is finite. This result is *a priori* surprising for this conservative system and is interpreted as follows. The energy absorption is not induced by a friction force in the plasma, but by a resonance effect due to *surfing* particles. These particles have a velocity $\mathbf{v} = \omega \mathbf{k}/k^2$ equal to the phase velocity of the wave associated with ϕ_{ext}. Power $\mathbf{F}_{\text{ext}} \cdot \mathbf{v}$ supplied to these particles is constant in time and strictly positive, to first order in the perturbation. So they pump out part of the exciting wave energy. This process of resonant energy absorption generally creates singularities. For an undamped harmonic oscillator such as a LC circuit for instance, the resonance frequency induces its own divergence of the response. Here the singularity corresponding to the phase velocity is smoothed by integration over all possible velocities distributed according to Maxwell-Boltzmann statistics and taking into account thermal fluctuations. This phenomenon is called Landau damping in plasma physics.

Plasmon modes   In the limit of large wavelengths $k \to 0$ with $\text{Im}\, z > 0$ fixed, we obtain from formulas (1.68) and (1.69) an asymptotic form of the susceptibility,

$$\chi(k,z) \simeq \frac{q^2 \rho k^2}{m} \frac{1}{z^2 - \omega_p^2},$$

where we have introduced the plasmon pulsation

$$\omega_p = \left(\frac{e^2}{m\epsilon_0} \right)^{1/2}.$$

Writing

$$\lim_{\epsilon \to 0^+} \frac{1}{z^2 - \omega_p^2} = \frac{1}{2\omega_p} \lim_{\epsilon \to 0^+} \left[\frac{1}{z - \omega_p} - \frac{1}{z + \omega_p} \right]$$

and using again equation (1.22), p. 14, we can separate the real and imaginary parts, to get[12]

$$\chi''(k,\omega) \simeq -\frac{q^2 \rho k^2 \pi}{2m\omega_p} \left[\delta(\omega - \omega_p) - \delta(\omega + \omega_p) \right] \quad \text{when } k \to 0, \ \omega \text{ fixed.} \qquad (1.76)$$

This behaviour is the signature of the presence of an undamped plasmon mode at frequency $\pm\omega_p$. Note that the existence of this mode is not an artefact due to mean-field approximation. In the limit of long wavelengths indeed, slices of non-neutral plasma are maintained at their equilibrium position by the bath, so that their

---

[12]See also in this context the exercise 1.4, p. 48.

dynamics is isomorphic to that of an undamped harmonic oscillator of frequency ω_p. For $k \neq 0$ the plasmon modes are damped by various processes including the surfing particles. For a plasma with two components such as an ionic salt, plasmons modes are always damped even in the limit $k \to 0$ as a result of collisions between positive and negative charges oscillating with opposite phase.

1.2.5 *Conductivity and Kubo formula*

◊ **Presentation** ◊

Let us consider a conductor submitted to a voltage inducing an electric current. At the microscopic level the latter is associated with the free motion of charged particles. The corresponding study of the conductivity is extremely difficult. It falls in fact within the very general framework of linear response in statistical mechanics. This issue is a field on its own extensively discussed in the literature. We present here a demonstration of Kubo formula without using the traditional tools of the many-body problem such as the density matrix. This formula shows that the response function at play is equal to a dynamical correlation function of the system at equilibrium. We discuss briefly what the expected behaviours are, and show how the exact microscopic expression of the conductivity can lead to simple phenomenological models.

Let H_0 be the quantum Hamiltonian for the particles constituting the conductive medium. The initial state $\mathcal{E}_0$ at $t_0 = -\infty$ is an equilibrium state at a given temperature T, with $\beta = 1/(k_B T)$. We operate in the canonical ensemble, considering that the system is infinitely extended without taking into account possible boundary effects. In addition the system is studied within the framework of quantum mechanics: the obtained expressions can easily be extended to the classical case. Thus, given an observable associated with a Hermitian operator A, the averaged equilibrium value $\langle A \rangle_0$ in state $\mathcal{E}_0$ is written

$$\langle A \rangle_0 = \frac{\text{Tr}\left[e^{-\beta H_0} A\right]}{\text{Tr}\, e^{-\beta H_0}}, \tag{1.77}$$

where Tr stands for the trace on the space of microscopic states available to the system. This trace can be computed using any basis of this space. In particular, if we introduce the orthonormal basis formed by eigenstates $|\psi_n\rangle$ of the Hamiltonian H_0, corresponding to energies E_n, we find

$$\langle A \rangle_0 = \frac{\sum_n e^{-\beta E_n} \langle \psi_n | A | \psi_n \rangle}{\sum_n e^{-\beta E_n}}. \tag{1.78}$$

Finally the initial current is zero, that is to say

$$\langle \mathbf{J} \rangle_0 = 0,$$

where $\mathbf{J}$ is the microscopic current operator.

One then submits the system to an external homogeneous electric field $\mathbf{E}_{ext}(t)$ depending on time and introduced adiabatically at $t_0 = -\infty$. This external field plays the role of forcing. At first we consider a generic time dependence, with the only condition that $\mathbf{E}_{ext}(t)$ decays exponentially fast to zero as $t \to -\infty$. In the presence of this external perturbation the Hamiltonian $H(t)$ is

$$H(t) = H_0 - \mathbf{P} \cdot \mathbf{E}_{ext}(t) \tag{1.79}$$

where $\mathbf{P}$ is the microscopic polarisation operator. We will examine the system response in current to the external forcing.

$$\Diamond \ \textbf{Analysis and solution} \ \Diamond$$

First, we will establish the Kubo formula, considering the evolution of the average value of any observable A. In general the system is submitted to a forcing described by the Hamiltonian

$$H(t) = H_0 - BF(t) \tag{1.80}$$

where B is some arbitrary observable, and $F(t)$ an arbitrary scalar function which tends exponentially fast to zero as $t \to -\infty$. We will later specify the formulas obtained in the specific case studied here.

The strategy is to first determine the evolved state $|\psi_n(t)\rangle$ under the action of $H(t)$ starting from an initial state that is an eigenstate $|\psi_n\rangle$ of the unperturbed Hamiltonian H_0. The evolution of the statistical average $\langle A \rangle_t$ is then simply obtained by weighting the quantum average $\langle \psi_n(t)|A|\psi_n(t)\rangle$ by the Boltzmann factor $e^{-\beta E_n}$ and summing over all possible quantum states, i.e.

$$\langle A \rangle_t = \frac{\sum_n e^{-\beta E_n} \langle \psi_n(t)|A|\psi_n(t)\rangle}{\sum_n e^{-\beta E_n}}. \tag{1.81}$$

Perturbative evolution of an eigenstate of the unperturbed Hamiltonian
It is possible to show that $|\psi_n(t)\rangle$ is solution of the integral equation:

$$|\psi_n(t)\rangle = e^{-iH_0(t-t_0)/\hbar}|\psi_n\rangle - \frac{1}{i\hbar}\int_0^{t-t_0} d\tau F(t-\tau)e^{-iH_0\tau/\hbar}B|\psi_n(t-\tau)\rangle. \tag{1.82}$$

We will actually establish the result in general terms in Chapter 3 (see Eq. (3.74), p. 155): One should then make the following substitutions $|\phi_0\rangle \to |\psi_n\rangle$, $|\phi(t)\rangle \to |\psi_n(t)\rangle$ and $W(t) \to -BF(t)$. However, the reader may also explicitly check[13] that expression (1.82) for $|\psi_n(t)\rangle$ is indeed a solution of Schrödinger equation

$$i\hbar\frac{\partial}{\partial t}|\psi_n(t)\rangle = H(t)|\psi_n(t)\rangle.$$

---

[13]It will be interesting to first perform the change variables $t' = t - \tau$ in the integral appearing in Eq. (1.82) before computing $(\partial/\partial t)|\psi_n(t)\rangle$.

For a weak forcing (1.82) results in a perturbative expansion in powers of $B\,F(t)$:

$$|\psi_n(t)\rangle = e^{-iH_0(t-t_0)/\hbar}|\psi_n\rangle$$

$$-\frac{1}{i\hbar}\int_0^{t-t_0}\mathrm{d}\tau F(t-\tau)e^{-iH_0\tau/\hbar}Be^{-iH_0(t-\tau-t_0)/\hbar}|\psi_n\rangle + \cdots \quad (1.83)$$

to first order in F.

The perturbative expansion (1.83) gives the evolution of the quantum average $\langle\psi_n(t)|A|\psi_n(t)\rangle$ of A in the state $|\psi_n(t)\rangle$. Setting the initial conditions at $t_0 = -\infty$ we obtain

$$\langle\psi_n(t)|A|\psi_n(t)\rangle \simeq \langle\psi_n|A|\psi_n\rangle$$

$$+\frac{1}{i\hbar}\int_0^{+\infty}\mathrm{d}\tau F(t-\tau)\langle\psi_n|\big[e^{-iH_0\tau/\hbar}Be^{iH_0\tau/\hbar}A$$

$$-Ae^{-iH_0\tau/\hbar}Be^{iH_0\tau/\hbar}\big]|\psi_n\rangle \quad (1.84)$$

to first order in F. To establish this result we used the fact that $|\psi_n\rangle$ is an eigenstate of H_0 and therefore that

$$e^{-iH_0(t-t_0)/\hbar}|\psi_n\rangle = e^{-iE_n(t-t_0)/\hbar}|\psi_n\rangle.$$

Statistical average over initial configurations Inserting expansion (1.84) in formula (1.81) and keeping only the linear terms in the forcing, we find

$$\langle A\rangle_t - \langle A\rangle_0 = \frac{1}{i\hbar}\int_0^{+\infty}\mathrm{d}\tau F(t-\tau)$$

$$\times \frac{\mathrm{Tr}\Big\{e^{-\beta H_0}\big[e^{-iH_0\tau/\hbar}Be^{iH_0\tau/\hbar}A - Ae^{-iH_0\tau/\hbar}Be^{iH_0\tau/\hbar}\big]\Big\}}{\mathrm{Tr}\,e^{-\beta H_0}} + \cdots . \quad (1.85)$$

Formula (1.85) therefore takes exactly the general form of linear response (1.4), p. 4. The response function of observable A to the applied forcing is

$$K_0(\tau) = -\frac{1}{i\hbar}\frac{\mathrm{Tr}\big\{[B,\ e^{-\beta H_0}]A_I(\tau)\big\}}{\mathrm{Tr}e^{-\beta H_0}}, \quad (1.86)$$

with the commutator

$$[B,\ e^{-\beta H_0}] = B\,e^{-\beta H_0} - e^{-\beta H_0}\,B \quad (1.87)$$

and the time-evolved operator

$$A_I(\tau) = e^{iH_0\tau/\hbar}\,A\,e^{-iH_0\tau/\hbar}. \quad (1.88)$$

To establish formula (1.86), we used the cyclic property of the trace, $\mathrm{Tr}(\mathcal{O}_1\mathcal{O}_2) = \mathrm{Tr}(\mathcal{O}_2\mathcal{O}_1)$, valid for any two traceable operators $\mathcal{O}_1$ and $\mathcal{O}_2$.

Kubo formula   It is customary to write the response function in terms of an operator $\dot{B}$ defined by[14]

$$\dot{B} = -\frac{1}{i\hbar}[H_0,\ B].$$

We then obtain the identity

$$-\frac{1}{i\hbar}[B,\ e^{-\beta H_0}] = e^{-\beta H_0} \int_0^\beta \mathrm{d}\beta_1\ e^{\beta_1 H_0}\dot{B}e^{-\beta_1 H_0}, \qquad (1.89)$$

by integration of the relation

$$\frac{\partial}{\partial\beta}(e^{\beta H_0}Be^{-\beta H_0}) = [H_0, e^{\beta H_0}Be^{-\beta H_0}].$$

Using identity (1.89) in expression (1.86) for the response function, we obtain the Kubo formula

$$K_0(\tau) = \left\langle \int_0^\beta \mathrm{d}\beta_1\ e^{\beta_1 H_0}\dot{B}e^{-\beta_1 H_0}\ A_I(\tau) \right\rangle_0. \qquad (1.90)$$

We can thus basically say that the response of A to a forcing of coupling B is given by the dynamic correlations of A and B at equilibrium. We find that the response function $K_0(\tau)$ is independent of F, and that it depends on the way the forcing is coupled to the system, as discussed on p. 4. In the framework of the N-body problem, Kubo formula is often established starting from Liouville equation, governing the density matrix evolution by the action of the forcing. This method is of course equivalent to the approach used here, which allows to avoid the explicit introduction of these concepts.

Frequency-dependent conductivity   We now specify the Kubo formula to the following situation. In the absence of external magnetic field and neglecting any relativistic effect, the Hamiltonian H_0 takes the form

$$H_0 = \sum_{i=1}^N \frac{\mathbf{p}_i^2}{2m_i} + V(\mathbf{r}_1, ..., \mathbf{r}_N),$$

where $\mathbf{p}_i = -i\hbar\boldsymbol{\nabla}_i$ is the momentum operator of the particle i canonically conjugate to its position $\mathbf{r}_i$, and $V(\mathbf{r}_1, ..., \mathbf{r}_N)$ the operator describing the interactions between particles depending only on their positions. The observable A is a vector and corresponds to the current

$$\mathbf{J} = \sum_{i=1}^N q_i \frac{\mathbf{p}_i}{m_i},$$

where q_i is the charge of the particle i while the perturbation $-B\,F(t)$ corresponds here to $-\mathbf{P}\cdot\mathbf{E}_{\mathrm{ext}}(t)$. We first determine the operator $\dot{\mathbf{P}}$. Using the microscopic expression $\mathbf{P} = \sum_{i=1}^N q_i\mathbf{r}_i$, an elementary calculation then yields

$$\dot{\mathbf{P}} = -\frac{1}{i\hbar}[H_0,\ \mathbf{P}] = \mathbf{J}.$$

---

[14]This operator is nothing but the time derivative of the evolved operator $B_I(\tau)$ at $\tau = 0$.

Now consider a monochromatic external field of the usual form $\mathbf{E}_{\text{ext}}(t) = \text{Re}(\mathbf{E}_z e^{-izt})$, and keeping a fixed direction in time. The relevant susceptibility is then the conductivity per unit volume $\sigma(z)$. Suppose that the response is isotropic, so that the induced current is collinear to the external field: $\sigma(z)$ then reduces to a scalar. Under this assumption we find, by inserting the expression (1.90) into general formula (1.8) for the susceptibility,

$$\sigma(z) = \frac{1}{3V} \int_0^\infty d\tau \, e^{iz\tau} \left\langle \int_0^\beta d\beta_1 \, e^{\beta_1 H_0} \mathbf{J} e^{-\beta_1 H_0} \cdot \mathbf{J}_I(\tau) \right\rangle_0 \qquad (1.91)$$

where V is the volume of the system. The conductivity is therefore expressed in terms of the dynamic auto-correlation function of the current at equilibrium.

Classical limit   In some conductive media such as ionic salts, quantum effects are small so that the conductivity can be computed with good accuracy within the framework of classical mechanics. The classic version of the Kubo formula is immediately obtained by using the trick

$$\int_0^\beta d\beta_1 \, e^{-(\beta-\beta_1)H_0} \mathbf{J} e^{-\beta_1 H_0} \to \beta e^{-\beta H_0} \mathbf{J}$$

which boils down to commute $\mathbf{J}$ and $e^{-\beta_1 H_0}$. We then obtain the classical conductivity

$$\sigma_{\text{cl}}(z) = \frac{\beta}{3V} \int_0^\infty d\tau \, e^{iz\tau} \langle \mathbf{J}(0) \cdot \mathbf{J}(\tau) \rangle_0, \qquad (1.92)$$

with

$$\mathbf{J}(\tau) = \sum_{i=1}^N q_i \, \mathbf{v}_i(\tau).$$

Here $\mathbf{v}_i(\tau)$ is the velocity of the particle i at time τ in the unperturbed evolution governed by H_0, from a given initial configuration at $\tau = 0$, identified by a point of coordinates $(\mathbf{r}_1, ..., \mathbf{r}_N; \mathbf{p}_1, ..., \mathbf{p}_N)$ in the $6N$ dimensions phase space of canonical variables. The statistical average over these initial configurations is an integral over this phase space with the Boltzmann factor $\exp[-\beta H_0(\mathbf{r}, \mathbf{p})]$.

Classical Kubo formula (1.92) can also be obtained from Liouville equation in a $6N$ dimensions phase space. Here we propose a simpler proof exploiting the interpretation of the response function $K_0(\tau)$ as the free evolution of $\langle \mathbf{J}_{\text{pulse}}(\tau) \rangle / V$ after applying an electric field strictly localised at time $t_0 = 0$ of the form

$$\mathbf{E}_{\text{ext}}(t) = \mathbf{F}_0 \delta(t).$$

For a set of initial conditions $\{\mathbf{r}_i, \mathbf{v}_i\}$ at time 0^-, the positions and velocities at time 0^+ after applying the above electric field become

$$\mathbf{r}_i(0^+) = \mathbf{r}_i \quad \text{and} \quad \mathbf{v}_i(0^+) = \mathbf{v}_i + \frac{q_i \mathbf{F}_0}{m_i}.$$

For this initial configuration, we get

$$\mathbf{J}_{\text{pulse}}(\tau) = \sum_i q_i \mathbf{v}_i \left(\tau | \{ \mathbf{r}_i; \mathbf{v}_i + \frac{q_i \mathbf{F}_0}{m_i} \} \right).$$

Here the velocities $\mathbf{v}_i$ at time τ can be obtained from the initial conditions $\{ \mathbf{r}_i; \mathbf{v}_i + \frac{q_i \mathbf{F}_0}{m_i} \}$ at 0^+ through the intrinsic evolution induced by H_0. We then get

$$\langle \mathbf{J}_{\text{pulse}}(\tau) \rangle = Z_{\text{cl}}^{-1} \int \prod_i \mathrm{d}\mathbf{r}_i \mathrm{d}\mathbf{p}_i e^{-\beta H_0(\{ \mathbf{r}_i, \mathbf{p}_i \})} \sum_i q_i \mathbf{v}_i \left(\tau | \{ \mathbf{r}_i; \mathbf{v}_i + \frac{q_i \mathbf{F}_0}{m_i} \} \right),$$

where Z_{cl} is the usual configuration integral

$$Z_{\text{cl}} = \int \prod_i \mathrm{d}\mathbf{r}_i \mathrm{d}\mathbf{p}_i e^{-\beta H_0(\{ \mathbf{r}_i, \mathbf{p}_i \})}.$$

By the change of variables

$$\mathbf{r}'_i = \mathbf{r}_i \quad ; \quad \mathbf{p}'_i = \mathbf{p}_i + q_i \mathbf{F}_0,$$

whose Jacobian is obviously equal to 1, and linearising the Gibbs factor $e^{-\beta H_0(\{ \mathbf{r}_i, \mathbf{p}_i \})}$ expressed in terms of the new variables,

$$\exp[-\beta H_0(\{ \mathbf{r}_i, \mathbf{p}_i \})] = \exp\left[-\beta \sum_i \frac{(\mathbf{p}'_i - q_i \mathbf{F}_0)^2}{2 m_i} - \beta V(\{ \mathbf{r}'_i \}) \right]$$

$$= \exp[-\beta H_0(\{ \mathbf{r}'_i, \mathbf{p}'_i \})] \left(1 + \beta \sum_i q_i \mathbf{v}'_i \cdot \mathbf{F}_0 + O(F_0^2) \right),$$

we get

$$\langle \mathbf{J}_{\text{pulse}}(\tau) \rangle = \beta Z_{\text{cl}}^{-1} \int \prod_i \mathrm{d}\mathbf{r}'_i \mathrm{d}\mathbf{p}'_i e^{-\beta H_0(\{ \mathbf{r}'_i, \mathbf{p}'_i \})} \sum_i q_i \mathbf{v}_i (\tau | \{ \mathbf{r}'_i; \mathbf{v}'_i \}) \sum_j q_j \mathbf{v}'_j \cdot \mathbf{F}_0$$

to first order in $\mathbf{F}_0$. The identity (1.5), established in the general case p. 5, becomes here

$$\frac{1}{V} \langle \mathbf{J}_{\text{pulse}}(\tau) \rangle = \mathbf{F}_0 K_0(\tau).$$

We finally get

$$K_0(\tau) = \frac{\beta}{3 V Z_{cl}} \int \prod_i \mathrm{d}\mathbf{r}_i \mathrm{d}\mathbf{p}_i e^{-\beta H_0(\{ \mathbf{r}_i, \mathbf{p}_i \})} \sum_i q_i \mathbf{v}_i (\tau | \{ \mathbf{r}_i; \mathbf{v}_i \}) \cdot \sum_j q_j \mathbf{v}_j$$

$$= \frac{\beta}{3V} \langle \mathbf{J}(0) \cdot \mathbf{J}(\tau) \rangle_0,$$

which gives indeed the classical Kubo formula (1.92) by applying the generic formula (1.8).

◊ Interpretation ◊

Determining the dynamical correlations of current at a microscopic level is extremely difficult. It is indeed necessary to take into account both the interactions and the quantum nature of particles. The problem becomes simpler in the classical case but it remains very difficult. One must then implement the arsenal of kinetic equations for the many-body problem which cannot be described here, even briefly! We just conclude with a few remarks on the fluctuation-dissipation theorem, as well as the phenomenological description of the correlation functions involved in Kubo formula.

Fluctuation-dissipation theorem Consider a real frequency $z = \omega$. The calculation of the power provided by the excitation field and averaged over a period

$$\overline{\mathcal{P}} = \frac{1}{T} \int_0^T dt \left\langle \frac{\partial H(t)}{\partial t} \right\rangle_t = -\frac{1}{T} \int_0^T dt \, \langle \mathbf{P} \rangle_t \cdot \frac{d\mathbf{E}_{\text{ext}}}{dt},$$

leads to

$$\overline{\mathcal{P}} = \frac{\sigma'(\omega)}{2} \, |\mathbf{E}_\omega|^2. \tag{1.93}$$

It is thus $\sigma'(\omega)$, the real part of the conductivity, that controls the power supplied, and is thus related to the dissipative processes at play in the system. This implies $\sigma'(\omega) > 0$. Note that formula (1.93), combined with the Kubo formula for conductivity, is a special case of the famous fluctuation-dissipation theorem. The dissipation is here determined by the dynamic fluctuations of the current via a proportionality factor that is simply the inverse temperature $\beta = 1/(k_B T)$.

Phenomenological model The dynamical correlations of the current can be described by a simple decreasing exponential $C \exp(-\gamma \tau)$. Constant C and the relaxation time γ^{-1} can then be fitted using various constraints, such as the value of the static conductivity $\sigma(0)$ or the short time behaviour of the response function[15]. Such a model can be interpreted within the usual Drude model. These phenomenological approaches are similar in spirit to those presented in the example of the dielectric section 1.2.2.

1.3   Exercises

■ Exercise 1.1. Response functions associated with linear operators

1. Give the response functions associated with operators

$$L_1 = \frac{d}{dt} + a \quad ; \quad L_2 = \frac{d^2}{dt^2} + b$$

where a and b are positive constants, i.e. the functions $K_i(t - t')$ solutions of $L_i K_i(t - t') = \delta(t - t')$ with $K_i(t - t') = 0$ for $t < t'$.

2. Compute susceptibilities $\chi_1(z)$ and $\chi_2(z)$ associated with L_1 and L_2. Discuss their poles and domains of analyticity. Same question if a and b are negative.

Solution p. 297.

---

[15] In the classical case, $K_0(0) = \rho q^2/m$ for a single type of identical particles moving with charge q and density ρ. Further information on the short time development of $K_0(\tau)$ can be obtained in terms of some static equilibrium correlations.

■ **Exercise 1.2. Response function for a RLC circuit**

Compute the response function of a RLC circuit using first the methods of Appendix C, that is to say the results (C.2) and (C.4), and then by performing the inverse Laplace transform of susceptibility

$$\chi(z) = \frac{1}{R - i(Lz - \frac{1}{Cz})}.$$

Solution p. 297.

■ **Exercise 1.3. Charged Brownian particle**

A charged particle is subject to an electric field $E(t)$ in one dimension[16] and a Brownian-type force due to the environment. The evolution of variable $r(t)$ representing the average value of the position of the particle is then simply governed by the equation:

$$m\ddot{r} + \gamma\dot{r} = E(t). \tag{1.94}$$

1. Compute the susceptibilities $\chi(z)$ and $\mu(z)$ associated to $\dot{r}(t)$ and $r(t)$, using $E(t) = E_z\, e^{-izt}$. Do these susceptibilities satisfy the K.K relation? Check the sum rule (1.27), p. 16 for $\chi(z)$.

2. Identify the response functions $V(\tau)$ and $R(\tau)$ associated respectively to $\dot{r}(t)$ and $r(t)$.

3. What term could be added to the equation of motion (1.94) to eliminate the constant part at large τ of the response function $R(\tau)$?

Solution p. 298.

■ **Exercise 1.4. Absorption line**

Imagine a susceptibility corresponding to a very narrow absorption line for a pulsation ω_0. More specifically, suppose that the imaginary part $\chi''(\omega)$ of the susceptibility $\chi(\omega)$ under consideration corresponds to a Dirac distribution centred at ω_0.

1. Why should we take $\chi''(\omega) = \sigma\delta(\omega - \omega_0) - \sigma\delta(\omega + \omega_0)$?

2. What do we then obtain for the real part of the susceptibility $\chi'(\omega)$ by applying Kramers-Kronig relations without precautions, that is to say without considering the singularities in $\pm\omega_0$?

---

[16] The generalisation to three dimensions is straightforward.

3. We propose now to recover the previous result by a more rigorous method. Show first that the identity

$$\lim_{\epsilon \to 0^+} \frac{1}{(x - i\epsilon)} = \mathrm{PP} \frac{1}{x} + i\pi\delta(x) \tag{1.95}$$

allows to define a susceptibility $\chi_\epsilon(z)$ analytic in $\mathbb{C}^+$ and $\mathbb{R}$, and satisfying $\chi_\epsilon''(\omega) \to \sigma\delta(\omega - \omega_0) - \sigma\delta(\omega + \omega_0)$ when $\epsilon \to 0^+$. Give expressions for $\chi_\epsilon(z)$, $\chi_\epsilon'(\omega)$ and $\chi_\epsilon''(\omega)$. Check then that the result of the previous question for $\chi'(\omega)$ is recovered in the limit $\epsilon \to 0^+$.

4. Obtain $\chi_\epsilon''(\omega)$ starting from $\chi_\epsilon'(\omega)$ using the Kramers-Kronig relations.

Solution p. 299.

■ **Exercise 1.5. Application of Kramers-Kronig relations in astrophysics**

1. Show that one can write the Kramers-Kronig relations (1.23), p. 15, as

$$\chi'(\omega) = \frac{2}{\pi} \mathrm{PP} \int_0^\infty d\omega' \, \frac{\omega' \chi''(\omega')}{(\omega'^2 - \omega^2)}, \tag{1.96}$$

$$\chi''(\omega) = -\frac{2\omega}{\pi} \mathrm{PP} \int_0^\infty d\omega' \, \frac{\chi'(\omega')}{(\omega'^2 - \omega^2)}.$$

2. In a simplified model for the interstellar medium, the vacuum is scattered with spherical grains. In this question we put a lower bound on the volume fraction occupied by the grains. To do so, consider the susceptibility $\chi(\omega) = \epsilon(\omega) - 1$ where $\epsilon(\omega)$ is the dielectric permittivity of the interstellar medium. Using Clausius-Mossotti relation, it is possible to show that

$$\chi'(0) = 4\pi\rho r^3 \frac{\epsilon_g - 1}{\epsilon_g + 2},$$

where r is the spherical grain radius, ρ the density and ϵ_g the static dielectric constant of the grain. In addition we remind that $\epsilon''(\omega)$ is an odd function and positive for $\omega > 0$.

Suppose that the function $\chi''(\omega)$ is experimentally measured between two pulsations ω_1 and ω_2. Show that these measurements allow to set a lower limit on the volume fraction occupied by the grain.

Solution p. 299.

■ Exercise 1.6. Sum rules

Establish the sum rules (1.28) and (1.29), i.e.

$$\frac{1}{\pi} \int_{-\infty}^{\infty} d\omega\, \omega \left[\chi''(\omega) - \frac{K_0(0)}{\omega} \right] = K_0'(0),$$

$$-\frac{1}{\pi} \int_{-\infty}^{\infty} d\omega\, \omega^2 \left[\chi'(\omega) + \frac{K_0'(0)}{\omega^2} \right] = K_0''(0).$$

To this end, we suppose that the first and second derivatives of the response function $K_0(\tau)$, denoted respectively $K_0'(\tau)$ and $K_0''(\tau)$ can also be seen as good response functions, without singularity in $\tau = 0$. Their susceptibilities are denoted $\tilde{\chi}(z)$ and $\tilde{\tilde{\chi}}(z)$, and will be connected to $\chi(z)$, the susceptibility associated with $K_0(\tau)$.

Solution p. 300.

■ Exercise 1.7. Response to noise

Consider an equation for a physical variable $x(t)$ given by $\mathcal{O}x(t) = \eta(t)$ with $x(t) \to 0$ when $t \to -\infty$. $\mathcal{O}$ is a linear differential operator and $\eta(t)$ is a random variable with mean zero and uncorrelated in time, i.e.

$$\langle \eta(t) \rangle = 0, \quad \langle \eta(t)\eta(t') \rangle = \alpha\delta(t - t').$$

The variable x can represent for example the position of a Brownian particle and η the influence of the environment on its movement.

1. Give the expression of $x(t)$ in terms of the causal response $K(t; t')$ associated with $\mathcal{O}$ and η. Then give an expression for $\langle x^2(t) \rangle$.

2. Show that if $\mathcal{O}$ corresponds to the differential operator associated with a damped harmonic oscillator, i.e.

$$\mathcal{O} = m\frac{d^2}{dt^2} + \gamma\frac{d}{dt} + m\omega^2$$

with $\gamma > 2m\omega$, then

$$\langle x^2(t) \rangle = \frac{\alpha}{2m\gamma\omega^2}.$$

Solution p. 300.

■ Exercise 1.8. Kramers-Kronig relations for a metal

We consider a metal of conductivity σ. In the presence of an external homogeneous electric field of single monochromatic component $\exp(-i\omega t)$, the metal behaves like a dielectric of relative dielectric constant $\varepsilon(\omega)$.

1. Specify the necessary conditions for a field to be created across a sample of finite size.

2. Let us recall that the current density $\mathbf{j}$ is given by $\mathbf{j} = (\partial \mathbf{P}/\partial t)$ where $\mathbf{P}$ is the polarisation. Show that in the static limit, i.e. $\omega \to 0$, $\varepsilon(\omega)$ behaves like $i\sigma/(\varepsilon_0\omega)$ where ε_0 is the vacuum dielectric constant. Conclude that $\varepsilon(z)$ has a simple pole at $z = 0$ with residue $i\sigma/\varepsilon_0$.

3. We assume that $\varepsilon(z) - 1$ is analytic in $\mathbb{C}^+$ (including the real axis) except at the origin, and that it decreases at least as $1/z$ for large z in $\mathbb{C}^+$. Find the modified Kramers-Kronig relations between the real part $\varepsilon'(\omega) - 1$ and imaginary part $\varepsilon''(\omega)$. Using these expressions, check that $\varepsilon'(\omega) - 1$ is finite in $\omega = 0$ while $\varepsilon''(\omega)$ diverges as $\sigma/(\varepsilon_0\omega)$.

4. Using a simple argument, find the asymptotic form of $\varepsilon(\omega) - 1$ for large ω (we will note n the number density of electrons, m their mass and q their charge). Conclude that $\varepsilon''(\omega)$ decays faster than $1/\omega^2$, and give a sum rule for the first moment of $\varepsilon''(\omega)$ in terms of n, q, m and ε_0.

5. Consider for example the simple model for $\varepsilon(\omega)$

$$\varepsilon''(\omega) = \frac{\sigma}{\varepsilon_0\omega(1 + \omega^2\tau^2)}.$$

Give the characteristic time τ in terms of the previous problem parameters.

6. What simple microscopic model leads to the expression of $\varepsilon(\omega)$ corresponding to the previous question? What improvements could be made to it?

Solution p. 301.

■ Exercise 1.9. Signal propagation in dielectric media

This exercise proposes to study wave propagation in a dispersive medium to show that causality implies that a signal cannot propagate faster than the speed of light in vacuum. To do so we recall that the electric permittivity $\epsilon(z)$ is built from a causal response function $G(\tau)$ given by the medium microscopic behaviour

$$\epsilon(z) = 1 + \int_0^\infty G(\tau)e^{iz\tau}\,\mathrm{d}\tau.$$

1. What conditions should one impose on $G(\tau)$ for $\epsilon(z)$ to be analytic and so that $\epsilon(z) \to 1$ when $z \to \infty$ in the upper half complex plane including the real axis?

2. Consider here the one-dimensional case for simplicity, with a domain corresponding to $x > 0$. Suppose that a source is turned on at $t = 0$ and $x = 0$, and that the field amplitude (for instance an electric field) in $x = 0$ is:

$$F(0, t) = \theta(t)\, f(t) \qquad \text{with} \qquad \forall\, t,\ |f(t)| < C,$$

for a given C. Show that $F(0, t)$ can be written as

$$F(0, t) = \int_{-\infty + i\gamma}^{\infty + i\gamma} dz\, g_z\, e^{-izt},$$

where γ is an arbitrary positive real number.

3. The propagation equation for each component of complex frequency $g_z(x, t) = g_z g(x) \exp(-izt)$ is given by:

$$\left(\frac{\partial^2}{\partial x^2} - \frac{n^2(z)}{c^2} \frac{\partial^2}{\partial t^2} \right) g_z(x, t) = 0,$$

where $n(z) = \sqrt{\epsilon(z)}$ is the refractive index (assuming that the permittivity μ is that of the vacuum for all z). By studying the domain of analyticity and the behaviour at large z of $n(z)$, give the general form of $F(x, t)$ and show that it is zero for $x > ct$.

Solution p. 302.

Chapter 2

Static Green's functions

A physical quantity is often related to a source by a linear partial differential equation (PDE). For example this quantity may be the electromagnetic field or the temperature respectively associated with sources of charge or heat. In this chapter we consider stationary situations where all quantities are time-independent. The physical quantity under consideration is therefore a static field, that is to say a function depending only on spatial coordinates. Here we present general methods for studying this type of problem.

First note that the sought physical quantity is not completely determined by the knowledge of the corresponding PDE. Uniqueness is in fact only ensured by imposing additional conditions. In particular, constraints at the system boundaries naturally appear in most physical situations. They generally have a clear physical meaning, and play a crucial role in determining the studied quantity. In electrostatics these boundary conditions (BC) amount for example to consider implicitly contributions of additional sources induced in the medium outside the system. It is generally quite difficult to find the type of BC ensuring both the existence and uniqueness of the solution of a given PDE: If the BC required for the existence are generally not too restrictive, uniqueness requires fairly restrictive BC. The question of boundary conditions is fundamental and it is therefore the first one to address in any particular situation.

Linearity of the PDE allows to determine the field of interest by using the superposition principle. It is then very natural to introduce the particular field created by a point source: This particular field is called a Green's function. The total field induced by an arbitrary distribution of sources then reduces to the superposition of individual fields created by point sources. This decomposition highlights the major role played by Green's functions in expressing the general solutions of linear PDEs. In this brief discussion we have not considered the issue of boundary conditions. It is discussed in detail for each of the particular cases presented in this chapter. Note here that for a given PDE there are a variety of Green's functions, associated with various boundary conditions. Remarkably the general solution of a PDE for a particular set of boundary conditions may be expressed in terms of a Green's function associated with other boundary conditions! This further enhances the importance of such functions.

In the first part of this chapter, we consider a generic linear PDE for which the corresponding Green's functions are introduced. Exploiting the superposition principle, we then show that a suitable integral over all point sources of any of these Green's functions is a solution of the PDE under consideration. The problem of boundary conditions is then discussed. We briefly present general properties of Green's functions, before writing them in terms of matrix elements of an operator. This operatorial interpretation paves the way for a natural extension of the concept of Green's function to situations where the goal is not to determine the field created by given sources. We then study successively two special PDE involving Laplace and Helmholtz operators in a three-dimensional space. These operators occur naturally in many areas of physics as a result of invariance properties of space. For each of them, we address the boundary condition problem, and express the solution of the corresponding PDE in terms of Green's functions. This study is then applied to low dimensions. In particular, we present a method specific to two dimensions based on conformal transformations. We finally conclude with PDE related to inhomogeneous operators involved in particular in quantum mechanics. The corresponding Green's functions are defined by extension of the aforementioned original concept. Their interest lies mainly in the determination of spectral properties of the involved operators as well as in the construction of perturbative expansions.

The second part of this chapter is devoted to examples, both illustrating and extending the general properties presented in the first part. We start with the standard problem of an electrostatic field created by charges in the vicinity of a conducting wall. The structure of the Green's functions obtained through a straightforward calculation allows us to recover the familiar picture associated with the method of image charges. It shades light on the systematic character of this method. Next we consider a situation encountered in fluid mechanics as well as in superconductivity: the field of interest is then solution of a Laplace equation with specific boundary conditions. It is possible to solve directly the PDE in the absence of explicit sources by decomposing the field in spherical harmonics. The following examples are related to quantum mechanics: the first one concerns the calculation of the density of states while the second deals with diffusion. We finally describe an application of conformal transformations to determine flow lines of the wind blowing on a wall.

2.1   General properties

2.1.1   *Definition and properties of Green's function*

Consider a system inside a domain $\mathcal{D}$ of arbitrary dimension including sources distributed with a given density $\rho(\mathbf{r})$, where $\mathbf{r}$ denotes a point of domain $\mathcal{D}$. Suppose these sources induce a field $\phi(\mathbf{r})$ uniquely determined by the system

$$\boxed{\begin{array}{c} \mathcal{O}\,\phi(\mathbf{r}) = \rho(\mathbf{r}) \\ \mathrm{BC}(\phi). \end{array}} \tag{2.1}$$

In the first equation of this system, $\mathcal{O}$ is a linear operator including partial derivatives with respect to the various components of $\mathbf{r}$. In addition, the notation $\mathrm{BC}(\phi)$ represents the boundary conditions (BC) specific to the situation under consideration. They include in general the value of $\phi(\mathbf{r})$ and/or its spatial derivatives on $\partial\mathcal{D}$, the boundary of domain $\mathcal{D}$. In the following we also call $\partial\mathcal{D}$ a surface by an abuse of language referring to the case where the domain $\mathcal{D}$ is three-dimensional. In full generality we allow the field $\phi(\mathbf{r})$ under consideration to take complex values.

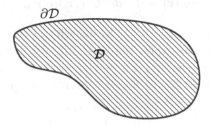

Fig. 2.1   Domain $\mathcal{D}$ and its boundary $\partial\mathcal{D}$

Since the PDE satisfied by ϕ is linear, it is tempting to invoke the superposition principle to decompose ϕ as a sum of fields created by adequate elementary sources. This attempt immediately fails because of difficulties caused by the boundary conditions. Which ones should we choose to unambiguously define elementary fields to ensure that the corresponding superposition satisfies the original boundary conditions on ϕ? There is in fact no general answer to this question which will then be handled on a case by case basis in the following.

◇ **Definition** ◇

Let us go back to the idea of superposition. By virtue of the identity

$$\rho(\mathbf{r}) = \int_{\mathcal{D}} d\mathbf{r}'\rho(\mathbf{r}')\delta(\mathbf{r}-\mathbf{r}') \tag{2.2}$$

valid for $\mathbf{r}$ strictly inside $\mathcal{D}$ (not on its boundary $\partial\mathcal{D}$), it is natural to introduce the set of all source points located in $\mathbf{r}'$ of $\mathcal{D}$. Each of these sources is associated with a purely local density which is nothing but $\delta(\mathbf{r}-\mathbf{r}')$. Each of them induces an elementary field, uniquely defined by given boundary conditions on $\partial\mathcal{D}$. This field is called **Green's function** of the operator $\mathcal{O}$, and is defined by the system

$$\boxed{\begin{array}{c} \mathcal{O}_{\mathbf{r}}\,G(\mathbf{r};\mathbf{r}') = \delta(\mathbf{r}-\mathbf{r}'), \qquad (2.3) \\ \mathrm{BC}(G). \end{array}}$$

Note that Green's function $G(\mathbf{r}; \mathbf{r}')$ depends on two positions that do not play the same role: $\mathbf{r}$ is the observation point where the basic field is evaluated, while $\mathbf{r}'$ is the position of the point source. Thus in the PDE satisfied by $G(\mathbf{r}; \mathbf{r}')$, the operator acts on the variable $\mathbf{r}$ as indicated by the notation $\mathcal{O}_\mathbf{r}$. In addition there are often several types of BC on G which may actually be different from those defining ϕ.

◇ Interest ◇

Let us now move forward with the idea of superposition. Consider the field $\phi_G(\mathbf{r})$ defined as the linear combination of the elementary fields $G(\mathbf{r}, \mathbf{r}')$, weighted by $\rho(\mathbf{r}')$ and summed over $\mathbf{r}'$, i.e.

$$\phi_G(\mathbf{r}) = \int_\mathcal{D} d\mathbf{r}' \, G(\mathbf{r}; \mathbf{r}') \, \rho(\mathbf{r}'). \tag{2.4}$$

Apply then the operator $\mathcal{O}_\mathbf{r}$ to $\phi_G(\mathbf{r})$. The linearity of this operator allows to pass it under the integral sign in expression (2.4). It also implies that this operator acts only on $G(\mathbf{r}; \mathbf{r}')$, the density $\rho(\mathbf{r}')$ playing the role of a simple multiplicative constant. Then using the PDE satisfied by $G(\mathbf{r}; \mathbf{r}')$, we eventually find

$$\mathcal{O}_\mathbf{r} \, \phi_G(\mathbf{r}) = \rho(\mathbf{r}). \tag{2.5}$$

As expected the field $\phi_G(\mathbf{r})$ satisfies the same PDE as the field $\phi(\mathbf{r})$ we are looking for. This clearly highlights the major interest of Green's functions in the resolution of general system (2.1). Once these functions are identified, it is indeed enough to compute a spatial integral to obtain any particular solution of the studied PDE.

Of course $\phi_G(\mathbf{r})$ does not reduce to $\phi(\mathbf{r})$ in general, because it does not satisfy the corresponding boundary conditions $BC(\phi)$. It typically differs by a surface integral✠ where the integration point runs through the domain boundary $\partial \mathcal{D}$, as we will later see in the case of Laplace operator. A comprehensive study of surface contributions is not possible, except in the simple case of homogeneous BC presented below.

✠ **Comment:**   Consider a field $\phi_\psi(\mathbf{r})$ defined by a surface integral over $\mathbf{r}' \in \partial \mathcal{D}$ of $G(\mathbf{r}; \mathbf{r}')$ or any of its spatial derivatives, weighted by an arbitrary function $\psi(\mathbf{r}')$. Exploiting again the linearity of operator $\mathcal{O}_\mathbf{r}$ and properties (2.3) defining $G(\mathbf{r}; \mathbf{r}')$, we find that the action of this operator on $\phi_\psi(\mathbf{r})$ leads to zero for $\mathbf{r}$ strictly inside $\mathcal{D}$. In other words $\phi_\psi(\mathbf{r})$ is a particular solution of the homogeneous PDE, i.e. with a vanishing density of sources. The sum $\phi_G(\mathbf{r}) + \phi_\psi(\mathbf{r})$ is then a solution of the original PDE, and it is conceivable that an appropriate adjustment of ϕ_ψ could give the sought field ϕ with the correct boundary conditions.

◇ Homogeneous boundary conditions ◇

These BC obey the following remarkable property. If two functions f_1 and f_2 defined on $\mathcal{D}$ satisfy these conditions, then $\alpha_1 f_1 + \alpha_2 f_2$ with α_1 and α_2 arbitrary smooth functions, also satisfies these BC.

Assume that the BC setting the field ϕ are homogeneous. Introduce then the homogeneous Green's function G_H defined by the same BC as in system (2.3). It is clear by homogeneity of these BC that the field ϕ_{G_H} given by superposition (2.4) also satisfies these BC. Then this field is actually the field ϕ we are looking for, i.e.

$$\phi(\mathbf{r}) = \int_D d\mathbf{r}'\ G_H(\mathbf{r};\mathbf{r}')\ \rho(\mathbf{r}'). \tag{2.6}$$

The homogeneous Green's function G_H therefore proves particularly useful to obtain the general solution of a PDE on ϕ by simple superposition of the corresponding elementary fields.

A simple example of homogeneous BC is given by the so-called homogeneous Dirichlet conditions, imposing that the solution vanishes at the domain boundary. Condition $BC(\phi)$ then reads $\phi(\mathbf{r}) = 0$ for $\mathbf{r} \in \partial D$. The homogeneous Green's function satisfies the same conditions $G_{DH}(\mathbf{r};\mathbf{r}') = 0$ for all $\mathbf{r} \in \partial D$. These homogeneous Dirichlet BC appear in many situations that we will address in more details later.

$$\Diamond\ \textbf{Usual properties}\ \Diamond$$

Some simple properties of Green's functions immediately arise from their definition. These properties are related to the problem symmetries determined both by the shape of domain D, the structure of operator $\mathcal{O}$ and the nature of boundary conditions. We list in the following the most frequently encountered properties.

Reciprocity Suppose that $\mathcal{O}$ is **Hermitian** on the set of functions defined on D and satisfying the given BC. It means that any functions $u(\mathbf{r})$ and $v(\mathbf{r})$ belonging to this set satisfy

$$\int_D d\mathbf{r}\ \left[u^*(\mathbf{r})\ \mathcal{O}_\mathbf{r}\ v(\mathbf{r}) - (\mathcal{O}_\mathbf{r}\ u(\mathbf{r}))^*\ v(\mathbf{r})\right] = 0,$$

where $u^*(\mathbf{r})$ stands for the complex conjugate of $u(\mathbf{r})$. We then have

$$\boxed{G(\mathbf{r}_1;\mathbf{r}_2) = G^*(\mathbf{r}_2;\mathbf{r}_1), \qquad \forall \mathbf{r}_1,\ \mathbf{r}_2 \in D.} \tag{2.7}$$

This relation is a particular example of the so-called reciprocity relations. One can prove it starting from the two equations

$$\mathcal{O}_\mathbf{r}\ G(\mathbf{r};\mathbf{r}_1) = \delta(\mathbf{r} - \mathbf{r}_1), \tag{2.8}$$

$$\mathcal{O}_\mathbf{r}\ G(\mathbf{r};\mathbf{r}_2) = \delta(\mathbf{r} - \mathbf{r}_2), \tag{2.9}$$

obtained by specifying the PDE satisfied by $G(\mathbf{r},\mathbf{r}')$ for two point sources $\mathbf{r}' = \mathbf{r}_1$ and $\mathbf{r}' = \mathbf{r}_2$ respectively. Multiply each side of equation (2.8) by $G^*(\mathbf{r};\mathbf{r}_2)$ on the one hand, and the complex conjugate of equation (2.9) by $G(\mathbf{r};\mathbf{r}_1)$ on the other hand, and then integrate each term over $\mathbf{r}$. We finally obtain, by subtracting the two resulting equations,

$$\int_D d\mathbf{r}\ \left[G^*(\mathbf{r};\mathbf{r}_2)\ \mathcal{O}_\mathbf{r}\ G(\mathbf{r};\mathbf{r}_1) - (\mathcal{O}_\mathbf{r}\ G(\mathbf{r};\mathbf{r}_2))^*\ G(\mathbf{r};\mathbf{r}_1)\right] = G^*(\mathbf{r}_1;\mathbf{r}_2) - G(\mathbf{r}_2;\mathbf{r}_1).$$

Since $\mathcal{O}$ is Hermitian, the left-hand side vanishes, leading to relation (2.7).

In the case where G is real, equation (2.7) becomes $G(\mathbf{r}_1; \mathbf{r}_2) = G(\mathbf{r}_2; \mathbf{r}_1)$. Points $\mathbf{r}_1$ and $\mathbf{r}_2$ then play symmetric roles: the elementary field created in $\mathbf{r}_1$ by $\mathbf{r}_2$ is identical to that created in $\mathbf{r}_2$ by $\mathbf{r}_1$.

Translation invariance If the problem symmetries involve translation invariance, then $G(\mathbf{r}; \mathbf{r}')$ depends only on the difference $\mathbf{r} - \mathbf{r}'$, as easily seen by expressing G in terms of the variables $\mathbf{r} - \mathbf{r}'$ and $\mathbf{r} + \mathbf{r}'$. Invariance under spatial translation by any vector $\mathbf{r}_0$ indeed implies identity $G(\mathbf{r} - \mathbf{r}'; \mathbf{r} + \mathbf{r}' + 2\mathbf{r}_0) = G(\mathbf{r} - \mathbf{r}'; \mathbf{r} + \mathbf{r}')$ for all $\mathbf{r}_0$, showing that G does not depend on the variable $\mathbf{r} + \mathbf{r}'$.

Translation invariance appears in various situations. The most common one is an infinitely large system (the domain $\mathcal{D}$ is then the whole space), with homogeneous Dirichlet BC and an operator $\mathcal{O}$ itself invariant by translation. Note that translation invariance can be restricted to particular spatial directions, especially in the case of semi-infinite systems.

Translation and rotation invariance Assume that the problem symmetries imply invariances both by rotation and translation. Based on the above discussion we have $G(\mathbf{r}; \mathbf{r}') = G(\mathbf{r} - \mathbf{r}')$. Moreover since G must remain unchanged under any rotation of centre $\mathbf{r}'$ and both arbitrary axis and angle, G does not depend on the angle $\mathbf{r} - \mathbf{r}'$ in a given coordinate system. G is then only a function of the relative distance $|\mathbf{r} - \mathbf{r}'|$. Note that if $\mathcal{O}$ is also Hermitian, reciprocity relation (2.7) combined with $G(\mathbf{r}_1; \mathbf{r}_2) = G(|\mathbf{r}_1 - \mathbf{r}_2|)$ implies that G is real.

Simultaneous invariance both by rotation and translation is typically observed for a system infinitely extended in all directions with an operator $\mathcal{O}$ presenting these invariance properties, and with homogeneous Dirichlet BC.

2.1.2 *Operatorial point of view*

Returning to general equation (2.1), we now assume that the boundary conditions are homogeneous. Since these BC define a unique solution ϕ for a given ρ, the inverse operator $\mathcal{O}^{-1}$ is well defined. This inverse operator depends on the type of chosen homogeneous BC. There are *a priori* several possible operators[1] $\mathcal{O}^{-1}$. We can write for a given set of BC

$$\phi(\mathbf{r}) = \mathcal{O}_\mathbf{r}^{-1}\, \rho(\mathbf{r}). \tag{2.10}$$

In this operatorial language, homogeneous Green's function $G_\mathrm{H}(\mathbf{r}, \mathbf{r}')$ is obtained by applying inversion formula (2.10) to a source point corresponding to a charge density $\delta(\mathbf{r} - \mathbf{r}')$, i.e.

$$G_\mathrm{H}(\mathbf{r}, \mathbf{r}') = \mathcal{O}_\mathbf{r}^{-1}\, \delta(\mathbf{r} - \mathbf{r}') \tag{2.11}$$

---

[1]In the following, for convenience of notation, we do not explicitly specify the dependence on the boundary conditions of $\mathcal{O}^{-1}$.

where $\mathcal{O}_{\mathbf{r}}^{-1}$ acts on position $\mathbf{r}$. It is easy to show that as $\mathcal{O}$ itself, its inverse $\mathcal{O}^{-1}$ is also linear. We then insert decomposition (2.2) of density $\rho(\mathbf{r})$ in inversion formula (2.10), and use the linearity property to get:

$$\phi(\mathbf{r}) = \int d\mathbf{r}' \rho(\mathbf{r}') \mathcal{O}_{\mathbf{r}}^{-1} \delta(\mathbf{r} - \mathbf{r}').$$

This expression coincides with expression (2.4) of the general solution with $G_{\mathrm{H}}(\mathbf{r}, \mathbf{r}') = \mathcal{O}_{\mathbf{r}}^{-1} \delta(\mathbf{r} - \mathbf{r}')$.

◊ **Spectral representation** ◊

Suppose now that operator $\mathcal{O}$ admits a complete set of orthonormal eigenfunctions $\psi_n(\mathbf{r})$ in the space of functions defined on $\mathcal{D}$, with the homogeneous boundary conditions under consideration. Noting λ_n the eigenvalue associated with $\psi_n(\mathbf{r})$, the action of $\mathcal{O}$ on $\psi_n(\mathbf{r})$ simply leads to

$$\mathcal{O} \, \psi_n(\mathbf{r}) = \lambda_n \, \psi_n(\mathbf{r}).$$

The set of eigenvalues defines the spectrum of $\mathcal{O}$. Note that some eigenvalues may be degenerate, i.e. $\lambda_n = \lambda_p$ with $n \neq p$. The spectrum moreover includes in general a discrete part for which n takes integer values, and a continuous part for which n is itself a continuous variable. The existence of an inverse operator $\mathcal{O}^{-1}$ prevents zero to be an eigenvalue. The kernel of $\mathcal{O}$, defined by $\mathcal{O}\psi = 0$, is nothing but the homogeneous PDE associated with general PDE (2.1). It then reduces to the identically vanishing function $\psi(\mathbf{r}) = 0$. Finally, $\psi_n(\mathbf{r})$ is also an eigenfunction of inverse operator $\mathcal{O}^{-1}$ with eigenvalue $1/\lambda_n$.

Since we assume that the set of functions $\psi_n(\mathbf{r})$ constitutes a complete basis, any function defined on $\mathcal{D}$ with the same BC can be uniquely decomposed as a linear combination of these functions. This remarkable property is equivalent to the so-called completeness relation

$$\sum_n \psi_n(\mathbf{r}) \, \psi_n^*(\mathbf{r}') = \delta(\mathbf{r} - \mathbf{r}'). \tag{2.12}$$

In this expression the summation over n must be understood as a symbolic notation representing the summation over the entire spectrum of $\mathcal{O}$: it includes in general a discrete sum as well as an integral over the continuous spectrum. Relation (2.12) is nothing but the decomposition of a density $\delta(\mathbf{r} - \mathbf{r}')$ corresponding to a source point located in $\mathbf{r}$, on the complete basis of eigenfunctions $\psi_n(\mathbf{r})$. Substituting $\delta(\mathbf{r} - \mathbf{r}')$ by its decomposition[2] in expression (2.11) and using the linearity of $\mathcal{O}_{\mathbf{r}}^{-1}$, we immediately find the spectral representation

---

[2]It is clear that Dirac distribution is poorly defined when $\mathbf{r}$ or $\mathbf{r}'$ belongs to $\partial\mathcal{D}$. It is not necessary to explicitly take into account this difficulty in the following: When it leads to ambiguities on the value of the quantities involved at the surface, it is implicitly agreed that these values are defined by a limit process starting from the inside of $\mathcal{D}$.

$$G_{\mathrm{H}}(\mathbf{r}; \mathbf{r}') = \sum_n \frac{\psi_n(\mathbf{r})\psi_n^*(\mathbf{r}')}{\lambda_n}. \qquad (2.13)$$

The knowledge of the spectrum of operator $\mathcal{O}$ thus gives access to the Green's function corresponding to the homogeneous BC considered via representation (2.13). Note that if $\mathcal{O}$ is Hermitian, all its eigenvalues λ_n are real: Reciprocity relations (2.7) are then obtained by taking the complex conjugate of representation (2.13).

◇ Extensions and alternative expressions ◇

We introduced the concept of Green's function for the determination of a field created by given sources. This concept can actually be extended to other physical situations, like in quantum mechanics, where the goal is not to compute a field generated by sources. By analogy this terminology is kept for quantities solutions of linear PDE of form (2.3). These quantities can no longer be interpreted in terms of elementary fields (electrostatic or other). They do have however a great interest in the analysis of the problem under consideration, as discussed in particular at section 2.1.7.

It is useful to write inversion formula (2.11) introducing concepts specific to vector spaces, both for simplification of algebraic manipulations, interpretation and further extensions. The vector space considered here is nothing but the space of functions defined on $\mathcal{D}$ with given homogeneous BC. This space is a Hilbert space of infinite dimension ✠.

Inversion formula (2.11) becomes in this context

$$G_{\mathrm{H}}(\mathbf{r}; \mathbf{r}') = \langle \mathbf{r} | \mathcal{O}^{-1} | \mathbf{r}' \rangle \qquad (2.14)$$

where we used Dirac notation. Thus Green's function $G_{\mathrm{H}}(\mathbf{r}, \mathbf{r}')$, created in an observation point $\mathbf{r}$ by a source localised in $\mathbf{r}'$, is nothing but the component along base vector $|\mathbf{r}\rangle$ of vector $\mathcal{O}^{-1}|\mathbf{r}\rangle$, image of $|\mathbf{r}\rangle$ by operator $\mathcal{O}^{-1}$, i.e. the matrix element of $\mathcal{O}^{-1}$ between $\langle \mathbf{r}|$ and $|\mathbf{r}\rangle$.

✠ **Comment:** We do not need here to give a precise mathematical definition of a Hilbert space of infinite dimension. Let us simply remember that it has properties very similar to those of finite vector spaces and refer the reader to Appendix D for a pragmatic reminder of concepts and properties required for operations performed below, in particular Dirac notation.

The technical interest of the concepts used here is well illustrated by the following manipulation. Let us write the tautology $\mathcal{O}^{-1} = \mathcal{O}^{-1}\mathcal{I}$ where $\mathcal{I}$ is the identity operator in the matrix element (2.14). We then replace $\mathcal{I}$ by the sum of orthogonal projectors along eigenvectors $|\psi_n\rangle$ of $\mathcal{O}$,

$$\mathcal{I} = \sum_n |\psi_n\rangle\langle\psi_n|. \qquad (2.15)$$

This identity is merely a rewriting of completeness relation (2.12). We then have

$$
\begin{aligned}
G_{\mathrm{H}}(\mathbf{r}; \mathbf{r}') &= \langle \mathbf{r} | \mathcal{O}^{-1} \mathcal{I} | \mathbf{r}' \rangle \\
&= \langle \mathbf{r} | \mathcal{O}^{-1} \sum_n |\psi_n\rangle \langle \psi_n | \mathbf{r}' \rangle \\
&= \sum_n \lambda_n^{-1} \langle \mathbf{r} | \psi_n \rangle \langle \psi_n | \mathbf{r}' \rangle,
\end{aligned}
\tag{2.16}
$$

which again gives spectral representation (2.13) *via* the identifications $\langle \mathbf{r}|\psi_n\rangle = \psi_n(\mathbf{r})$ and $\langle \psi_n|\mathbf{r}'\rangle = \psi_n^*(\mathbf{r}')$. We will give in section 2.1.6 examples of extensions and other interpretations of Green's functions based on matrix notation (2.14).

2.1.3 *Laplace operator*

Let us now specify our study to the case of the Laplace operator Δ, defined by

$$
\Delta = \sum_{i=1}^{d} \frac{\partial^2}{\partial x_i^2}.
\tag{2.17}
$$

Here x_i are Cartesian coordinates of a space of dimension d. The infinitesimal volume element is denoted $\mathrm{d}\mathbf{r} = \prod_{i=1}^{d} \mathrm{d}x_i$.

The Laplace operator is the simplest operator that can be constructed from the spatial partial derivatives $\partial/\partial x_i$, which is invariant under the natural symmetries of translation and rotation of the empty space: It is nothing but the scalar square of the gradient operator, i.e. $\Delta = \nabla^2$. This operator therefore appears naturally in physics. Thus the general form of the PDE in system (2.1) specified in the case $\mathcal{O} = -\Delta$ is **Poisson equation**[3]

$$
\boxed{-\Delta\phi(\mathbf{r}) = \rho(\mathbf{r}).}
\tag{2.18}
$$

This is one of the most common partial differential equations in physics. In electrostatic for instance ϕ/ϵ_0 is the electric potential and ρ reduces to the charge density.

We will begin the study of the Laplace operator by examining the problem of boundary conditions ensuring the uniqueness for Poisson equation. We then show how the use of the operator Green's functions allows to determine an integral equation satisfied by each solution of Poisson equation. This analysis provides a more precise knowledge of Green's functions of the Laplacian, as well as an expression of the solution of Poisson equation depending on the type of BC considered.

---

[3] Here the choice of a minus sign has no deeper meaning. This sign is introduced so that the PDE (2.18) thus defined coincides with the Poisson equation in electrostatics up to a factor $1/\epsilon_0$ in the source term.

◇ Boundary conditions ◇

As emphasised in the general case of an arbitrary operator $\mathcal{O}$, the uniqueness of solution ϕ of PDE (2.18) for given sources requires the introduction of appropriate boundary conditions $\mathrm{BC}(\phi)$ at the boundary $\partial\mathcal{D}$ defining the closed domain of interest $\mathcal{D}$. It is therefore crucial to determine the nature of these boundary conditions.

Physical argument   It is useful to remind the phenomena observed in electrostatics before moving on to a rigorous mathematical analysis of this problem. Imagine charges in a box whose walls are made of a conductive material kept at a fixed potential. Experiments show that an electrostatic equilibrium is reached with the appearance of influence charges in the conductive material. The total electrostatic field $\mathbf{E} = -\boldsymbol{\nabla}\phi$ takes a specific value at any point. This suggests that values of ϕ on the boundary $\partial\mathcal{D}$ are sufficient to ensure uniqueness. In a similar manner if the walls are now insulating, and if we drop surface charges onto them, the field $\mathbf{E}$ inside the box is again fully determined. The presence of surface charges inducing here a discontinuity of the component of $\mathbf{E}$ normal to the wall, it is now the conditions on the normal component of $\boldsymbol{\nabla}\phi$ which seem sufficient to ensure uniqueness.

These considerations suggest to carefully consider situations where ϕ or its derivative normal to $\partial\mathcal{D}$ are fixed on $\partial\mathcal{D}$. It will be our strategy after presenting two integral formulae called Green's formulae which will be used several times in the following.

Green identities   Let u and v be two complex-valued functions defined in domain $\mathcal{D}$ and square-integrable, i.e. such that integrals $\int_{\mathcal{D}} \mathrm{d}\mathbf{r} |u(\mathbf{r})|^2$ and $\int_{\mathcal{D}} \mathrm{d}\mathbf{r} |v(\mathbf{r})|^2$ are convergent. Then the pair of functions (u,v) satisfies the **first Green identity**:

$$\boxed{\int_{\mathcal{D}} \mathrm{d}\mathbf{r}\, [u(\mathbf{r})\Delta v(\mathbf{r}) + \boldsymbol{\nabla} u(\mathbf{r}) \cdot \boldsymbol{\nabla} v(\mathbf{r})] = \oint_{\partial\mathcal{D}} \mathrm{d}\Sigma\, u(\mathbf{r})\, [\mathbf{n} \cdot \boldsymbol{\nabla} v(\mathbf{r})].} \qquad (2.19)$$

In the surface integral appearing in the second member, the position $\mathbf{r}$ runs through the surface of $\partial\mathcal{D}$, $\mathbf{n}$ is the unit vector normal to the surface at the point considered and directed from the inside to the outside of the domain, and $\mathrm{d}\Sigma$ is the infinitesimal surface element (see Figure 2.2).

The proof of formula (2.19) is immediate from Green-Ostrogradski theorem. It states that a volume integral over $\mathcal{D}$ of the divergence of a vector is equal to the flux of this vector through the surface of $\partial\mathcal{D}$. Formula (2.19) is simply the expression of this identity for the vector $u\boldsymbol{\nabla}v$ of divergence $(u\Delta v + \boldsymbol{\nabla}u \cdot \boldsymbol{\nabla}v)$.

The **second Green identity** is the anti-symmetric part in u and v of the first Green identity,

$$\boxed{\int_{\mathcal{D}} \mathrm{d}\mathbf{r}\, [u(\mathbf{r})\Delta v(\mathbf{r}) - v(\mathbf{r})\Delta u(\mathbf{r})] = \oint_{\partial\mathcal{D}} \mathrm{d}\Sigma\, \mathbf{n} \cdot [u(\mathbf{r})\boldsymbol{\nabla}v(\mathbf{r}) - v(\mathbf{r})\boldsymbol{\nabla}u(\mathbf{r})].} \qquad (2.20)$$

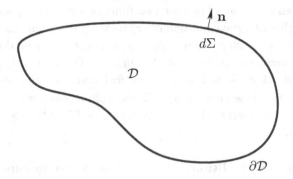

Fig. 2.2   In this figure **n** is the unit vector normal to the surface $\partial\mathcal{D}$ at the point considered and pointing outwards.

This identity is simply established by subtracting term by term the relations obtained by applying the first Green identity to the pairs (u, v) and (v, u) respectively.

Identification of natural boundary conditions   Let us return to the choice of boundary conditions. Suppose then that ϕ_1 and ϕ_2 are two solutions of (2.18) in $\mathcal{D}$ and introduce their difference $\alpha = \phi_1 - \phi_2$. It clearly satisfies the **Laplace equation**

$$\boxed{\Delta\alpha(\mathbf{r}) = 0.} \tag{2.21}$$

This PDE is a special case of Poisson equation (2.18) without source. Application of the first Green identity with $u = \alpha^*$ and $v = \alpha$ then gives

$$\int_{\mathcal{D}} d\mathbf{r}\, \boldsymbol{\nabla}\alpha^*(\mathbf{r}) \cdot \boldsymbol{\nabla}\alpha(\mathbf{r}) = \oint_{\partial\mathcal{D}} d\Sigma\, \alpha^*(\mathbf{r})\, [\mathbf{n} \cdot \boldsymbol{\nabla}\alpha(\mathbf{r})]. \tag{2.22}$$

So if $\alpha^*(\mathbf{r})[\mathbf{n} \cdot \boldsymbol{\nabla}\alpha(\mathbf{r})] = 0$ at any point on boundary $\partial\mathcal{D}$ we have

$$\int_{\mathcal{D}} d\mathbf{r}\, |\boldsymbol{\nabla}\alpha(\mathbf{r})|^2 = 0. \tag{2.23}$$

We therefore have that $\boldsymbol{\nabla}\alpha(\mathbf{r}) = 0$ at any point $\mathbf{r}$ of $\mathcal{D}$, and $\alpha(\mathbf{r})$ then reduces to a constant. This means that the two solutions ϕ_1 and ϕ_2 of Poisson equation differ only by a constant. These considerations allow us to formulate different choices of boundary conditions that guarantee uniqueness of the solution of PDE (2.18).

Dirichlet boundary conditions   These BC are defined by imposing the value of field $\phi(\mathbf{r})$ at any point $\mathbf{r}$ on boundary $\partial\mathcal{D}$, i.e.

$$\boxed{\textbf{Dirichlet:} \quad \phi(\mathbf{r}) = D(\mathbf{r}) \quad \text{for} \quad \mathbf{r} \in \partial\mathcal{D},} \tag{2.24}$$

where D is a given function. Then for two functions ϕ_1 and ϕ_2 solutions of PDE (2.18) satisfying this BC, the condition $\alpha^*(\mathbf{r})[\mathbf{n} \cdot \boldsymbol{\nabla}\alpha(\mathbf{r})] = 0$ on the boundary $\partial\mathcal{D}$ is also satisfied. Thus the difference $\alpha = \phi_1 - \phi_2$ is constant inside domain $\mathcal{D}$ and by continuity of the involved functions at the surface $\partial\mathcal{D}$, this constant is necessarily zero. We conclude that ϕ_1 and ϕ_2 are identical and the solution of PDE (2.18) is unique if it exists. Note that in general the uniqueness does not guarantee the existence. This is well illustrated by the discussion p. 67 on the so-called Neumann-Green's functions.

Neumann boundary conditions   These BC are defined by imposing the normal component value of the field gradient at any point $\mathbf{r}$ of the surface $\partial\mathcal{D}$, as well as the field value at a given point $\mathbf{r}_0$ of $\mathcal{D}$, i.e.

$$\boxed{\textbf{Neumann:}\quad \mathbf{n} \cdot \boldsymbol{\nabla}\phi(\mathbf{r}) = N(\mathbf{r}) \quad \text{for} \quad \mathbf{r} \in \partial\mathcal{D} \quad \text{and} \quad \phi(\mathbf{r}_0) = c,} \tag{2.25}$$

where N is a given function and c is a given complex number. For two functions ϕ_1 and ϕ_2 solutions of PDE (2.18) with this BC, the condition $\alpha^*(\mathbf{r})[\mathbf{n} \cdot \boldsymbol{\nabla}\alpha(\mathbf{r})] = 0$ on boundary $\partial\mathcal{D}$ is again satisfied, and therefore the difference $\alpha = \phi_1 - \phi_2$ is constant inside domain $\mathcal{D}$. This constant is actually zero because $\alpha(\mathbf{r})$ vanishes at $\mathbf{r} = \mathbf{r}_0$, implying uniqueness $\phi_1 = \phi_2$ but of course not necessarily the existence of the solution.

Other boundary conditions   There are naturally other boundary conditions which guarantee the uniqueness of the solution of PDE (2.18), in particular the so-called mixed conditions combining Dirichlet and Neumann BC on complementary parts of surface $\partial\mathcal{D}$. The reader may also refer to exercise 3.9 in Chapter 3 p. 209, presenting the so-called Robin boundary conditions involved in diffusion problems. In practice, the most common are Dirichlet or Neumann boundary conditions.

◇ **Green's functions and integral equations** ◇

A Green's function of the Laplacian satisfies the PDE (2.18) with $\rho(\mathbf{r}) = \delta(\mathbf{r} - \mathbf{r}')$, i.e.

$$-\Delta_{\mathbf{r}} G(\mathbf{r}; \mathbf{r}') = \delta(\mathbf{r} - \mathbf{r}'). \tag{2.26}$$

Let us now return to the field ϕ, defined as a solution of Poisson equation (2.18). We will show that ϕ satisfies an integral equation involving any Green's function of the Laplacian, solution of PDE (2.26). To do so let us rename the integration variable in the second Green's formula (2.20) $\mathbf{r}'$ and set $u(\mathbf{r}') = \phi(\mathbf{r}')$ and $v(\mathbf{r}') = G(\mathbf{r}'; \mathbf{r})$ where $\mathbf{r}$ is here a fixed point. We get

$$\int_{\mathcal{D}} d\mathbf{r}' \left[\phi(\mathbf{r}')\Delta_{\mathbf{r}'} G(\mathbf{r}'; \mathbf{r}) - G(\mathbf{r}'; \mathbf{r})\Delta_{\mathbf{r}'}\phi(\mathbf{r}') \right]$$

$$= \oint_{\partial\mathcal{D}} d\Sigma' \, \mathbf{n}' \cdot \left[\phi(\mathbf{r}')\boldsymbol{\nabla}_{\mathbf{r}'} G(\mathbf{r}'; \mathbf{r}) - G(\mathbf{r}'; \mathbf{r})\boldsymbol{\nabla}_{\mathbf{r}'}\phi(\mathbf{r}') \right].$$

According to PDE (2.18) and (2.26) respectively,

$$-\Delta_{\mathbf{r}'}\phi(\mathbf{r}') = \rho(\mathbf{r}') \quad \text{and} \quad -\Delta_{\mathbf{r}'}G(\mathbf{r}';\mathbf{r}) = \delta(\mathbf{r}' - \mathbf{r}),$$

the previous identity becomes

$$\boxed{\begin{aligned}
\phi(\mathbf{r}) = &\int_{\mathcal{D}} d\mathbf{r}'\rho(\mathbf{r}')G(\mathbf{r}';\mathbf{r}) \\
&- \oint_{\partial\mathcal{D}} d\Sigma'\,\mathbf{n}' \cdot [\phi(\mathbf{r}')\boldsymbol{\nabla}_{\mathbf{r}'}G(\mathbf{r}';\mathbf{r}) - G(\mathbf{r}';\mathbf{r})\boldsymbol{\nabla}_{\mathbf{r}'}\phi(\mathbf{r}')].
\end{aligned}} \quad (2.27)$$

Since the right-hand side depends on values of field $\phi(\mathbf{r})$ and its gradient on the boundary $\partial\mathcal{D}$, equation (2.27) is interpreted as an integral equation satisfied by $\phi(\mathbf{r})$. The main advantage of this equation is its validity for any Green's function solution of PDE (2.26) in any domain $\mathcal{D}$! Thus there is an integral equation for each Green's function. To go further we must take into account the BC satisfied by field $\phi(\mathbf{r})$. This is what we shall do in the following after introducing various Green's functions of the Laplacian.

As already stated, the specification of boundary conditions amounts to implicitly introduce additional sources at the domain boundary $\partial\mathcal{D}$. So each Green's function can be seen as a dressed elementary field, created both by a point source and implicit sources induced by the corresponding BC. Then the interpretation of each term of expression (2.27) is not straightforward because the volume integral also implicitly contains surface contributions. Note moreover that the volume term reduces to the field ϕ_G defined by the simple superposition (2.4), p. 56 only if G is symmetric in the exchange of source and observation points.

$$\Diamond \textbf{ Green's functions } \Diamond$$

The discussion about boundary conditions ensuring the uniqueness of solution of Poisson equation applies of course to Green's functions, solutions of Poisson equation for the particular source $\rho(\mathbf{r}) = \delta(\mathbf{r} - \mathbf{r}')$. We should however take into account the dependence of the Green's functions on two arguments $\mathbf{r}$ and $\mathbf{r}'$. More specifically, it is very useful to introduce several types of Green's functions, each defined by specific BC.

Green's functions for the Dirichlet case They are defined by PDE (2.26) with the boundary conditions $G(\mathbf{r};\mathbf{r}') = D(\mathbf{r};\mathbf{r}')$, for all $\mathbf{r} \in \partial\mathcal{D}$ and $\mathbf{r}' \in \mathcal{D}$. These BC define a whole set of Green's functions generated by all possible choices for the function $D(\mathbf{r};\mathbf{r}')$.

Green's functions for homogeneous Dirichlet problem A special case is given by the choice $D(\mathbf{r};\mathbf{r}') = 0$, which defines Green's function G_{HD} corresponding to the homogeneous Dirichlet BC:

Green's function	$-\Delta_{\mathbf{r}} G_{HD}(\mathbf{r}; \mathbf{r}') = \delta(\mathbf{r} - \mathbf{r}')$
for homogeneous Dirichlet problem	$G_{HD}(\mathbf{r}; \mathbf{r}') = 0 \quad \forall \mathbf{r} \in \partial \mathcal{D}, \quad \forall \mathbf{r}' \in \mathcal{D}$

As illustrated by integral formula (2.6) on p. 57, homogeneous Green's function turns out to be useful for solving many problems where BC on field ϕ are also of homogeneous Dirichlet type. We will see later that G_{HD} also provides a representation of any field ϕ defined by inhomogeneous Dirichlet BC for which the given function $D(\mathbf{r})$ in BC (2.24) is not identically vanishing.

Green's function $G_{HD}(\mathbf{r}_1; \mathbf{r}_2)$ is symmetric in the exchange of observation point $\mathbf{r}_1$ and source point $\mathbf{r}_2$, i.e.

$$\boxed{G_{HD}(\mathbf{r}_1; \mathbf{r}_2) = G_{HD}(\mathbf{r}_2; \mathbf{r}_1).} \tag{2.28}$$

As proposed in exercise 2.4, p. 112, the reader can establish this remarkable property by setting $G_{HD}(\mathbf{r}; \mathbf{r}_1) = u(\mathbf{r})$ and $G_{HD}(\mathbf{r}; \mathbf{r}_2) = v(\mathbf{r})$ in the second Green's formula (2.20). In addition since the Laplacian operator is Hermitian in the space of functions defined on $\mathcal{D}$ with homogeneous Dirichlet BC, G_{HD} also satisfies reciprocal relation (2.7). We deduce from symmetry (2.28) that G_{HD} is real. Note also that these properties can be derived from spectral representation (2.13). It is indeed possible to choose a set of real eigenfunctions for the Laplacian, which immediately implies that G_{HD} is real and symmetric.

An explicit calculation of G_{HD} is possible for domains with enough symmetries (see example of section 2.2.1, p. 90, or exercise 2.8, p. 114). For a domain of arbitrary shape, such a determination is a very difficult problem. The origin of this difficulty appears clearly in the physical context of electrostatics. G_{HD} is indeed nothing but the potential created by a point charge within a domain delimited by conducting walls maintained at zero potential. Surface charges are induced on these walls so that the total electrostatic field identically vanishes in the conductor. Green's function G_{HD} is then simply the sum of Coulomb potential created in vacuum both by point charges and induced charges. Its determination is then strictly equivalent to finding the induced charges distribution, a highly difficult problem for a surface $\partial \mathcal{D}$ without simple symmetries ✠.

✠ **Comment:**   This difficulty can be found in the study of the spectrum of Laplacian operator, which formally gives access to G_{HD} via representation (2.13), p. 60. Indeed, the determination of eigenvalues λ_n and corresponding eigenfunctions remains an open problem, except for simple surfaces $\partial \mathcal{D}$ such as a sphere or a cube. Note that these eigenvalues coincide in quantum mechanics, up to the factor $\hbar^2/(2m)$, with the energy levels of a free particle of mass m confined in a box of boundary $\partial \mathcal{D}$. This apparently simple problem is in fact rich and complex. In particular, unlike in the case of simple geometries where the energies are evenly spaced, the energy levels turn out to be randomly distributed and correlated: this is an aspect of quantum chaos [27] and its relation with random matrix theory [63].

Green's functions for the Neumann case It would seem *a priori* natural to proceed in the same way for Neumann boundary conditions, and to define Green's functions for the Neumann problem as the solutions of PDE (2.26) with boundary conditions $\mathbf{n} \cdot \boldsymbol{\nabla} G_N(\mathbf{r}; \mathbf{r}') = N(\mathbf{r}; \mathbf{r}')$ and $G_N(\mathbf{r}_0; \mathbf{r}') = c(\mathbf{r}')$ for all $\mathbf{r} \in \partial \mathcal{D}$, $\mathbf{r}' \in \mathcal{D}$ and a given $\mathbf{r}_0$. It turns out however that many of these solutions do not exist, i.e. the above system of equations has no solution for some choices. It can easily be shown by applying Green-Ostrogradski theorem to the flux of $\boldsymbol{\nabla}_{\mathbf{r}} G(\mathbf{r}; \mathbf{r}')$ across surface $\partial \mathcal{D}$, leading to the identity[4]:

$$\oint_{\partial \mathcal{D}} d\Sigma \, \mathbf{n} \cdot \boldsymbol{\nabla}_{\mathbf{r}} G(\mathbf{r}; \mathbf{r}') = -1, \qquad (2.29)$$

valid for any Green's function of the Laplacian, independently of the chosen BC. Since $\mathbf{n} \cdot \boldsymbol{\nabla}_{\mathbf{r}} G_N(\mathbf{r}; \mathbf{r}') = N(\mathbf{r}; \mathbf{r}')$, G_N could only exist if function N satisfies the condition

$$\oint_{\partial \mathcal{D}} d\Sigma \, N(\mathbf{r}; \mathbf{r}') = -1. \qquad (2.30)$$

This is a necessary condition for the existence of G_N but it may not be a sufficient condition. So there is no Green's function satisfying homogeneous Neumann BC where $N(\mathbf{r}, \mathbf{r}') = 0$. On the other hand, we can *a priori* impose inhomogeneous condition where $N(\mathbf{r}, \mathbf{r}')$ reduces to

$$N(\mathbf{r}; \mathbf{r}') = -1/s \qquad \forall \mathbf{r} \in \partial \mathcal{D}, \qquad (2.31)$$

where $s = \oint_{\partial \mathcal{D}} d\Sigma$ is the measure of surface $\partial \mathcal{D}$, assumed to be finite. We then define the special Green's functions for Neumann problem:

$$
\boxed{
\begin{array}{l|l}
\textbf{Special} & -\Delta_{\mathbf{r}} G_{\bar{N}}(\mathbf{r}; \mathbf{r}') = \delta(\mathbf{r} - \mathbf{r}') \\
\textbf{Neumann} & \mathbf{n} \cdot \boldsymbol{\nabla}_{\mathbf{r}} G_{\bar{N}}(\mathbf{r}; \mathbf{r}') = -\dfrac{1}{s} \quad \forall \mathbf{r} \in \partial \mathcal{D}, \quad \forall \mathbf{r}' \in \mathcal{D} \\
\textbf{Green's functions} & \quad\quad G_{\bar{N}}(\mathbf{r}_0; \mathbf{r}') = c(\mathbf{r}') \quad \forall \mathbf{r}' \in \mathcal{D}
\end{array}
} \qquad (2.32)
$$

As we shall later see $G_{\bar{N}}$ is called on to play an important role in situations where the physical boundary conditions on field ϕ are of Neumann type (2.25).

Unlike the case of G_{HD} the functions $G_{\bar{N}}$ do not have general symmetry properties[5]. Their determination is just as complicated as the calculation of G_{HD} for domains $\mathcal{D}$ of any shape, so that an explicit calculation of $G_{\bar{N}}$ is only possible for simple geometries.

---

[4]In the context of electrostatics this identity is obviously Gauss theorem.
[5]See however exercise 2.5 on p. 112.

◇ **Homogeneous Green's function for an infinite size system** ◇

When the domain $\mathcal{D}$ is the space $\mathbb{R}^d$ as a whole, boundary $\partial\mathcal{D}$ is sent to infinity and the Green's function of the homogeneous Dirichlet problem for infinite system, denoted G_∞, is defined by

$$
\boxed{
\begin{aligned}
-\Delta_{\mathbf{r}} G_\infty(\mathbf{r}; \mathbf{r}') &= \delta(\mathbf{r} - \mathbf{r}') \\
G_\infty(\mathbf{r}; \mathbf{r}') &\to 0 \quad \text{for} \quad |\mathbf{r}| \to \infty \quad \text{with} \quad \mathbf{r}' \quad \text{fixed.}
\end{aligned}
}
\tag{2.33}
$$

As shown in the overview section, invariances by translation and rotation of the Laplace operator and selected BC imply that $G_\infty(\mathbf{r}; \mathbf{r}')$ is a function of $|\mathbf{r} - \mathbf{r}'|$. In the following, we explicitly compute $G_\infty(|\mathbf{r} - \mathbf{r}'|)$ in the three-dimensional case. The case of lower dimensions will be discussed in section 2.1.5, while higher dimensions are treated in exercise 2.2, p. 111.

Calculation of G_∞ in three dimensions   Linearity of the PDE satisfied by G_∞ combined with translational invariance guide us towards using Fourier transform to determine the Green's function. Consider the Fourier transform of G_∞ defined by

$$
\widehat{G}_\infty(\mathbf{k}) = \int \mathrm{d}\mathbf{r}\; e^{-i\mathbf{k}\cdot(\mathbf{r}-\mathbf{r}')}\; G_\infty(|\mathbf{r} - \mathbf{r}'|).
$$

In Fourier space partial derivatives with respect to spatial coordinates become simple multiplications by powers of $i\mathbf{k}$. Therefore $\widehat{G}_\infty(\mathbf{k})$ satisfies the purely algebraic equation

$$
\mathbf{k}^2\, \widehat{G}_\infty(\mathbf{k}) = 1,
\tag{2.34}
$$

whose solution is immediate. We find $\widehat{G}_\infty(\mathbf{k}) = 1/\mathbf{k}^2$ up to a subtle point discussed later, then $G_\infty(|\mathbf{r} - \mathbf{r}'|)$ is readily obtained by inverse Fourier transform, leading to

$$
G_\infty(\mathbf{r} - \mathbf{r}') = \frac{1}{(2\pi)^3} \int \mathrm{d}\mathbf{k}\; e^{i\mathbf{k}\cdot(\mathbf{r}-\mathbf{r}')}\frac{1}{\mathbf{k}^2}.
\tag{2.35}
$$

The integral over $\mathbf{k}$ can be done by taking as Oz axis the direction of the fixed vector $\mathbf{r} - \mathbf{r}'$ and introducing the corresponding spherical coordinates

$$
\begin{aligned}
G_\infty(|\mathbf{r} - \mathbf{r}'|) &= \frac{1}{(2\pi)^3} \int_0^\infty \mathrm{d}k\, k^2 \int_0^\pi \mathrm{d}\theta \sin\theta \int_0^{2\pi} \mathrm{d}\phi\; e^{ik|\mathbf{r}-\mathbf{r}'|\cos\theta}\frac{1}{k^2} \\
&= \frac{1}{2\pi^2} \int_0^\infty \mathrm{d}k \frac{\sin(k|\mathbf{r} - \mathbf{r}'|)}{k|\mathbf{r} - \mathbf{r}'|} \\
&= \frac{1}{2\pi^2|\mathbf{r} - \mathbf{r}'|} \int_0^\infty \mathrm{d}\xi \frac{\sin\xi}{\xi}.
\end{aligned}
$$

This last integral can be elegantly evaluated just by applying Kramers-Kroenig relation (1.23) to function e^{iz}. One eventually gets

$$
\boxed{
G_\infty(|\mathbf{r} - \mathbf{r}'|) = \frac{1}{4\pi|\mathbf{r} - \mathbf{r}'|}.
}
\tag{2.36}
$$

We thus recover, up to a factor q/ϵ_0, the well-known expression for Coulomb potential in three dimensions created by a point charge q in vacuum.

Discussion of boundary conditions It turns out that expression (2.36) for G_∞ meets the homogeneous Dirichlet BC of system (2.33). However, we have at no time imposed any constraint whatsoever on the value of G_∞ at infinity! It is therefore natural to ask why the result was selected among countless other possibilities of boundary conditions. To clarify this paradox, let us return to equation (2.34). This equation admits for solutions, in the sense of distributions,

$$\frac{1}{\mathbf{k}^2} + \widehat{D}_h(\mathbf{k}),$$

where $\widehat{D}_h(\mathbf{k})$ is a distribution of support localised in $\mathbf{k} = \mathbf{0}$. This distribution can be written as a linear combination of $\delta(\mathbf{k})$ and all its derivatives. Its inverse Fourier transform, $D_h(\mathbf{r} - \mathbf{r}')$, is simply a harmonic function solution of Laplace equation (2.21) *in the whole space*. This function is a linear combination of spherical harmonics[6] of the form $R^l\, Y_l^m(\theta, \varphi)$ with l an integer, m an integer satisfying $-l \leq m \leq l$, and where $\mathbf{R} = \mathbf{r} - \mathbf{r}'$ is the relative position vector characterised by angles θ and φ in spherical coordinates. It is then clear that only the identically vanishing function D_h has the correct decaying behaviour at infinity. This justifies the identification of $\widehat{G}_\infty(\mathbf{k})$ to $1/\mathbf{k}^2$.

Spectral viewpoint Note that G_∞ can also be easily computed by spectral considerations. It is convenient to introduce a cubic finite domain of side L, and the corresponding Green's function G_{HD}^L for the homogeneous Dirichlet problem. Eigenfunctions of the Laplace operator involved in this method are stationary waves, linear combinations of plane waves $e^{\pm i\mathbf{k}\cdot\mathbf{r}}$ for the corresponding eigenvalue $-\mathbf{k}^2$ with $\mathbf{k}$ quantised. In spectral representation (2.13) of $G_{\mathrm{HD}}^L(\mathbf{r}; \mathbf{r}')$, the limit $L \to \infty$ for fixed positions r and r' then gives exactly integral expression (2.35) of $G_\infty(|\mathbf{r} - \mathbf{r}'|)$. So let us emphasise that expression (2.35)

$$G_\infty(\mathbf{r} - \mathbf{r}') = \frac{1}{(2\pi)^3} \int d\mathbf{k}\, e^{i\mathbf{k}\cdot\mathbf{r}} \frac{1}{\mathbf{k}^2} e^{-i\mathbf{k}\cdot\mathbf{r}'},$$

has a natural interpretation within this spectral point of view[✠].

> ✠ **Comment:** This discussion also helps to understand why it is natural to compute G_∞ using Fourier transform. The Laplace operator indeed commutes with the operator generating translations, which is simply the operator ∇. This commutation property is obvious since $\Delta = \nabla^2$. Plane waves are actually eigenfunctions of the translation operator. To borrow a language used in quantum mechanics, Laplace and translation operators are simultaneously diagonalised by Fourier transform.

---

[6]Appendix G contains reminders about spherical harmonics.

Asymptotic behaviour   In high dimensions $(d > 3)$, $G_\infty(|\mathbf{r} - \mathbf{r}'|)$ decays as $1/|\mathbf{r} - \mathbf{r}'|^{d-2}$ at large relative distances (see exercise 2.2 p. 111). In low dimension $(d < 3)$, $G_\infty(|\mathbf{r} - \mathbf{r}'|)$ diverges as $\ln(|\mathbf{r} - \mathbf{r}'|)$ for $d = 2$ and as $|\mathbf{r} - \mathbf{r}'|$ for $d = 1$, as we will see in section 2.1.5. All these asymptotic behaviours are non-integrable: G_∞ is said to be long range. These slow decays and these divergences even in low dimension have major consequences as we shall later see.

◇ **Solution of Poisson equation** ◇

Dirichlet BC on ϕ   Let us return to integral equation (2.27),

$$\phi(\mathbf{r}) = \int_{\mathcal{D}} d\mathbf{r}' \rho(\mathbf{r}') G(\mathbf{r}'; \mathbf{r}) - \oint_{\partial \mathcal{D}} d\Sigma' \, \mathbf{n}' \cdot [\phi(\mathbf{r}')\boldsymbol{\nabla}_{\mathbf{r}'} G(\mathbf{r}'; \mathbf{r}) - G(\mathbf{r}'; \mathbf{r})\boldsymbol{\nabla}_{\mathbf{r}'}\phi(\mathbf{r}')],$$

and remind that it is valid for any Green's function of Laplace operator, regardless of its BC and those defining ϕ. Suppose we impose Dirichlet BC (2.24) on ϕ. It is clear that the knowledge of a Green's function will not be sufficient to determine ϕ from the integral equation above, because the value of $\boldsymbol{\nabla}_{\mathbf{r}'}\phi(\mathbf{r}')$ on the domain boundary $\partial \mathcal{D}$ is not known *a priori*. In fact, it makes sense to use here the Green's function for the homogeneous Dirichlet problem G_{HD} vanishing on $\partial \mathcal{D}$. The above expression then reduces to ⌖

$$\boxed{\phi(\mathbf{r}) = \int_{\mathcal{D}} d\mathbf{r}' \, \rho(\mathbf{r}') G_{\mathrm{HD}}(\mathbf{r}; \mathbf{r}') - \oint_{\partial \mathcal{D}} d\Sigma' \, D(\mathbf{r}')\mathbf{n}' \cdot \boldsymbol{\nabla}_{\mathbf{r}'} G_{\mathrm{HD}}(\mathbf{r}'; \mathbf{r}),} \qquad (2.37)$$

where $D(\mathbf{r}')$ is the value of ϕ on $\partial \mathcal{D}$ imposed by BC (2.24) and where we used symmetry (2.28) of function $G_{\mathrm{HD}}(\mathbf{r}; \mathbf{r}')$. Thus the knowledge of G_{HD} gives immediate access to ϕ for any density ρ.

⌖ **Comment:**   In expression (2.37) the volume term vanishes for $\mathbf{r}$ in $\partial \mathcal{D}$. So the surface integral must reduce to $D(\mathbf{r})$, again for $\mathbf{r}$ in $\partial \mathcal{D}$. It therefore means that for $\mathbf{r}$ and $\mathbf{r}'$ in $\partial \mathcal{D}$

$$-\mathbf{n}' \cdot \boldsymbol{\nabla}_{\mathbf{r}'} G_{\mathrm{HD}}(\mathbf{r}'; \mathbf{r}) \qquad (2.38)$$

is necessarily the Dirac distribution on the surface $\partial \mathcal{D}$. This result has a simple electrostatic interpretation where the above expression is the component of the field $\mathbf{E}$ normal to the surface. As shown on p. 66 indeed, the presence of a point charge (located here in $\mathbf{r}$) induces a polarisation cloud on the conducting wall at zero potential. This cloud is localised in the charge vicinity. When this charge is close to the wall, it is completely screened by the polarisation cloud so that the electrostatic field vanishes except at point $\mathbf{r}$. Note also that the integral of the normal component of the electric field is given by Gauss theorem (2.29). So expression (2.38) indeed matches with the Dirac distribution. The reader can then check this property explicitly for a particular geometry, as proposed in exercise 2.15 on p. 120.

Neumann boundary conditions on ϕ Suppose now that Neumann BC (2.25) are imposed. The case of Dirichlet BC suggests choosing homogeneous Neumann BC for G. But it has been previously shown that such a Green's function does not exist! We can however consider the special Green's function $G_{\bar{N}}$, corresponding to definition (2.32). The introduction of $G_{\bar{N}}$ in the integral equation (2.27) leads to

$$\phi(\mathbf{r}) = \overline{\phi}_{\partial \mathcal{D}} + \int_{\mathcal{D}} d\mathbf{r}' \, \rho(\mathbf{r}')G_{\bar{N}}(\mathbf{r}';\mathbf{r}) + \oint_{\partial \mathcal{D}} d\Sigma' \, G_{\bar{N}}(\mathbf{r}';\mathbf{r})N(\mathbf{r}'), \qquad (2.39)$$

where

$$\overline{\phi}_{\partial \mathcal{D}} = \frac{1}{s} \oint_{\partial \mathcal{D}} d\Sigma' \, \phi(\mathbf{r}')$$

is the average of ϕ on the domain surface area $\partial \mathcal{D}$ and $N(\mathbf{r}')$ is the value of $\mathbf{n}' \cdot \boldsymbol{\nabla}_{\mathbf{r}'}\phi(\mathbf{r}')$ on $\partial \mathcal{D}$ imposed by BC (2.25). The knowledge of $G_{\bar{N}}$ determines surface and volume integrals in expression (2.39). The average $\overline{\phi}_{\partial \mathcal{D}}$, which acts as a constant, is then adjusted by imposing the value of $\phi(\mathbf{r}_0)$ specified in BC (2.25). This completes the full determination of ϕ from $G_{\bar{N}}$.

Use of G_∞ As mentioned earlier, determining Green's functions G_{DH} or $G_{\bar{N}}$ can quickly become difficult. The boundary conditions defining the Green's functions are however not necessarily taken on the domain boundary $\partial \mathcal{D}$. We can in fact introduce another boundary on a larger domain **including** the domain of interest $\mathcal{D}$. It is particularly advantageous in this context to use the Green's function of the infinite homogeneous system.

Using this Green's function we can use its symmetries and its dependence only on the difference $\mathbf{r} - \mathbf{r}'$ to see that integral equation (2.27), p. 65, becomes

$$\phi(\mathbf{r}) = \int_{\mathcal{D}} d\mathbf{r}' G_\infty(|\mathbf{r} - \mathbf{r}'|)\rho(\mathbf{r}') + \oint_{\partial \mathcal{D}} d\Sigma' \, G_\infty(|\mathbf{r} - \mathbf{r}'|)\mathbf{n}' \cdot \boldsymbol{\nabla}_{\mathbf{r}'}\phi(\mathbf{r}')$$

$$\qquad (2.40)$$

$$+ \oint_{\partial \mathcal{D}} d\Sigma' \, \phi(\mathbf{r}') \, \mathbf{n}' \cdot \boldsymbol{\nabla}_{\mathbf{r}} G_\infty(|\mathbf{r} - \mathbf{r}'|).$$

The interpretation of this result is more illuminating than for general expression (2.27). Indeed, since G_∞ is the bare elementary field created by a single source point in the entire space, the various contributions to this formula can be simply interpreted. The volume integral is the total field created by the original sources distributed within $\mathcal{D}$ with density ρ. The first surface integral,

$$\oint_{\partial \mathcal{D}} d\Sigma' \, G_\infty(|\mathbf{r} - \mathbf{r}'|)\mathbf{n}' \cdot \boldsymbol{\nabla}_{\mathbf{r}'}\phi(\mathbf{r}'),$$

is the total field created by induced sources located on $\partial \mathcal{D}$, and distributed with the surface density $\mathbf{n}' \cdot \boldsymbol{\nabla}_{\mathbf{r}'}\phi(\mathbf{r}')$. The second surface integral contains a derivative of elementary field G_∞. The quantity $\mathbf{n}' \cdot \boldsymbol{\nabla}_{\mathbf{r}} G_\infty(|\mathbf{r} - \mathbf{r}'|)$ is in fact the elementary field created by a bare source located at $\mathbf{r}'$ and with density $\mathbf{n}' \cdot \boldsymbol{\nabla}_{\mathbf{r}} \delta(\mathbf{r} - \mathbf{r}')$. It is indeed solution of the corresponding PDE (2.18) in the whole space with homogeneous Dirichlet boundary conditions at infinity. Returning to the terminology of electrostatics, this localised distribution has a zero net charge, i.e.

$$\int_{\mathbb{R}^d} d\mathbf{r} \; \mathbf{n}' \cdot \boldsymbol{\nabla}_{\mathbf{r}} \delta(\mathbf{r} - \mathbf{r}') = 0,$$

and only its dipole (or first moment)

$$\int_{\mathbb{R}^d} d\mathbf{r} \; (\mathbf{r} - \mathbf{r}') \; \mathbf{n}' \cdot \boldsymbol{\nabla}_{\mathbf{r}} \delta(\mathbf{r} - \mathbf{r}') = -\mathbf{n}'$$

is non-zero, all other higher order multipoles (moments) vanishing identically. Therefore it describes a dipole strictly localised in $\mathbf{r}'$ of intensity $-\mathbf{n}'$, so that the surface integral

$$\oint_{\partial \mathcal{D}} d\Sigma' \; \phi(\mathbf{r}') \; \mathbf{n}' \cdot \boldsymbol{\nabla}_{\mathbf{r}} G_\infty(|\mathbf{r} - \mathbf{r}'|)$$

is nothing but the field created by the assembly of all these dipoles distributed with the surface density $-\phi(\mathbf{r}')$.

One should note that though formula (2.40) is specific to the use of G_∞, it turns out to be very useful in various cases. It is clearly not completely explicit since it involves the values of ϕ and $\boldsymbol{\nabla}\phi$ on $\partial \mathcal{D}$ which cannot be imposed simultaneously under specific BC. The knowledge of G_∞ however makes this expression for ϕ particularly interesting. First a subsequent explicit resolution is sometimes possible by imposing appropriate BC as shown in the example of section 2.2.2 on p. 93. Formula (2.40) also provides information on the asymptotic behaviour and the importance of finite size effects. Note in particular that even when the boundaries are sent to infinity, surface terms may give non-zero contributions to $\phi(\mathbf{r})$ at a fixed point $\mathbf{r}$ as a result of the long range behaviour of G_∞.

2.1.4 *Helmholtz operator*

The Helmholtz operator, defined by

$$\mathcal{O} = -\Delta + m^2 \tag{2.41}$$

with m a real positive constant, appears in many fields such as statistical mechanics of plasmas or particle physics. It is also the simplest extension of the Laplace operator that preserves both the invariance by rotation and translation of empty space[7]. The general form of the PDE associated with Helmholtz operator is:

$$\boxed{-\Delta\phi(\mathbf{r}) + m^2\phi(\mathbf{r}) = \rho(\mathbf{r}).} \tag{2.42}$$

The methods introduced previously to study Poisson equation are readily applicable to the analysis of this PDE. We will then resume the presentation of the previous section, sketching the proofs and highlighting the specific properties of Helmholtz operator.

---

[7]The discussion on the operator $-\Delta - m^2$, whose properties are very different from the Helmholtz operator, is included in the general discussion in section 2.1.6.

◇ **Analysis of boundary conditions** ◇

In the case of the Laplace operator we have shown that Dirichlet boundary conditions (2.24), p. 63 uniquely determine the solution of Poisson equation. We will show that these BC also determine uniquely the solution of PDE (2.42). To this end let us consider two solutions ϕ_1 and ϕ_2 of this PDE. Their difference $\alpha = \phi_1 - \phi_2$ is such that

$$\Delta\alpha(\mathbf{r}) = m^2\alpha(\mathbf{r}) \,,$$

so that first Green's formula (2.19), p. 62, applied to $u = \alpha^*$ and $v = \alpha$ gives

$$\int_{\mathcal{D}} d\mathbf{r} \left[m^2|\alpha(\mathbf{r})|^2 + |\boldsymbol{\nabla}\alpha(\mathbf{r})|^2 \right] = \oint_{\partial\mathcal{D}} d\Sigma \, \alpha^*(\mathbf{r}) \left[\mathbf{n} \cdot \boldsymbol{\nabla}\alpha(\mathbf{r})\right]. \qquad (2.43)$$

If ϕ_1 and ϕ_2 fulfil the same Dirichlet BC (2.24) then α is identically zero on the surface $\partial\mathcal{D}$. The volume integral of the left-hand side of identity (2.43) must therefore be zero. Since the corresponding integrand cannot be negative at any point[8], the difference α is identically zero on any domain $\mathcal{D}$. This shows that Dirichlet BC (2.24) indeed guarantee uniqueness of the solution of PDE (2.42).

For the Laplace operator with Neumann BC (2.25) one must specify in particular the value of ϕ at a given point. This is not necessary for Helmholtz operator, for which it is sufficient to impose values of $\mathbf{n} \cdot \boldsymbol{\nabla}\phi$ on boundary $\partial\mathcal{D}$, i.e.

$$\mathbf{n} \cdot \boldsymbol{\nabla}\phi(\mathbf{r}) = N(\mathbf{r}) \qquad \text{for} \quad \mathbf{r} \in \partial\mathcal{D} \,. \qquad (2.44)$$

Indeed, by exploiting again identity (2.43) and that $\mathbf{n} \cdot \boldsymbol{\nabla}\alpha$ vanishes on the boundary $\partial\mathcal{D}$, we find that the difference $\alpha = \phi_1 - \phi_2$ between two solutions is necessarily identically zero at any point in the domain. We shall call conditions (2.44) simple Neumann BC.

We should keep in mind that the above BC do not guarantee *a priori* the existence of a solution. Moreover, if they occur naturally in most physical situations, other BC leading to uniqueness of the solution are also possible.

◇ **Green's functions of Helmholtz operator** ◇

Green's functions $G(\mathbf{r}, \mathbf{r}')$ are solutions of PDE

$$\left(-\Delta_{\mathbf{r}} + m^2\right) G(\mathbf{r}; \mathbf{r}') = \delta(\mathbf{r} - \mathbf{r}') \qquad (2.45)$$

with particular boundary conditions. As in the Laplacian case, Green's function G_{DH} for homogeneous Dirichlet problem is called upon to play an important role. This Green's function $G_{\mathrm{DH}}(\mathbf{r}, \mathbf{r}')$ is again real and symmetric in the exchange of observation point $\mathbf{r}$ and source point $\mathbf{r}'$. These properties can be established by repeating the demonstrations introduced in the Laplacian case and noting that Helmholtz operator is also Hermitian.

---

[8]It is here that the analysis for operator $-\Delta - m^2$ is different (see p. 86).

Note that unlike in the case of Laplace equation, it is possible to construct simple Green's functions satisfying homogeneous Neumann BC: the boundary function N then identically vanishes. Green's function G_{NH} so defined is real and symmetric, just like G_{HD} and for similar reasons.

As in the case of the Laplacian, it is particularly useful to introduce Green's functions for the homogeneous Dirichlet problem of an infinite system $G_\infty(\mathbf{r}; \mathbf{r}')$ vanishing when $|\mathbf{r}| \to \infty$ for a fixed $\mathbf{r}'$. Rotation and translation invariances imply that $G_\infty(\mathbf{r}; \mathbf{r}')$ depends only on $|\mathbf{r} - \mathbf{r}'|$. Here we explicitly compute $G_\infty(|\mathbf{r} - \mathbf{r}'|)$ in the three-dimensional case. The case of dimensions $d < 3$ is discussed in section 2.1.5. Once again linearity and translation invariances naturally lead us to take the Fourier transform on each side of PDE (2.45). This leads to an algebraic equation for $\widehat{G}_\infty(\mathbf{k})$ whose solution reduced to

$$\widehat{G}_\infty(\mathbf{k}) = \frac{1}{\mathbf{k}^2 + m^2} \, . \tag{2.46}$$

Its inverse transform reads

$$G_\infty(\mathbf{r} - \mathbf{r}') = \frac{1}{(2\pi)^3} \int d\mathbf{k} \, e^{i\mathbf{k}\cdot(\mathbf{r}-\mathbf{r}')} \frac{1}{\mathbf{k}^2 + m^2}.$$

The integral over $\mathbf{k}$ is computed using spherical coordinates already introduced to evaluate inverse Fourier transform (2.35). The angular integrations are elementary and lead to

$$G_\infty(\mathbf{r} - \mathbf{r}') = \frac{i}{4\pi^2|\mathbf{r} - \mathbf{r}'|} \int_0^{+\infty} dk \frac{k}{k^2 + m^2} \left(e^{-ik|\mathbf{r}-\mathbf{r}'|} - e^{ik|\mathbf{r}-\mathbf{r}'|} \right)$$

$$= -\frac{i}{4\pi^2|\mathbf{r} - \mathbf{r}'|} \int_{-\infty}^{+\infty} dk \frac{k}{k^2 + m^2} e^{ik|\mathbf{r}-\mathbf{r}'|}. \tag{2.47}$$

This integral can be carried out by applying the residue theorem to function $ze^{iz|\mathbf{r}-\mathbf{r}'|}/(z^2 + m^2)$ whose poles are located on the imaginary axis in $z_\pm = \pm im$. In the upper half complex plane, the integrand decreases exponentially fast when $\mathrm{Im}\, z \to +\infty$. It is therefore appropriate to introduce the closed contour obtained by supplementing the real axis with a semicircle in the upper half complex plane (see Figure 2.3). By Jordan's Lemma, the integral over the semicircle is zero in the limit of infinite radius. The only relevant residue is the one of pole $z_+ = im$ so that we eventually find

$$\boxed{G_\infty(\mathbf{r} - \mathbf{r}') = \frac{e^{-m|\mathbf{r}-\mathbf{r}'|}}{4\pi|\mathbf{r} - \mathbf{r}'|}.} \tag{2.48}$$

For $m = 0$ we recover expression (2.36) of G_∞ for the Laplacian. For $m \neq 0$, the long range of G_∞ observed in the Laplacian case disappears: The presence of term $e^{-m|\mathbf{r}-\mathbf{r}'|}$ induces an exponentially fast decay of G_∞ when $|\mathbf{r} - \mathbf{r}'| \to \infty$. This mechanism called screening plays a fundamental role in plasma physics. Then up to the factor q/ϵ_0, Green's function (2.48) is then equal to the Debye potential created by a point charge q immersed in a medium containing free charges, and $1/m$ is the

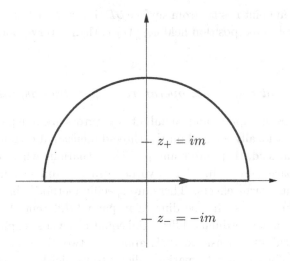

Fig. 2.3 Contour used to compute the integral (2.47) using residue theorem.

Debye length. In particle physics this Green's function can be identified with the Yukawa potential between two elementary entities generated by the exchange[9] of particles of mass proportional to m.

◇ General expression of the solution ◇

As in the case of the Laplacian, let us apply the second Green's formula (2.20), p. 62, after renaming the integration variable $\mathbf{r}'$, and $u(\mathbf{r}') = \phi(\mathbf{r}')$ and $v(\mathbf{r}') = G(\mathbf{r}';\mathbf{r})$, where $\mathbf{r}$ is still a fixed point. Using PDE (2.42) and (2.45) respectively satisfied by $\phi(\mathbf{r}')$ and $G(\mathbf{r}';\mathbf{r})$, we find that the expression of ϕ in terms of any Green's function G is identical to expression (2.27) obtained for the Laplacian.

Again, if ϕ satisfies Dirichlet BC, it is wise to use homogeneous Dirichlet Green's function G_{DH} in general expression (2.27), giving exactly formula (2.37), p. 70. If ϕ satisfies simple Neumann BC (2.44), the use of homogeneous Neumann Green's function G_{NH} leads to formula

$$\phi(\mathbf{r}) = \int_{\mathcal{D}} d\mathbf{r}' \, \rho(\mathbf{r}') G_{NH}(\mathbf{r};\mathbf{r}') + \oint_{\partial\mathcal{D}} d\Sigma' \, G_{NH}(\mathbf{r}';\mathbf{r}) N(\mathbf{r}'). \qquad (2.49)$$

As in the Laplacian case, an explicit determination of Green's functions G_{HD} or G_{NH} is not an easy task for a domain $\mathcal{D}$ with no particular symmetries. It may then be more advantageous to consider the version (2.40) of general formula (2.27) in terms of Green's function G_∞ for an infinite system. In particular we immediately see that since G_∞ is now short range, boundary effects on $\phi(\mathbf{r})$ will be extremely

---

[9]This exchange process can describe the strong interaction for example. Note that electromagnetic interactions are generated by exchanges of massless photons, resulting in the long-range decay of Coulomb interactions as $1/|\mathbf{r} - \mathbf{r}'|$.

small if observation point $\mathbf{r}$ is far from surface $\partial \mathcal{D}$, i.e. at a distance large compared to $1/m$. The simple superposition field $\phi_{G_\infty}(\mathbf{r})$ is then a very good approximation for $\phi(\mathbf{r})$.

2.1.5 *Laplace and Helmholtz operators in low dimensions*

Previous properties of Laplace and Helmholtz operators are independent of dimension. We have occasionally focused on the three-dimensional case for obvious practical reasons. That said, there are many physical situations where everything happens as if the space were reduced to two or even one dimension, for symmetry reasons or by confinement effects. There are specific methods that are very useful for these lower dimensions. In one dimension partial differential equations (2.18) and (2.42) become in fact ordinary differential equations whose explicit solutions are elementary. Difficulties intrinsic to PDE appear in two dimensions. Nevertheless the method of conformal transformations allows to explicitly compute the Green's functions of interest in geometries frequently encountered in practice, by introducing an equivalent problem in a domain of very simple geometry. We conclude this section with a brief description of the Green's functions for a plane and a line both infinite.

$$\Diamond \ \textbf{Segment of length } L \ \Diamond$$

In one dimension domain $\mathcal{D}$ is a segment of length L. We choose for convenience the origin at the segment centre so that the boundaries describing $\partial \mathcal{D}$ reduce to two points of respective abscissa $x_- = -L/2$ and $x_+ = L/2$ (see Figure 2.4).

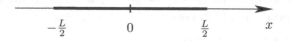

Fig. 2.4   Segment of length L.

Helmholtz operator   Let us first consider Helmholtz operator. Integration of the differential equation

$$\left[-\frac{d^2}{dx^2} + m^2 \right] \phi(x) = \rho(x) \tag{2.50}$$

is elementary using the method of variation of constants described in Appendix C, p. 272. Here, two independent functions ϕ_1 and ϕ_2 solutions of homogeneous equation (2.50) with a vanishing second member are $\phi_1(x) = e^{mx}$ and $\phi_2(x) = e^{-mx}$ whose Wronskian $\phi_1 \phi_2' - \phi_1' \phi_2$ is the constant $-2m$. We then find the general solution of

(2.50),

$$\phi(x) = \left[c_1 - \frac{1}{2m} \int_{-L/2}^{x} \mathrm{d}x' \, \rho(x') e^{-mx'} \right] e^{mx}$$

$$+ \left[c_2 + \frac{1}{2m} \int_{-L/2}^{x} \mathrm{d}x' \, \rho(x') e^{mx'} \right] e^{-mx} , \quad (2.51)$$

where c_1 and c_2 are integration constants determined by the BC imposed on ϕ.

Green's functions are easily computed using formula (2.51) with $\rho(x') = \delta(x' - x_0)$ and appropriate BC. We thus obtain for homogeneous Dirichlet BC

$$G_{\mathrm{HD}}(x; x_0) = -\frac{1}{2m} \operatorname{sh}(m|x - x_0|) + \frac{1}{2m} \coth(mL) \operatorname{ch}(m(x - x_0))$$
$$- \frac{1}{2m \operatorname{sh}(mL)} \operatorname{ch}(m(x + x_0)), \quad (2.52)$$

and

$$G_{\mathrm{NH}}(x; x_0) = -\frac{1}{2m} \operatorname{sh}(m|x - x_0|) + \frac{1}{2m} \coth(mL) \operatorname{ch}(m(x - x_0))$$
$$+ \frac{1}{2m \operatorname{sh}(mL)} \operatorname{ch}(m(x + x_0)), \quad (2.53)$$

for homogeneous Neumann BC We check that in agreement with general properties, $G_{\mathrm{HD}}(x, x_0)$ and $G_{\mathrm{NH}}(x, x_0)$ are real and symmetric in the exchange of x and x_0. They are obviously not translation invariant because of the edges.

For a large domain ($L \gg 1/m$), if x and x_0 stay away from both edges ($|x - \pm L/2| \gg 1/m$ and $|x_0 - \pm L/2| \gg 1/m$), then boundary effects on G_{HD} and G_{NH} become exponentially small, in agreement with the general prediction: these two functions then tend to $e^{-m|x-x_0|}/(2m)$ which is nothing but Green's function G_∞ for an infinite line as we shall see later. Finally, the reader may check that inserting Green's functions (2.52) and (2.53) in formulas (2.37) and (2.49) respectively gives indeed general solution (2.51), where constants c_1 and c_2 are selected by the corresponding BC on ϕ.

Laplacian For the Laplace operator the problem becomes even simpler, as it only requires integrating twice density $\rho(x)$ to get

$$\phi(x) = c_1 + c_2 x + \int_{-L/2}^{x} \mathrm{d}x' \, \rho(x') \, (x' - x). \quad (2.54)$$

The homogeneous Dirichlet Green's function simply becomes

$$G_{\mathrm{HD}}(x; x_0) = -\frac{1}{2}|x - x_0| - \frac{x x_0}{L} + \frac{L}{4}, \quad (2.55)$$

whereas there is no homogeneous Neumann Green's function. Note that we can find expression (2.55) by taking the limit $m \to 0$ of formula (2.52) for Helmholtz operator while keeping L, x and x_0 fixed.

◇ Conformal transformations in two dimensions ◇

In this section we present specific methods for dimension 2 based on conformal transformations. We will show that the interest of these transformations is to replace a two-dimensional domain by another one of simpler geometry. In addition the Laplacian operator and its Green's functions have remarkable transformation properties.

Definition First let us recall the definition of conformal transformations in arbitrary dimension. A transformation from coordinates $\{\xi^i\}$ to coordinates $\{\eta^i\}$ is conformal if matrix elements g_{ij} and $\widehat{g}_{ij}$, corresponding respectively to metrics for coordinates $\{\xi^i\}$ and $\{\eta^i\}$

$$\widehat{g}_{ij} = f(\xi^k)\, g_{ij}$$

with $f(\xi^i)$ a scalar function[10]. It is easy to identify translations, rotations and dilations in Euclidean space as part of this group. The peculiarity of dimension 2 is that the group of conformal transformations is infinite-dimensional, unlike with other dimensions. We restrict ourselves to this dimension 2 from now on.

Conformal transformations and analytic functions Let us now determine the two-dimensional conformal transformations preserving orientation. Euclidean space $\mathbb{R}^2$ is described by Cartesian coordinates (x, y). In these coordinates the metric g is simply the identity matrix $\mathcal{I}$, and the elementary line segment reduces to $ds^2 = dx^2 + dy^2$. Consider the change of coordinates

$$x \mapsto x'(x, y) \qquad \text{and} \qquad y \mapsto y'(x, y).$$

The metric $\widehat{g}$ in coordinates $(x',\ y')$ is defined by rewriting the line element as

$$ds^2 = \widehat{g}_{x'x'}\mathrm{d}x'^2 + 2\widehat{g}_{x'y'}\mathrm{d}x'\mathrm{d}y' + \widehat{g}_{y'y'}\mathrm{d}y'^2.$$

It is given by the matrix relation (see Appendix F):

$$\widehat{g} = \begin{pmatrix} \frac{\partial x}{\partial x'} & \frac{\partial x}{\partial y'} \\ \frac{\partial y}{\partial x'} & \frac{\partial y}{\partial y'} \end{pmatrix}^T \mathcal{I} \begin{pmatrix} \frac{\partial x}{\partial x'} & \frac{\partial x}{\partial y'} \\ \frac{\partial y}{\partial x'} & \frac{\partial y}{\partial y'} \end{pmatrix},$$

where A^T is the transpose of matrix A.

---

[10]See Appendix F for reminders on changes of coordinates.

It is then possible to show[✠] from this equation that $(x, y) \mapsto (x', y')$ is a conformal transformation, i.e. $\hat{g} = f(x, y) \, \mathcal{I}$, if and only if

$$\frac{\partial x'}{\partial x} = \frac{\partial y'}{\partial y},$$

$$\frac{\partial x'}{\partial y} = -\frac{\partial y'}{\partial x} \qquad (2.56)$$

with

$$f(x, y) = \left[\left(\frac{\partial x'}{\partial x} \right)^2 + \left(\frac{\partial x'}{\partial y} \right)^2 \right]^{-1}.$$

Equations (2.56) are nothing but the Cauchy-Riemann conditions, imposing that function

$$\mathcal{F}(x, y) = x'(x, y) + i y'(x, y)$$

is an analytic function of a single complex variable $z = x + iy$, i.e. $\mathcal{F}(x, y) = \mathcal{F}(z)$.

✠ **Comment:** Let us sketch some steps of this demonstration. Setting

$$A_1 = \left(\frac{\partial x}{\partial x'} \right)^2 - \left(\frac{\partial y}{\partial y'} \right)^2$$

and

$$A_2 = \left(\frac{\partial x}{\partial y'} \right)^2 - \left(\frac{\partial y}{\partial x'} \right)^2,$$

we first obtain the condition

$$A_1 = A_2 \qquad (2.57)$$

and

$$\left(\frac{\partial x}{\partial x'} \right) \left(\frac{\partial x}{\partial y'} \right) + \left(\frac{\partial y}{\partial x'} \right) \left(\frac{\partial y}{\partial y'} \right) = 0. \qquad (2.58)$$

The latter implies

$$A_1 \left(\frac{\partial x}{\partial y'} \right)^2 = -A_2 \left(\frac{\partial y}{\partial y'} \right)^2. \qquad (2.59)$$

We must then inspect conditions (2.57), (2.58) and (2.59). First of all, one can convince himself it is not possible to have $A_i \neq 0$ (with $i = 1, 2$). Conditions $A_1 = A_2 = 0$, Eq. (2.58) together with orientation preservation then lead to the result.

Furthermore, since

$$\left(\frac{\partial x'}{\partial x} \right)^2 + \left(\frac{\partial x'}{\partial y} \right)^2 = \left| \frac{d\mathcal{F}}{dz} \right|^2,$$

we get $\hat{g} = |d\mathcal{F}/dz|^{-2} \, \mathcal{I}$, which requires $d\mathcal{F}/dz \neq 0$. In summary, a two-dimensional coordinate transformation is a conformal transformation[11] if and only if $\mathcal{F}(z)$ is analytic in z and $d\mathcal{F}/dz \neq 0$. In other words, any analytic function $\mathcal{F}(z)$ is associated with a conformal transformation. This is the reason why the group of two-dimensional conformal transformations is of infinite dimension.

Geometric interpretation It is worth noting that conformal transformations have a very simple local geometric interpretation. Indeed, in the vicinity of a given point (x_0, y_0), a conformal transformation reduces to composing a rotation of angle $\arg(d\mathcal{F}/dz(z_0))$ and a homothety of scaling factor $|d\mathcal{F}/dz(z_0)|$. Therefore this transformation preserves local angles (e.g. angles of the domain boundary in Figure 2.5). Rotation angle and scaling factor depend on the point (x_0, y_0) under

---

[11]When the transformation reverses orientation, $\mathcal{F}(\bar{z})$ is an anti-holomorphic function.

consideration. They are constant throughout the space only if $\mathcal{F}(z) = az + b$ with a and b complex constants.

Consider the image of a domain $\mathcal{D}$ by a conformal transformation associated with the function $\mathcal{F}(z)$ analytic on $\mathcal{D}$ and such that $d\mathcal{F}/dz \neq 0$ for $(x, y) \in \mathcal{D}$. The transformation is bijective, i.e. each point of the image domain has a single antecedent in $\mathcal{D}$. In addition the image domain has the same topology as $\mathcal{D}$. When the boundary $\partial\mathcal{D}$ (here a line) has angular points, it is often useful to introduce a conformal transformation becoming singular at these points. Then the transformation no longer preserves angles in the vicinity of such singular points. This is then used to transform an edge into a smooth surface using an appropriate choice for $\mathcal{F}(z)$: This is the principle of Schwartz-Cristoffel transformation. We give a simple illustration in section 2.2.5, p. 106.

Transformation of Laplace operator and Dirac distribution   Laplace operator and Dirac distribution become under a conformal change of coordinates $\mathcal{F}(z)$

$$\frac{\partial^2}{\partial x^2} + \frac{\partial^2}{\partial y^2} = \left|\frac{d\mathcal{F}}{dz}\right|^2 \left(\frac{\partial^2}{\partial x'^2} + \frac{\partial^2}{\partial y'^2}\right), \tag{2.60}$$

and

$$\delta(x - x_0)\delta(y - y_0) = \left|\frac{d\mathcal{F}}{dz}\right|^2 \delta(x' - x_0')\delta(y' - y_0'). \tag{2.61}$$

Note that these formulae are applications of general results (F.5) and (F.2) of Appendix F. In other words Laplace operator and Dirac distribution are transformed exactly in the same way.

Dirichlet Green's functions   The above results suggest using conformal transformations to calculate Green's functions of Laplace operator in domains of geometry without simple symmetry. Let us first consider Dirichlet Green's function G_D for a given boundary $\mathcal{D}$. The idea is to first find a conformal transformation,

$$\mathbf{r} = (x, y) \mapsto \mathbf{r}' = (x', y') \quad \text{with} \quad x' = \operatorname{Re}\mathcal{F}(z) \quad \text{and} \quad y' = \operatorname{Im}\mathcal{F}(z) ,$$

mapping domain $\mathcal{D}$ onto a domain $\mathcal{D}'$ of simpler geometry as illustrated in Figure 2.5. In the new coordinates we define Green's function G_D' as the unique solution of

$$-\Delta_{\mathbf{r}'} G_D'(\mathbf{r}'; \mathbf{r}'_0) = \delta(\mathbf{r}' - \mathbf{r}'_0)$$

with Dirichlet BC[12]

$$G_D'(\mathbf{r}'; \mathbf{r}'_0)\big|_{\mathbf{r}' \in \partial\mathcal{D}'} = D'(\mathbf{r}') ,$$

---

[12] For simplicity we limited ourselves to the case where the function $D(\mathbf{r}; \mathbf{r}_0)$, defined p. 65 for the general case, depends only on $\mathbf{r}$.

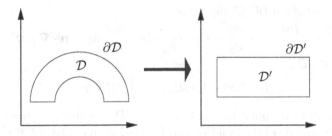

Fig. 2.5 Conformal transformation in the complex plane mapping a domain $\mathcal{D}$ of complicated boundary $\partial\mathcal{D}$ onto a domain $\mathcal{D}'$ with a rectangular boundary $\partial\mathcal{D}'$.

where $\partial\mathcal{D}'$ is the image of $\partial\mathcal{D}$ by the conformal transformation, and $D'(\mathbf{r}') = D(\mathbf{r})$. This function G'_{D} is then by construction easier to determine. In addition the function G_{D} defined by

$$\boxed{G_{\mathrm{D}}(\mathbf{r};\mathbf{r}_0) = G'_{\mathrm{D}}(\mathbf{r}';\mathbf{r}'_0),}$$

is the Dirichlet Green's function of the original problem. Indeed, it satisfies the BC and taking into account the properties of transformations (2.61) of the Laplacian and the Dirac distribution, we find

$$-\Delta_r G_{\mathrm{D}}(\mathbf{r};\mathbf{r}_0) = -\left|\frac{\mathrm{d}\mathcal{F}}{dz}\right|^2 \Delta_{\mathbf{r}'} G'_{\mathrm{D}}(\mathbf{r}';\mathbf{r}'_0),$$

$$= \left|\frac{\mathrm{d}\mathcal{F}}{dz}\right|^2 \delta(x'-x'_0)\delta(y'-y'_0),$$

$$= \delta(x-x_0)\delta(y-y_0).$$

For $D(\mathbf{r}) = 0$ we obtain homogeneous Dirichlet Green's function G_{HD}. As shown in section 2.1.3, the knowledge G_{HD} gives access to any field ϕ defined by Dirichlet BC. When BC on ϕ are of Neumann type, using G_{HD} in the general formula (2.27), p. 65 does not provide an explicit expression for ϕ. The expression then obtained can nevertheless help at a later resolution.

Harmonic functions In the absence of sources, the field ϕ is a harmonic function solution of Laplace equation $\Delta\phi = 0$. It is sometimes wiser to look directly for the solution of the PDE by the method of conformal transformations, without using Green's functions. So suppose we want to solve $\Delta\phi = 0$ in $\mathcal{D}$ either with Dirichlet or Neumann BC. Since the field ϕ is harmonic in $\mathcal{D}$, it can be seen as the real part of an analytic function of the form

$$\mathcal{A}(z) = \phi(x,y) + i\psi(x,y).$$

Let us interpret at first the BC on ϕ in terms of BC on $\mathcal{A}$. It is clear that Dirichlet BC (2.24) are written as

$$\boxed{\text{Dirichlet}: \qquad \mathrm{Re}\,\mathcal{A}(z) = D(x,y) \qquad \text{for} \qquad z = (x,y) \in \partial\mathcal{D}.} \qquad (2.62)$$

In the case of Neumann BC (2.25), since

$$d\psi = \frac{\partial \psi}{\partial x}\,dx + \frac{\partial \psi}{\partial y}\,dy = -\frac{\partial \phi}{\partial y}\,dx + \frac{\partial \phi}{\partial x}\,dy = \mathbf{n} \cdot \boldsymbol{\nabla}\phi\,d\Sigma,$$

we have for $z \in \partial\mathcal{D}$

$$\operatorname{Im}\mathcal{A}(z) - \operatorname{Im}\mathcal{A}(z_i) = \int_{\mathcal{C}} d\Sigma N(\mathbf{r}),$$

where z_i is an arbitrary point on the boundary $\partial\mathcal{D}$, $\mathcal{C}$ a contour going from z_i to z (see Figure 2.6) and $N(\mathbf{r})$ the function involved in Neumann BC (2.25). Introducing

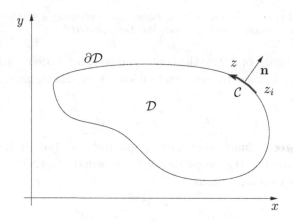

Fig. 2.6　Integration contour $\mathcal{C}$ going from a fixed point z_i on the boundary to any boundary point z.

$$\mathcal{N}(x, y) = \int_{\mathcal{C}} d\Sigma N(\mathbf{r}),$$

Neumann BC become:

$$\boxed{\text{Neumann}: \quad \begin{aligned} &\operatorname{Im}\mathcal{A}(z) = \operatorname{Im}\mathcal{A}(z_i) + \mathcal{N}(x, y) \quad\quad \text{for} \quad z = (x, y) \in \partial\mathcal{D}, \\ &\operatorname{Re}\mathcal{A}(z_0) = c \quad\quad \text{for} \quad \text{fixed} \quad z_0 = (x_0, y_0). \end{aligned}} \quad (2.63)$$

Note that the arbitrary constant $\operatorname{Im}\mathcal{A}(z_i)$ only appears in the imaginary part of $\mathcal{A}(z)$ and therefore not in $\phi(x, y)$. Since the BC are written in terms of $\mathcal{A}(z)$, one has to perform a change of coordinates $z' = \mathcal{F}(z)$, defined by the analytic function $\mathcal{F}(z)$, onto a simpler domain. The problem thus reduces to the determination of field $\phi'(\mathbf{r}') = \phi(\mathbf{r})$ solution of Laplace equation $\Delta'\phi' = 0$ on domain $\mathcal{D}'$ with BC induced by those on ϕ. Of course the field $\phi'(\mathbf{r}')$ is itself the real part of an analytic function $\mathcal{A}'(z')$ satisfying the BC deduced from conditions (2.62) or (2.63). Solution of Laplace equation $\Delta\phi = 0$ in domain $\mathcal{D}$ finally reads:

$$\boxed{\phi(x, y) = \operatorname{Re}\mathcal{A}(z) \quad\quad \text{with} \quad\quad \mathcal{A}(z) = \mathcal{A}'(z') = \mathcal{A}'\left(\mathcal{F}(z)\right).}$$

We present in section 2.2.5, p. 106, an application of these methods to a hydrodynamic problem.

◇ **Infinite line and plane** ◇

Helmholtz operator on a plane Let us start with Helmholtz operator in two dimensions, where domain $\mathcal{D}$ is the infinite plane. The corresponding homogeneous Dirichlet Green's function admits $\widehat{G}_\infty(\mathbf{k}) = 1/(\mathbf{k}^2 + m^2)$ as Fourier transform, an expression valid in all dimensions (see p. 74). Here in two dimensions we have

$$
\begin{aligned}
G_\infty(|\mathbf{r} - \mathbf{r}_0|) &= \frac{1}{(2\pi)^2} \int d^2\mathbf{k} \, e^{i\mathbf{k}\cdot(\mathbf{r}-\mathbf{r}_0)} \frac{1}{\mathbf{k}^2 + m^2} \\
&= \frac{1}{(2\pi)^2} \int_0^\infty dk \, k \int_0^{2\pi} d\theta \, e^{ik|\mathbf{r}-\mathbf{r}_0|\cos\theta} \frac{1}{\mathbf{k}^2 + m^2} \\
&= \frac{1}{2\pi} \int_0^\infty dk \, \frac{k \, J_0(k|\mathbf{r} - \mathbf{r}_0|)}{\mathbf{k}^2 + m^2},
\end{aligned}
\tag{2.64}
$$

where J_0 is the zero order Bessel function of first kind. The last integral is expressed in terms of the third kind Bessel function of order zero, denoted K_0 and also called Hankel function of imaginary argument[13]. We then find:

$$
\boxed{G_\infty(|\mathbf{r} - \mathbf{r}_0|) = \frac{1}{2\pi} K_0(m|\mathbf{r} - \mathbf{r}_0|) \ .}
\tag{2.65}
$$

The asymptotic behaviour of $G_\infty(|\mathbf{r}-\mathbf{r}_0|)$ at large relative distance is obtained from the asymptotic expansion of function K_0 at large argument and reads

$$
G_\infty(|\mathbf{r} - \mathbf{r}_0|) \simeq \frac{e^{-m|\mathbf{r}-\mathbf{r}_0|}}{2(2\pi m|\mathbf{r} - \mathbf{r}_0|)^{1/2}} \quad \text{when} \quad |\mathbf{r} - \mathbf{r}_0| \to \infty.
$$

We thus see that $G_\infty(|\mathbf{r}-\mathbf{r}_0|)$ decreases exponentially fast. As in three dimensions the presence of m induces a screening effect and G_∞ is short ranged. We find a logarithmic behaviour when the relative distance $|\mathbf{r} - \mathbf{r}_0|$ is small compared to $1/m$

$$
G_\infty(|\mathbf{r} - \mathbf{r}_0|) \sim -\frac{1}{2\pi} \ln(m|\mathbf{r} - \mathbf{r}_0|) \ ,
$$

from the expansion of function K_0 at small arguments. Since the screening should not be relevant at short-distance we infer that this behaviour should come from the Green's function of the Laplacian, similarly to what is observed in the three-dimensional case (see p. 74).

Helmholtz operator on the line Let us consider now the one-dimensional case where $\mathcal{D}$ is an infinite line. The inverse transform of $\widehat{G}_\infty$ now reads

$$
G_\infty(|x - x_0|) = \frac{1}{2\pi} \int_{-\infty}^\infty dk \, e^{ik(x-x_0)} \frac{1}{k^2 + m^2}.
$$

The calculation of this integral is performed by the method of residues[14], leading to

$$
\boxed{G_\infty(|x - x_0|) = \frac{1}{2m} e^{-m|x-x_0|}.}
\tag{2.66}
$$

---

[13]The reader can find definitions and other properties of Bessel functions in books [26] or [4].
[14]This calculation is similar to that of integral (2.47), p. 74.

As previously announced G_∞ is indeed the common limit of Green's functions G_{HD} (2.52) and G_{HN} (2.53) for systems of finite size L when $L \to \infty$ at fixed positions. As in two and three dimensions, $G_\infty(|x - x_0|)$ decreases exponentially fast for distances large compared to $1/m$. At short distance $G_\infty(|x - x_0|)$ behaves, up to some constants, as $-|x - x_0|/2$, just like the homogeneous Dirichlet Green's function for the Laplacian (2.55) of a large finite size system (i.e. $L \gg x, x_0$).

Laplacian Regarding the Laplacian on an infinite line or plane, it turns out that there is no Green's function which does not diverge at infinity. In particular there is no homogeneous Dirichlet Green's function. This is a consequence of the non-existence of inverse Fourier transform of function $1/k^2$ in one and two dimensions, due to its non-integrable singular behaviour at $\mathbf{k} = 0$. However, the functions

$$-\frac{1}{2\pi} \ln\left(|\mathbf{r} - \mathbf{r}_0|/\ell\right) \tag{2.67}$$

with an arbitrary constant ℓ and

$$-\frac{|x - x_0|}{2} \tag{2.68}$$

are Green's functions of Laplace operator in two and one dimensions respectively. In the two-dimensional case the reader can apply a test function, as proposed in exercise 2.3 p. 111, to check explicitly that function (2.67) indeed satisfies the corresponding PDE in the sense of distributions. In one dimension the verification is immediate by applying successively identities $(\mathrm{d}/\mathrm{d}x)|x| = \theta(x) - \theta(-x)$ and $(\mathrm{d}/\mathrm{d}x)\theta(x) = \delta(x)$ where $\theta(x)$ is the Heaviside function. Note that Green's functions (2.67) and (2.68) appear naturally in the short distance behaviour of Green's functions G_∞ associated with Helmholtz operators. They are also identical to the Helmholtz asymptotic forms when $m \to 0^+$ up to diverging constants.

Green's functions (2.67) and (2.68) have a simple physical interpretation. Up to infinite constants due to the presence of non-localised sources, they actually correspond to the electrostatic potential created in three dimensions respectively by a line and a plane uniformly charged. As an exercise, the reader can compute the potential difference between two observation points $\mathbf{r}_1$ and $\mathbf{r}_2$,

$$\int \mathrm{d}^3\mathbf{r}' \left[\frac{\rho(\mathbf{r}')}{4\pi|\mathbf{r}_1 - \mathbf{r}'|} - \frac{\rho(\mathbf{r}')}{4\pi|\mathbf{r}_2 - \mathbf{r}'|} \right],$$

respectively for linear $\rho(\mathbf{r}') = \delta(x' - x_0)$ and planar $\rho(\mathbf{r}') = \delta(x' - x_0)\delta(y' - y_0)$ charge distributions (see Figure 2.7). On can check that these differences indeed reduce to Green's functions (2.67) and (2.68), up to some constants. Note that the electric field created by previous charge distributions is finite, and it is obviously given by the gradient of these Green's functions.

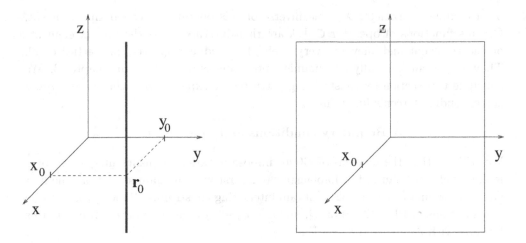

Fig. 2.7 Distribution of linear and planar sources involved in the construction of Green's functions in two and one dimensions respectively.

2.1.6 *Inhomogeneous operators*

We studied in detail the Green's functions for Laplace and Helmholtz operators, which appear frequently in stationary problems. There are however many cases where the field we want to study satisfies the inhomogeneous PDE

$$[-\Delta + f(\mathbf{r})]\,\phi(\mathbf{r}) = \rho(\mathbf{r}) \tag{2.69}$$

where the operator

$$\mathcal{O} = -\Delta + f(\mathbf{r})$$

now involves, in addition to the Laplacian, a real function $f(\mathbf{r})$ given by characteristics of the system under consideration. This type of PDE may appear for example in the study of inhomogeneous plasmas or in quantum mechanics. In the latter case operator $\mathcal{O}$ and PDE (2.69) are respectively related to the Hamiltonian and the stationary Schrödinger equation. Examples of sections 2.2.3 and 2.2.4, presented in the second part of this chapter, are applications of general properties discussed below, respectively to density of states calculation and diffusion in quantum mechanics.

Let us return to PDE (2.69). For functions $f(\mathbf{r})$ which cannot be reduced to simple constants[15], a comprehensive and systematic investigation of the properties of $\mathcal{O}$ and its Green's functions becomes very difficult. The main reason is the lack of symmetry (translation and rotation invariances for instance) of this operator due to the presence of $f(\mathbf{r})$. Here we just state some simple general results.

In a first step we address the problem of boundary conditions and uniqueness of the solution of PDE (2.69). We then introduce the operator $\mathcal{O} + \lambda$, which depends

---

[15]In this case, and if this constant is positive, $\mathcal{O}$ is obviously Helmholtz operator.

on a complex parameter λ. The inverse of this operator is called the resolvent. Green's functions of operator $\mathcal{O} + \lambda$ are then interpreted as the matrix elements of the resolvent and they are very useful for studying spectral properties of $\mathcal{O}$. They also occur naturally in dynamical problems as we will see in Chapter 3. We conclude this section by constructing perturbative expansions of the resolvent and corresponding Green's functions.

$\Diamond$ **Boundary conditions and uniqueness** $\Diamond$

We will see that the analysis of BC to impose in order to ensure uniqueness of ϕ is more delicate than in the Laplacian case. First we can again start from the first Green's formula (2.19), p. 62 to obtain interesting constraints. So let ϕ_1 and ϕ_2 be two solutions of PDE (2.69) on $\mathcal{D}$, and $\alpha = \phi_1 - \phi_2$ their difference. Applying this formula with $u = \alpha^*$ and $v = \alpha$ leads to

$$\int_{\mathcal{D}} d\mathbf{r} \left[f(\mathbf{r}) |\alpha(\mathbf{r})|^2 + |\boldsymbol{\nabla}\alpha(\mathbf{r})|^2 \right] = \oint_{\partial\mathcal{D}} d\Sigma \, \alpha^*(\mathbf{r}) \left[\mathbf{n} \cdot \boldsymbol{\nabla}\alpha(\mathbf{r}) \right] . \tag{2.70}$$

If f is a strictly positive real-valued function, $f(\mathbf{r}) > 0$ for all $\mathbf{r} \in \mathcal{D}$, then as in the case of Helmholtz operator, we see that Dirichlet BC (2.24), or simple Neumann BC (2.44) are sufficient to imply that difference α is identically zero in $\mathcal{D}$ and so uniqueness is ensured.

If $f(\mathbf{r})$ takes positive and negative real values, then identity (2.70) does not allow to conclude about uniqueness. Indeed, if Dirichlet or Neumann BC imply that the volume integral is zero, the integrand does not necessarily vanish because its sign is not constant. In fact since the difference $\alpha = \phi_1 - \phi_2$ is solution of PDE (2.69) without a second member,

$$[-\Delta + f(\mathbf{r})] \, \alpha(\mathbf{r}) = 0 ,$$

with homogeneous Dirichlet or Neumann BC, the problem of uniqueness is related to the presence of eigenvalue $\lambda = 0$ in the spectrum of operator $\mathcal{O}$ with the homogeneous version of the BC chosen for ϕ. Thus for Dirichlet BC on ϕ, we consider the spectrum of the operator with homogeneous Dirichlet BC. If $\lambda = 0$ is not an eigenvalue of $\mathcal{O}$ with these BC, then $\alpha(\mathbf{r})$ is necessarily identically zero and uniqueness is guaranteed. If $\lambda = 0$ is an eigenvalue of $\mathcal{O}$ with chosen BC, then $\alpha(\mathbf{r})$ is an eigenfunction associated with $\lambda = 0$, and it is not identically zero. In other words, given a particular solution of PDE (2.69) with specified Dirichlet BC, we can construct an infinite number of other solutions by adding any eigenfunction of $\mathcal{O}$ with eigenvalue $\lambda = 0$ satisfying homogeneous Dirichlet BC. The argument is the same for Neumann BC on ϕ, leading to the same result.

Once uniqueness guaranteed within given BC, Green's functions corresponding to these BC are well defined. Manipulations of second Green's formula (2.20) similar to that introduced for the Laplacian leads to a general expression for the solution of PDE (2.69) in terms of these Green's functions, which happens to be identical to formula (2.27), p. 65. Again, a field ϕ created by an arbitrary source with given BC is completely set by the knowledge of appropriate Green's functions. For Dirichlet BC in particular, ϕ is explicitly obtained from formula (2.37) in terms of homogeneous Dirichlet Green's function.

$\lozenge$ Resolvent and Green's functions $\lozenge$

Resolvent Let us now introduce the operator

$$(\mathcal{O} + \lambda) \tag{2.71}$$

where λ is a given complex number with $\text{Im} \, \lambda \neq 0$. The corresponding generic PDE reads

$$[\mathcal{O} + \lambda] \, \phi(\mathbf{r}) = \rho(\mathbf{r}) \tag{2.72}$$

or

$$[-\Delta + f(\mathbf{r}) + \lambda] \, \phi(\mathbf{r}) = \rho(\mathbf{r}). \tag{2.73}$$

The BC ensuring uniqueness are again obtained by manipulating the first Green's formula. Noting again $\alpha = \phi_1 - \phi_2$ the difference between two possible solutions of PDE (2.73), we find from the imaginary part of the first Green's formula

$$(\text{Im} \, \lambda) \int_{\mathcal{D}} d\mathbf{r} \, |\alpha(\mathbf{r})|^2 = \text{Im} \left\{ \oint_{\partial \mathcal{D}} d\Sigma \, \alpha^*(\mathbf{r}) \, [\mathbf{n} \cdot \boldsymbol{\nabla}\alpha(\mathbf{r})] \right\}. \tag{2.74}$$

If BC are of Neumann or Dirichlet type, the surface integral on the right-hand side also vanishes, so that, as $\text{Im}(\lambda) \neq 0$, the volume integral of the left side is also zero. As $|\alpha(\mathbf{r})|^2$ keeps a constant sign, $\alpha(\mathbf{r})$ is identically zero, ensuring uniqueness of the solution. The operator $\mathcal{O} + \lambda$ is then an invertible operator in the space of functions satisfying the considered BC and its inverse,

$$\boxed{\textbf{Resolvent :} \quad \frac{1}{\mathcal{O} + \lambda}}$$

is called the resolvent operator associated with $\mathcal{O}$.

Green's functions Let us fix homogeneous Dirichlet or Neumann BC on the surface $\partial \mathcal{D}$ of the domain under consideration. Green's function $G_\lambda(\mathbf{r}; \mathbf{r}')$ of operator $\mathcal{O} + \lambda$ is solution of the PDE

$$[-\Delta_{\mathbf{r}} + f(\mathbf{r}) + \lambda] \, G_\lambda(\mathbf{r}; \mathbf{r}') = \delta(\mathbf{r} - \mathbf{r}') \tag{2.75}$$

with these boundary conditions. As shown in the general case, p. 60, $G_\lambda(\mathbf{r}; \mathbf{r}')$ can be interpreted as the matrix element of the resolvent between bra $\langle \mathbf{r} |$ and ket $| \mathbf{r}' \rangle$, i.e.

$$G_\lambda(\mathbf{r}; \mathbf{r}') = \langle \mathbf{r} | \frac{1}{\mathcal{O} + \lambda} | \mathbf{r}' \rangle. \tag{2.76}$$

If operator $\mathcal{O}$ allows a complete basis of orthonormal eigenfunctions ψ_n of eigenvalues λ_n for the chosen BC then, since the ψ_n are obviously also eigenfunctions of the resolvent with eigenvalues $1/(\lambda_n + \lambda)$, spectral representation (2.13), p. 60 is written here as

$$G_\lambda(\mathbf{r}; \mathbf{r}') = \sum_n \frac{1}{\lambda + \lambda_n} \, \psi_n(\mathbf{r})\psi_n^*(\mathbf{r}'). \qquad (2.77)$$

Even if this spectral representation is useful, it does not yet explicitly give G_λ, as it is still necessary to determine eigenvalues and eigenfunctions of $\mathcal{O}$! Note that each term in this representation is well defined, because $(\lambda + \lambda_n)$ cannot vanish since λ has a non-zero imaginary part and all λ_n are real in agreement with the Hermitian character of $\mathcal{O}$.

◊ Interest of Green's functions ◊

As for Helmholtz operator, the knowledge of the homogeneous Green's functions gives access to the general solution of PDE (2.73) *via* formulae identical to expressions (2.37) on p. 70 and (2.49), p. 75, depending upon the nature of imposed BC on ϕ. As we shall see in the following chapter, this type of PDE with complex parameter occurs naturally in solving dynamical problems by Laplace or Fourier transform with respect to time.

Another major benefit of Green's functions G_λ appears in their spectral representation (2.77). If these quantities, now seen as functions of λ, are in general analytic outside the real axis ($\text{Im}\,\lambda \neq 0$), they should display singularities on the real axis at $\lambda = -\lambda_n$ where λ_n is an eigenvalue of $\mathcal{O}$. As we shall see in section 2.2.3, p. 99, within an example specific to quantum mechanics where $\mathcal{O}$ reduces to a Hamiltonian, the knowledge of analytical properties of $G_\lambda(\mathbf{r}, \mathbf{r}')$ provides information on the spectrum of the original operator $\mathcal{O}$. This property justifies the term of "resolvent" for operator $1/(\mathcal{O} + \lambda)$.

◊ Perturbative expansion ◊

For an arbitrary function $f(\mathbf{r})$, an explicit determination of G_λ remains a rather difficult problem. In this context it is useful to write PDE (2.73) as:

$$[-\Delta + \lambda]\,\phi(\mathbf{r}) = \rho(\mathbf{r}) - f(\mathbf{r})\phi(\mathbf{r}) \qquad (2.78)$$

where operator $\mathcal{O}$ has been separated into a so-called free part[16] for which Green's functions $G_\lambda^{(0)}(\mathbf{r}, \mathbf{r}')$ are more accessible, and a part proportional to $f(\mathbf{r})$. Suppose that $\phi(\mathbf{r})$ satisfies homogeneous BC. The idea is then to start by considering the whole right-hand side of PDE (2.78) as a source, even if it depends on ϕ. This immediately leads to the integral equation

$$\phi(\mathbf{r}) = \int_{\mathcal{D}} d\mathbf{r}' \, G_\lambda^{(0)}(\mathbf{r}; \mathbf{r}')\big[\rho(\mathbf{r}') - f(\mathbf{r}')\phi(\mathbf{r}')\big],$$

---

[16]This name comes from quantum mechanics, where the free term refers to the Hamiltonian of a free particle.

where $G_\lambda^{(0)}(\mathbf{r}; \mathbf{r}')$ is the Green's function satisfying the same homogeneous BC as ϕ. Introducing

$$\phi_0(\mathbf{r}) = \int_{\mathcal{D}} d\mathbf{r}' \, \rho(\mathbf{r}') G_\lambda^{(0)}(\mathbf{r}; \mathbf{r}'),$$

the previous equation becomes:

$$\boxed{\phi(\mathbf{r}) = \phi_0(\mathbf{r}) - \int_{\mathcal{D}} d\mathbf{r}' \, G_\lambda^{(0)}(\mathbf{r}; \mathbf{r}') f(\mathbf{r}') \phi(\mathbf{r}').}$$

This integral equation gives rise to an expansion of ϕ in powers of f, through an iteration procedure. This leads to:

$$\phi(\mathbf{r}) = \phi_0(\mathbf{r}) - \int_{\mathcal{D}} d\mathbf{r}' \, G_\lambda^{(0)}(\mathbf{r}; \mathbf{r}') f(\mathbf{r}') \phi_0(\mathbf{r}')$$

$$+ \int_{\mathcal{D}^2} d\mathbf{r}' d\mathbf{r}'' \, G_\lambda^{(0)}(\mathbf{r}; \mathbf{r}') f(\mathbf{r}') G_\lambda^{(0)}(\mathbf{r}'; \mathbf{r}'') f(\mathbf{r}'') \phi(\mathbf{r}'').$$

If $f(\mathbf{r})$ can be seen as a perturbation of the free part, this line of reasoning leads at the first order in f to:

$$\boxed{\phi(\mathbf{r}) \simeq \phi_0(\mathbf{r}) - \int_{\mathcal{D}} d\mathbf{r}' \, G_\lambda^{(0)}(\mathbf{r}; \mathbf{r}') f(\mathbf{r}') \phi_0(\mathbf{r}').}$$

The above reasoning can also be applied to build a perturbative expansion of the resolvent associated with operator $\mathcal{O}$ in the vicinity of the free resolvent. To do so let us write[*] the tautology

$$(\mathcal{O} + \lambda)(\mathcal{O} + \lambda)^{-1} = \mathcal{I},$$

$$(-\Delta + \lambda)(\mathcal{O} + \lambda)^{-1} + f(\mathcal{O} + \lambda)^{-1} = \mathcal{I}, \tag{2.79}$$

and multiply to the left both sides of the above equation by $(-\Delta + \lambda)^{-1}$ to get:

$$\frac{1}{\mathcal{O} + \lambda} = \frac{1}{-\Delta + \lambda} - \frac{1}{-\Delta + \lambda} f \frac{1}{\mathcal{O} + \lambda}.$$

By taking the matrix elements of this operatorial identity between bra $\langle \mathbf{r}|$ and ket $|\mathbf{r}'\rangle$, we find the equation

> [*] **Comment:** Note that equation (2.79) deals with operators that do not commute with each other. The order of the operators is therefore important. Furthermore, in equation (2.79), f must be understood as the operator associated with $f(\mathbf{r})$. This operator is diagonal in the basis of $|\mathbf{r}\rangle$ and its matrix elements are $f(\mathbf{r})$, so that
>
> $$f = \int_{\mathcal{D}} d\mathbf{r}_1 |\mathbf{r}_1\rangle f(\mathbf{r}_1) \langle \mathbf{r}_1|.$$

$$\boxed{G_\lambda(\mathbf{r}; \mathbf{r}') = G_\lambda^{(0)}(\mathbf{r}; \mathbf{r}') - \int_{\mathcal{D}} d\mathbf{r}_1 \, G_\lambda^{(0)}(\mathbf{r}; \mathbf{r}_1) \, f(\mathbf{r}_1) \, G_\lambda(\mathbf{r}_1; \mathbf{r}').} \tag{2.80}$$

This identity can be written graphically as shown in Figure 2.8.

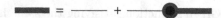

Fig. 2.8   Representation of (2.80) : Bold and thin lines correspond respectively to G_λ and $G_\lambda^{(0)}$, and the circle to $-f$.

The expected perturbative expansion is finally obtained though an iteration procedure if identity (2.80) similar to that introduced above, eventually leading to

$$G_\lambda(\mathbf{r}; \mathbf{r}') = G_\lambda^{(0)}(\mathbf{r}; \mathbf{r}') - \int_{\mathcal{D}} d\mathbf{r}_1\, G_\lambda^{(0)}(\mathbf{r}; \mathbf{r}_1)\, f(\mathbf{r}_1)\, G_\lambda^{(0)}(\mathbf{r}_1; \mathbf{r}')$$
$$+ \int_{\mathcal{D}^2} d\mathbf{r}_1 d\mathbf{r}_2\, G_\lambda^{(0)}(\mathbf{r}; \mathbf{r}_1)\, f(\mathbf{r}_1)\, G_\lambda^{(0)}(\mathbf{r}_1; \mathbf{r}_2) f(\mathbf{r}_2)\, G_\lambda^{(0)}(\mathbf{r}_2; \mathbf{r}')$$
$$+ \cdots \tag{2.81}$$

We can again give an interpretation of this perturbative expansion in terms of diagrams as shown in Figure 2.9.

Fig. 2.9   Representation of Eq. (2.81) using the same conventions as for Figure 2.8.

2.2   Applications and examples

2.2.1   *Origin of the method of image charges*

◊ **Presentation** ◊

In electrostatics, it is common to encounter a situation where an infinite plane, taken e.g. at $x = 0$ is the edge of a conductor kept at zero potential (see Figure 2.10). The conductor fills the half-space $x < 0$ while the other half-space $x > 0$ is the vacuum. The most suitable Green's function to compute the electrostatic potential created by a charge distribution outside the conductor is defined by

$$-\Delta_\mathbf{r} G_{\mathrm{HD}}(\mathbf{r}; \mathbf{r}') = \delta(\mathbf{r} - \mathbf{r}')$$

in the half-space defined by $x > 0$, with homogeneous Dirichlet BC, $G_{\mathrm{HD}}(\mathbf{r}; \mathbf{r}') = 0$, on the conductive wall at $x = 0$ on the one hand, and at infinity ($|\mathbf{r}| \to \infty$) in the vacuum on the other hand. The electrostatic potential created by an arbitrary distribution localised in the half-plane $x > 0$ is then given by the general formula (2.37), p. 70, up to the factor $1/\epsilon_0$. The surface term vanishes here since $D(\mathbf{r}) = 0$. We will determine the Green's function G_{DH} and explain the origin of the method of images.

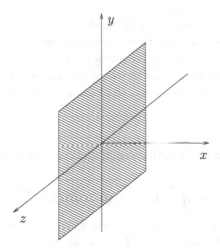

Fig. 2.10   Planar conductor at $x = 0$.

◊ Analysis and solution ◊

In order to determine G_{HD}, we could follow the calculation of Green's function in the infinite system presented p. 68. Here the system is still invariant under any translation along axis Oy and Oz, but this invariance is broken along the Ox axis by the presence of the conductive wall at $x = 0$. A method for computing G_{HD} would be to first perform a Fourier transform along Oy and Oz. This method is proposed in exercise 2.7 on p. 113. There is however another effective and systematic approach exploiting spectral formula (2.13), p. 60 as described below.

Spectrum determination   It is easy to obtain the basis of eigenfunctions of the Laplace operator with homogeneous Dirichlet BC in $x = 0$: Functions[✠]

$$\psi_{k_x,k_y,k_z}(\mathbf{r}) = \frac{2}{\sqrt{(2\pi)^3}}\, e^{i(k_y y + k_z z)}\, \sin(k_x x) \tag{2.82}$$

with $k_x > 0$ form an orthonormal basis of the space of functions defined in the half-space $x > 0$ and vanishing at $x = 0$. These functions satisfy the completeness relation

$$\int_0^\infty \mathrm{d}k_x \int_{-\infty}^\infty \mathrm{d}k_y \int_{-\infty}^\infty \mathrm{d}k_z\; \psi_{k_x,k_y,k_z}(\mathbf{r})\psi_{k_x,k_y,k_z}^*(\mathbf{r}') = \delta(\mathbf{r} - \mathbf{r}')\,, \tag{2.83}$$

for $\mathbf{r}$ and $\mathbf{r}'$ in domain $(x,\ x' > 0)$, together with the orthonormality condition

$$\int_{x>0} \mathrm{d}\mathbf{r}\; \psi_{k_x,k_y,k_z}(\mathbf{r})\psi_{k_x',k_y',k_z'}^*(\mathbf{r}) = \delta(k_x - k_x')\delta(k_y - k_y')\delta(k_z - k_z') \tag{2.84}$$

for $k_x > 0$ and $k_x' > 0$. This result is obtained by noting that expression (2.82) is a superposition of plane waves of wave vector (k_x, k_y, k_z) and $(-k_x, k_y, k_z)$ respectively.

> ✠ **Comment:**   Strictly speaking, we should introduce a finite domain and identify the corresponding eigenfunctions with homogeneous Dirichlet BC. We then send all the boundaries other than the conductive wall in $x = 0$ to infinity. It turns out that this procedure is equivalent to constructing eigenfunctions with periodic BC in directions Oy and Oz. This equivalence justifies the use of the set of functions (2.82).

Formula of the images   Inserting eigenfunctions (2.82) in the spectral formula (2.13) we find

$$G_{\mathrm{HD}}(\mathbf{r};\mathbf{r}') = \int_0^\infty dk_x \int_{-\infty}^\infty dk_y \int_{-\infty}^\infty dk_z \, \frac{\psi_{k_x,k_y,k_z}(\mathbf{r})\psi^*_{k_x,k_y,k_z}(\mathbf{r}')}{\mathbf{k}^2}. \qquad (2.85)$$

By replacing $\sin(k_x x)$ and $\sin(k_x x')$ by their expressions in terms of plane waves in integral (2.85), we obtain a sum of four terms. Since the integration domain is restricted to $k_x > 0$, it makes sense to combine these terms by pair to obtain an integral in k_x on the entire real axis. We then recognise the integral over $\mathbf{k}$ involved in expression (2.35) p. 68 for $G_\infty(r - r')$ and $G_\infty(\mathbf{r} - \mathbf{r}'_{\mathrm{im}})$ with $\mathbf{r}'_{\mathrm{im}} = (-x', y', z')$. One eventually finds

$$\boxed{G_{\mathrm{HD}}(\mathbf{r};\mathbf{r}') = G_\infty(\mathbf{r} - \mathbf{r}') - G_\infty(\mathbf{r} - \mathbf{r}'_{\mathrm{im}}).} \qquad (2.86)$$

Thus we find the famous formula of the images: the Green's function $G_{\mathrm{HD}}(\mathbf{r};\mathbf{r}')$ is, up to a factor q/ϵ_0, the potential created by a point charge q at $\mathbf{r}'$ and an effective opposite point charge $-q$ located at $\mathbf{r}'_{\mathrm{im}}$.

Note that it is just as easy to obtain the Green's function satisfying homogeneous Neumann BC $(\partial/\partial x)G_{NH}(\mathbf{r};\mathbf{r}') = 0$ for $x = 0$.   A physical realisation of these BC is obtained by replacing the conductor in the half-space $x < 0$ by a dielectric of constant ε_1, and the vacuum in the half-space $x > 0$ by another dielectric of constant ε_2. In the limit $\varepsilon_1 \ll \varepsilon_2$, everything happens as if the E_x component of the electric field in the region $x > 0$ became zero at $x = 0$ because of matching conditions $\varepsilon_1 E_x^{(1)} = \varepsilon_2 E_x^{(2)}$. We now use a basis of eigenfunctions similar to the previous one, obtained by the simple substitution $\sin(k_x x) \to \cos(k_x x)$ in the formula (2.82), in order to determine G_{HN}. We thus find

$$\boxed{G_{\mathrm{HN}}(\mathbf{r};\mathbf{r}') = G_\infty(\mathbf{r} - \mathbf{r}') + G_\infty(\mathbf{r} - \mathbf{r}'_{\mathrm{im}}).}$$

The image charge of q is now equal to q.

◇ Interpretation ◇

We have shown that the method of images arises very naturally from the spectral representation of Green's functions in terms of eigenfunctions of the Laplace operator satisfying suitable BC. The reader can compare the efficiency of this method to that of a direct calculation by Fourier transform. The method of images is quite general: it is used in the exercises 2.9, p. 115 and 2.15, p. 120, as well as in Chapter 3 for d'Alembert operator in the example of section 3.2.2 regarding Fraunhofer diffraction.

Screening effect If the image charge is fictitious, the resulting screening effect is observed in real conductors! In other words, at a large distance, the electrostatic potential produced by the charge q decreases faster than in the whole vacuum. In fact as a consequence of the rearrangement of free charges in the conductor, a localised charge distribution of sign opposite to q appears near the plane $x = 0$. At mesoscopic level, the corresponding surface charge density is simply determined by component E_x of the electric field in vacuum at $x = 0^+$, the electric field identically vanishing in the conductor. Note that this prediction from macroscopic electrostatic is in perfect agreement with a microscopic analysis of polarisation process inside the conductor. Such an analysis can be carried out within Debye theory, which is a mean-field approach quite similar to the Vlasov approximation presented in Chapter 1. Here we have to determine the charge distribution induced in the conductor by an external point charge, *via* Poisson-Boltzmann equation. This mean field theory confirms the prediction from the macroscopic approach.

2.2.2 *Ball with uniform motion in a fluid*

◊ **Presentation** ◊

There are many situations in hydrodynamics which reduce to the study of ideal and incompressible fluids. The flow is often stationary and irrotational, so that the velocity field $\mathbf{u}(\mathbf{r})$ satisfies

$$\boldsymbol{\nabla} \times \mathbf{u} = \mathbf{0}.$$

It then derives from a potential ϕ, that is to say $\mathbf{u}(\mathbf{r}) = \boldsymbol{\nabla}\phi(\mathbf{r})$. The equation describing the fluid incompressibility,

$$\boldsymbol{\nabla} \cdot \mathbf{u}(\mathbf{r}) = 0 \,,$$

then implies Laplace equation

$$\Delta\phi(\mathbf{r}) = 0. \tag{2.87}$$

Let us specify the BC that $\phi(\mathbf{r})$ must satisfy at the boundary $\partial\mathcal{D}$ of the domain where the fluid lies. A part $\partial\mathcal{D}_P$ of this border consists of the surface of an impervious material. On this wall, the velocity $\mathbf{u}(\mathbf{r})$ is necessarily tangent to it as shown in Figure 2.11, so that BC reads

$$\mathbf{n} \cdot \mathbf{u}(\mathbf{r}) = 0 \text{ for } \mathbf{r} \in \partial\mathcal{D}_P$$

where $\mathbf{n}$ is the vector normal to the wall, directed from the fluid toward the exterior material. Boundary $\partial\mathcal{D}_P$ is generally not closed such as for the pipe section shown in Figure 2.12. It is then necessary to supplement it by a fictitious surface $\partial\mathcal{D}_F$ on which the normal component of the velocity must be known, so that $\partial\mathcal{D} = \partial\mathcal{D}_P \cup \partial\mathcal{D}$ defines a closed domain. The velocity potential $\phi(\mathbf{r})$ is then the solution of Laplace equation without sources (2.87) with Neumann BC

$$\mathbf{n} \cdot \boldsymbol{\nabla}\phi(\mathbf{r}) = N(\mathbf{r}) \text{ pour } \mathbf{r} \in \partial\mathcal{D}. \tag{2.88}$$

Boundary function $N(\mathbf{r})$, given for a particular problem, satisfies the condition

$$\oint_{\partial\mathcal{D}} \mathrm{d}\Sigma \, N(\mathbf{r}) = 0,$$

induced by the conservation of matter.

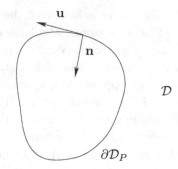

Fig. 2.11   The fluid velocity **u** is tangential to the boundary $\partial\mathcal{D}_P$.

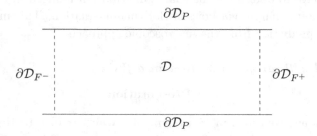

Fig. 2.12   In this example of flow in a pipe, the cylindrical wall $\partial\mathcal{D}_P$ is supplemented by two fictitious walls $\partial\mathcal{D}_{F\pm}$ so that the total surface $\partial\mathcal{D} = \partial\mathcal{D}_P \cup \partial\mathcal{D}_{F+} \cup \partial\mathcal{D}_{F-}$ defines a closed domain.

Analogy with superconductivity   The previous situation is similar to that for a material in a superconducting phase, submitted to a magnetic field with an amplitude below the critical value. We then observe the Meissner effect, which can be summarised as follows: the magnetic field lines do not penetrate into the material, and the total magnetic field is identically zero[17]. Given the continuity conditions of the magnetic field at the surface separating the two media, the component normal to the superconducting wall is zero. Magnetic field lines are therefore tangential to the superconductor surface. The magnetic field **B** outside the superconductor satisfies the same equations as previous velocity field **u**, namely $\nabla \cdot \mathbf{B} = 0$ and $\nabla \times \mathbf{B} = \mathbf{0}$ assuming the absence of current density in the domain under consideration. We have homogeneous Neumann BC at the superconductor walls.

Ball in uniform motion   We will consider the example of a ball immersed in a perfect incompressible fluid. Assume that an external operator imposes on the ball a uniform motion at a constant speed **V**. This motion induces a flow in the fluid

---

[17]It therefore appears a screening effect of the magnetic field, which can be described phenomenologically by assigning to the superconductor a diamagnetic susceptibility equal to -1. We refer the reader to the works mentioned in the bibliographical notes for a detailed description of superconductivity.

initially at rest. It is convenient to work in the Galilean reference frame linked to the ball, since the fluid velocity **u** is then stationary in the irrotational regime (see Figure 2.13). In practice for a real fluid, this regime will be observed only if the velocity **V** is sufficiently small, so that no turbulence appears in the wake of the ball. In its reference frame, the ball is at rest and sees a flow whose velocity is $-\mathbf{V}$ at infinity. Note that this problem is similar to that of a superconducting sphere subjected to a uniform external magnetic field.

◊ **Analysis and solution** ◊

The wall is composed of the sphere $S_R = \partial\mathcal{D}_P$ of radius R. It is complemented by a large sphere whose radius will be sent to infinity. We will denote this limit by $S_\infty = \partial\mathcal{D}_F$. The boundary of the closed domain is here $\partial\mathcal{D} = S_R \cup S_\infty$. The problem geometry suggests to use spherical coordinates shown in Figure 2.13. Neumann BC

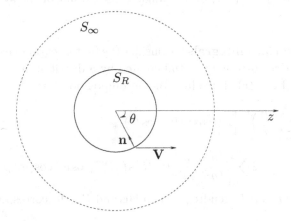

Fig. 2.13 Ball of radius R (bounded by the sphere S_R) moving in a fluid at constant velocity **V** along the z-axis. The surface S_∞ is a fictitious spherical wall at infinity.

(2.88) then reads

$$\mathbf{n} \cdot \boldsymbol{\nabla}\phi = 0 \quad \text{on} \quad S_R \quad \text{and} \quad \mathbf{n} \cdot \boldsymbol{\nabla}\phi = -V\cos\theta \quad \text{on} \quad S_\infty.$$

It is useful in the following to decompose $\phi(\mathbf{r})$ as

$$\phi(\mathbf{r}) = -Vr\cos\theta + \psi(\mathbf{r}) \, ,$$

where the first term reduces to the constant velocity field of the flow without ball $\mathbf{u}_0(\mathbf{r}) = -\mathbf{V}$. The field $\psi(\mathbf{r})$ also satisfies Laplace equation

$$\Delta\psi(\mathbf{r}) = 0,$$

with Neumann BC

$$\mathbf{n} \cdot \boldsymbol{\nabla}\psi = -V\cos\theta \quad \text{on} \quad S_R \quad \text{and} \quad \mathbf{n} \cdot \boldsymbol{\nabla}\psi = 0 \quad \text{on} \quad S_\infty. \tag{2.89}$$

Let us now move to the determination of $\psi(\mathbf{r})$. As explained in the first part of this chapter, there is no homogeneous Neumann Green's function for Laplace operator. Rather than introducing the special Neumann Green's functions, which are not easily accessible anyway, it is more efficient to work with the integral equation (2.40), p. 71 for $\psi(\mathbf{r})$, based on Green's function G_∞ of the infinite system. Here since no source ρ is present, the sole integrals to consider are the ones defined on surfaces S_R and S_∞. In fact, as we shall verify *a posteriori*, field $\psi(\mathbf{r})$ decreases sufficiently fast at infinity for the contribution of S_∞ to vanish. We then find, taking into account the boundary condition on S_R,

$$\psi(\mathbf{r}) = -V \oint_{S_R} d\Sigma'\, G_\infty(|\mathbf{r} - \mathbf{r}'|)\, \cos\theta' + \oint_{S_R} d\Sigma'\, \psi(\mathbf{r}')\, \mathbf{n}' \cdot \boldsymbol{\nabla}_{\mathbf{r}} G_\infty(|\mathbf{r} - \mathbf{r}'|), \quad (2.90)$$

with $G_\infty(|\mathbf{r} - \mathbf{r}'|) = 1/(4\pi|\mathbf{r} - \mathbf{r}'|)$. We will solve this integral equation using the decomposition of $1/|\mathbf{r} - \mathbf{r}'|$ in terms of Legendre polynomials and exploiting the independence of $\psi(\mathbf{r}) = \psi(r, \theta)$ on the angle φ as a result of the system cylindrical symmetry.

Calculation of surface integrals   Consider the first integral in expression (2.90), analogous to the electrostatic potential created by a distribution of surface charges and now denoted by $\psi_c(\mathbf{r})$. Introduce the decomposition✠

$$\frac{1}{|\mathbf{r} - \mathbf{r}'|} = \frac{1}{r} \sum_{n=0}^{+\infty} \left(\frac{R}{r}\right)^n \left[P_n(\cos\theta) P_n(\cos\theta') \right.$$

$$\left. + 2 \sum_{m=1}^{n} \frac{(n-m)!}{(n+m)!} P_n^m(\cos\theta) P_n^m(\cos\theta') \cos(m(\varphi - \varphi')) \right] \quad (2.91)$$

for $\mathbf{r}' \in S_R$, where P_n are Legendre polynomials and P_n^m the corresponding Legendre functions.

Since

$$d\Sigma' = R^2 d\Omega' = R^2 \sin\theta' d\theta' d\varphi',$$

the integration over φ' of terms $\cos(m(\varphi - \varphi'))$ on $[0, 2\pi[$ vanishes for $m \geq 0$. Moreover, as $\cos\theta' = P_1(\cos\theta')$, integration in $\cos\theta'$ on $[-1, 1]$ is straightforward by orthogonality of P_1 and P_n for $n \neq 1$.

> ✠ **Comment:**   Relation
> (2.91) arises from the decomposition of the Green's function $G_\infty(|\mathbf{r} - \mathbf{r}'|) = 1/(4\pi|\mathbf{r} - \mathbf{r}'|)$ in spherical harmonics, with $r' = R < r$. It can easily be obtained from result (2.128), p. 115 of exercise 2.8. In addition, Appendix G contains reminders about Legendre polynomials.

One finally gets

$$\psi_c(\mathbf{r}) = -\frac{VR^3}{3r^2} \cos\theta. \quad (2.92)$$

To simply compute the second integral, denoted from now on $\psi_d(\mathbf{r})$, note that it is similar to the electrostatic potential created by a surface distribution of dipoles, as already noticed in section 2.1.3. It is particularly useful to see each strictly localised elementary dipole as the limit of an extended dipole along direction **n** and consisting of two opposite surface charges, i.e.

$$\mathbf{n}' \cdot \boldsymbol{\nabla_r} G_\infty(|\mathbf{r} - \mathbf{r}'|) = \lim_{\delta \to 0^+} (2\delta)^{-1} \left[G_\infty(|\mathbf{r} - \mathbf{r}' + \mathbf{n}'\delta|) - G_\infty(|\mathbf{r} - \mathbf{r}' - \mathbf{n}'\delta|) \right].$$

Inserting this identity into the expression for $\psi_d(\mathbf{r})$, we get

$$\psi_d(\mathbf{r}) = R^2 \lim_{\delta \to 0^+} \frac{1}{2\delta} \left[\int_{-1}^{1} d(\cos\theta') \, \psi(R, \cos\theta') \int_0^{2\pi} d\varphi' \, G_\infty(|\mathbf{r} + (R+\delta)\mathbf{n}'|) \right.$$
$$\left. - \int_{-1}^{1} d(\cos\theta') \, \psi(R, \cos\theta') \int_0^{2\pi} d\varphi' \, G_\infty(|\mathbf{r} + (R-\delta)\mathbf{n}'|) \right].$$

For a finite value of δ, each of these integrals has the same form as that defining $\psi_c(\mathbf{r})$, and it is easily calculated by the same method based on decomposition (2.91). We then find, after taking the limit $\delta \to 0^+$,

$$\psi_d(\mathbf{r}) = \frac{1}{2} \sum_{n=0}^{+\infty} n(R/r)^{n+1} P_n(\cos\theta) \int_{-1}^{1} d(\cos\theta') \, \psi(R, \cos\theta') \, P_n(\cos\theta'). \quad (2.93)$$

Solution of the integral equation   Replacing $\psi_c(\mathbf{r})$ and $\psi_d(\mathbf{r})$ by formulas (2.92) and (2.93) respectively in

$$\psi(\mathbf{r}) = \psi_c(\mathbf{r}) + \psi_d(\mathbf{r}) ,$$

we obtain an expression of $\psi(\mathbf{r})$ valid at any point $\mathbf{r}$ of the domain, which depends only on $\psi(R, \cos\theta')$ on the surface S_R. We derive a similar expression for $\boldsymbol{\nabla}\psi$. Its insertion in boundary condition $\mathbf{n} \cdot \boldsymbol{\nabla}\psi = -V\cos\theta$ at $r = R$ allows to determine the unique decomposition of $\psi(R, \cos\theta')$ on the orthogonal basis of $P_n(\cos\theta')$. Using again $P_1(\cos\theta') = \cos\theta'$, we find

$$\psi(R, \cos\theta') = -\frac{VR}{2} \cos\theta'.$$

Expression (2.93) for $\psi_d(\mathbf{r})$ then only contains the term $n = 1$, which eventually gives

$$\boxed{\psi(r, \theta) = -\frac{VR^3 \cos\theta}{2r^2}.} \quad (2.94)$$

We check that the potential $\psi(\mathbf{r})$ decreases sufficiently fast at infinity, justifying *a posteriori* the omission of the contribution from the fictitious surface S_∞ in formula (2.90). Note that $\psi(\mathbf{r})$ has a dipolar structure similar to the electrostatic potential created by a localised dipole.

The velocity field in the reference frame of the ball is finally obtained by inserting formula (2.94) in

$$\mathbf{u}(\mathbf{r}) = -\mathbf{V} + \boldsymbol{\nabla}\psi(\mathbf{r}).$$

It is represented in Figure 2.14. For a superconducting sphere submitted to a weak homogeneous magnetic field, we recover the magnetic field lines associated with of the Meissner effect.

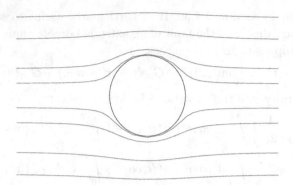

Fig. 2.14   Velocity field of the fluid in the reference frame of the ball.

$$\Diamond \ \textbf{Interpretation} \ \Diamond$$

This example illustrates the interest of Green's functions with boundary conditions at infinity for solving problems with boundaries. Since the Green's function is proportional to the Coulomb potential created by a point charge in vacuum, surface contributions allow for fruitful electrostatic interpretations. However in the present case without source, i.e. $\rho(\mathbf{r}) = 0$, a direct solution of Laplace equation is also possible without using any Green's function, as briefly described below.

Solution by decomposing in spherical harmonics   The field $\psi(\mathbf{r})$ solution of Laplace equation is a harmonic function. Moreover, it depends only on r and θ due to the problem cylindrical symmetry. Its most general form can be written as an infinite linear combination of spherical harmonics[18]

$$\psi(r, \theta) = \sum_{l=0}^{+\infty} (A_l r^l + B_l r^{-l-1}) P_l(\cos \theta).$$

The boundary condition at infinity requires $A_l = 0$ for all $l \neq 0$. Since the velocity is the gradient of ψ, we can set the constant $A_0 = 0$. The boundary condition on S_R then reads

$$\sum_l (l+1) B_l R^{-l-2} P_l(\cos \theta) = -V \cos \theta.$$

As the P_l form an orthogonal basis while $P_1(\cos \theta) = \cos \theta$, B_l vanishes for all $l \neq 1$ and $B_1 = -VR^3/2$. We then recover expression (2.94) obtained previously for $\psi(\mathbf{r})$. Properties of Legendre polynomials $P_l(\cos \theta)$ are the key to explicitly solve the problem by both methods.

---

[18]Appendix G contains reminders on spherical harmonics.

2.2.3 *Density of states of a quantum particle*

◊ **Presentation** ◊

As described in the general part of this chapter, knowledge of the spectrum of operators at play gives access to Green's functions. Conversely, a direct calculation of Green's functions can determine spectral properties. This inversed strategy will be illustrated in this example from quantum mechanics, where the formalism of Green's functions provides useful information on the spectrum of the Hamiltonian, particularly in condensed matter physics.

To fix ideas consider a quantum particle of Hamiltonian

$$H = -\frac{\hbar^2}{2m}\Delta_{\mathbf{r}} + V(\mathbf{r}) ,$$

enclosed in a domain $\mathcal{D}$. An eigenfunction $\psi_n(\mathbf{r})$ of H, with eigenvalue E_n, is solution of PDE

$$[H - E_n]\psi_n(\mathbf{r}) = 0 , \tag{2.95}$$

with homogeneous Dirichlet BC on the domain boundary $\partial\mathcal{D}$. This equation takes the general form (2.72), p. 87 with $\mathcal{O} = H$, $\lambda = -E_n$ and $\rho(\mathbf{r}) = 0$. For this value of λ, the operator $[\lambda + H]$ cannot be inverted: $\psi_n(\mathbf{r})$ would otherwise be identically zero, since it is solution of PDE (2.95) with correct BC! We can therefore say that $\lambda = -E_n$ must be a singularity of Green's function $G_\lambda(\mathbf{r}, \mathbf{r}')$ associated with resolvent $[\lambda + H]^{-1}$. We will first explain the nature of this singularity and show how to infer the density of states of the particle. We then discuss one-dimensional examples. This second step will provide a playground to apply some general techniques.

◊ **Analysis and solution** ◊

Spectral representation (2.77) of $G_\lambda(\mathbf{r}; \mathbf{r}')$ here becomes

$$G_\lambda(\mathbf{r}, \mathbf{r}') = \sum_n \frac{\psi_n(\mathbf{r})\psi_n^*(\mathbf{r}')}{\lambda + E_n}. \tag{2.96}$$

Here $\lambda = -E_n$ clearly appears as a simple pole of $G_\lambda(\mathbf{r}; \mathbf{r}')$ seen as an analytic function of the complex variable λ. Residue of this pole boils down to the product $\psi_n(\mathbf{r})\psi_n^*(\mathbf{r}')$ or the sum of such products if the eigenenergy E_n is degenerate. Thus an *a priori* knowledge of $G_\lambda(\mathbf{r}; \mathbf{r}')$ for λ in the complex plane gives access to the spectrum of H by identification of both the positions and the residues of simple poles on the real axis.

Expression for the density of states Let us now show, by applying the previous observations, how the behaviour of $G_\lambda(\mathbf{r}; \mathbf{r}' = \mathbf{r})$ when λ approaches the real axis allows to determine the density of states $\rho(E)$ defined by

$$\rho(E) = \text{Tr}\{\delta(H - E\mathcal{I})\} = \sum_n \delta(E_n - E),$$

where $\mathcal{I}$ is the identity and where the second equality is the expression of the trace in the eigenbasis $\{\psi_n\}$ of the Hamiltonian. Let us set $\lambda = -E + i\epsilon$ with $\epsilon > 0$ and define

$$G_E^+(\mathbf{r}; \mathbf{r}) = \lim_{\epsilon \to 0^+} G_{-E+i\epsilon}(\mathbf{r}; \mathbf{r}).$$

Using identity (A.1), p. 265,

$$\lim_{\epsilon \to 0^+} \frac{1}{E_n - E + i\epsilon} = \text{PP}\left(\frac{1}{E_n - E}\right) - i\pi\delta(E_n - E) ,$$

spectral representation (2.96) becomes in the limit under consideration

$$G_E^+(\mathbf{r}; \mathbf{r}) = \text{PP}\left(\sum_n \frac{|\psi_n(\mathbf{r})|^2}{E_n - E}\right) - i\pi \sum_n |\psi_n(\mathbf{r})|^2 \, \delta(E_n - E). \tag{2.97}$$

Let us take the imaginary part of each member of this equality. Since the first term in the right-hand side of equation (2.97) is real, only the second term contributes. Moreover the sum over n reduces to $\rho(E)$ after integration over $\mathbf{r}$ on the domain $\mathcal{D}$, as a result of normalisation of eigenfunctions $\psi_n(\mathbf{r})$. We finally obtain the expression

$$\rho(E) = -\frac{1}{\pi} \int_{\mathcal{D}} \mathrm{d}\mathbf{r} \; \text{Im}\big[G_E^+(\mathbf{r}; \mathbf{r})\big]. \tag{2.98}$$

Note that it is also possible to define $G_E^-(\mathbf{r}; \mathbf{r})$ by taking the limit of $G_{-E+i\epsilon}(\mathbf{r}; \mathbf{r})$ when $\epsilon \to 0^-$. We then obtain a formula analogous to (2.98) expressing $\rho(E)$ in terms of $G_E^-(\mathbf{r}, \mathbf{r})$ with a prefactor $1/\pi$ instead of $-1/\pi$. This implies that a portion of the real axis corresponding to the continuous part of the spectrum is necessarily a branch cut for $G_\lambda(\mathbf{r}; \mathbf{r})$. It is indeed discontinuous since $G_E^+(\mathbf{r}; \mathbf{r})$ and $G_E^-(\mathbf{r}; \mathbf{r})$ have opposite signs.

Free particle in one dimension Consider a free particle on an infinite line whose Hamiltonian H_0 is simply given by

$$H_0 = -\frac{\hbar^2}{2m} \frac{\mathrm{d}^2}{\mathrm{d}x^2}.$$

Let us determine Green's function $G_\lambda(x; x')$ from the ordinary differential equation

$$[-\frac{\hbar^2}{2m} \frac{\mathrm{d}^2}{\mathrm{d}x^2} + \lambda]G_\lambda(x; x') = \delta(x - x') \tag{2.99}$$

with $G_\lambda(\pm\infty; x') = 0$. This equation is a one-dimensional Helmholtz equation with complex parameter. Similarly to the calculation leading to formula (2.66), p. 83, it is convenient to perform a Fourier transform leading to

$$G_\lambda(x; x') = \frac{1}{2\pi} \int_{-\infty}^{\infty} \mathrm{d}k \, \frac{e^{ik(x-x')}}{\lambda + \hbar^2 k^2/(2m)}. \tag{2.100}$$

In this formula the integrant, seen as an analytic function of the variable k, has two simple poles at $k_\pm(\lambda) = \pm(-2m\lambda/\hbar^2)^{1/2}$. The function $(-Z)^{1/2}$ is here defined by the choice of determination

$$(-Z)^{1/2} = \sqrt{|Z|}e^{i(\arg Z + \pi)/2}$$

with a branch cut on the negative real axis and $\arg Z \in]-\pi, \pi[$. If $(x - x')$ is positive, it is appropriate to complete the real axis of integration over k by a large semi-circle in the upper half complex plane. Indeed, the integrant then satisfies Jordan's lemma on this semicircle so that application of Cauchy theorem to the corresponding closed contour involves only the pole $k_+(\lambda)$. Similarly, if $(x - x')$ is negative, by closing in the lower half complex plane now, only the pole $k_-(\lambda)$ contributes. We eventually get

$$G_\lambda(x; x') = i \left(\frac{-m}{2\hbar^2 \lambda} \right)^{1/2} \exp\left[i|x - x'|(-2m\lambda/\hbar^2)^{1/2} \right], \qquad (2.101)$$

which can be seen as the analytic continuation of the formula (2.66).

When $\lambda = -E + i\epsilon$ approaches the positive real axis either from above or from below, $G_\lambda(x; x')$ tends to the same real value

$$G_E^+(x; x') = G_E^-(x; x') = \sqrt{\frac{m}{2\hbar^2 |E|}} \, e^{-|x - x'|\sqrt{2m|E|}/\hbar^2} \text{ for } E < 0. \qquad (2.102)$$

Application of formula (2.98) gives then $\rho(E) = 0$ for $E < 0$. When $\lambda = -E + i\epsilon$ approaches the negative real axis we find

$$G_E^\pm(x; x') = \mp i \sqrt{\frac{m}{2\hbar^2 E}} \, e^{\mp i|x - x'|\sqrt{2mE/\hbar^2}} \text{ for } E > 0. \qquad (2.103)$$

Since these two values are complex conjugate, $G_\lambda(x; x')$ is not continuous when crossing the negative real axis, which is actually a branch cut starting from the branching point $\lambda = 0$, in agreement with the existence of a continuum of positive eigenvalues E_n.

A direct application of formula (2.98) leads to an infinite density of states for $E > 0$ since $G_\lambda(x; x)$ is a constant. This singularity is simply due to the infinite size of the domain under consideration. We can repeat this study for a segment of length L. Then, as already noted for the Helmholtz operator in one dimension p. 77, boundary effects become negligible when L is sufficiently large. Expression (2.101) is asymptotically correct far from the edges so that formula (2.98) reads for $E > 0$

$$\rho(E) \simeq L \sqrt{\frac{m}{2\pi^2 \hbar^2 E}} \text{ when } L \to \infty. \qquad (2.104)$$

We thus retrieve the well-known expression which can be obtained directly through a simple counting argument based on eigenenergy quantification

$$E_n = n^2 \pi^2 \hbar^2 / (2mL^2).$$

Particle in a localised well   If we submit the previous particle to an attractive potential $V(x) = -U\delta(x)$ with $U > 0$, its Hamiltonian becomes

$$H = H_0 + V(x) = -\frac{\hbar^2}{2m}\frac{d^2}{dx^2} + V(x).$$

After considering the case of H_0 with a continuous spectrum, we will now illustrate how determining Green's function $G_\lambda(x; x')$ associated with $[\lambda + H]^{-1}$ also gives access to the discrete part of the spectrum of H. Henceforth we denote $G_\lambda^{(0)}(x; x')$ the previous Green's function associated with $[\lambda + H_0]^{-1}$.

Let us start from identity (2.80), proven in p. 89 and written here as

$$G_\lambda(x; x') = G_\lambda^{(0)}(x; x') - \int_{-\infty}^{\infty} dx'' \, G_\lambda^{(0)}(x; x'') \, V(x'') \, G_\lambda(x''; x').$$

Since $V(x'') = -U\delta(x'')$, this integral equation for $G_\lambda(x; x')$ takes the simple form

$$G_\lambda(x; x') = G_\lambda^{(0)}(x; x') + U \, G_\lambda^{(0)}(x; 0) \, G_\lambda(0; x'). \qquad (2.105)$$

Setting $x = 0$ in this relation, we immediately compute $G_\lambda(0; x')$ by solving a basic algebraic equation. Using the value found in the formula (2.105), we find

$$G_\lambda(x; x') = G_\lambda^{(0)}(x; x') + \frac{UG_\lambda^{(0)}(x; 0) \, G_\lambda^{(0)}(0; x')}{1 - iU\left[(-m/(2\hbar^2\lambda)\right]^{1/2}}, \qquad (2.106)$$

with the free Green's function $G_\lambda^{(0)}(x; x')$ given by formula (2.101). This completes the exact and explicit determination of $G_\lambda(x; x')$.

We will now focus on the singularities of expression (2.106). The negative real axis in λ clearly remains a cut for $G_\lambda(x; x')$, since it is a cut for $G_\lambda^{(0)}(x; x')$. The spectrum of H therefore has a continuous part of energy $E > 0$ like that of H_0. For real and positive λ, $G_\lambda^{(0)}(x; x')$ is analytic. Due to the presence of the denominator

$$1 - iU\left(-m/(2\hbar^2\lambda)\right)^{1/2}$$

in the formula (2.106), a simple pole of $G_\lambda(x; x')$ appears for the value λ_P cancelling the denominator, i.e.

$$\lambda_P = \frac{mU^2}{2\hbar^2}. \qquad (2.107)$$

This isolated simple pole corresponds to a localised eigenstate of energy $E_0 = -\lambda_P = -mU^2/(2\hbar^2)$: it is the ground state of H which might as well be obtained by direct resolution of the eigenvalue equation. Using formulas (2.102) for $G_{\lambda_P}^{(0)}(x; 0)$ and $G_{\lambda_P}^{(0)}(0; x')$ the reader can check that the residue of pole λ_P gives the product $\psi_0(x)\psi_0^*(x')$.

◊ Interpretation ◊

This simple example illustrates the mechanisms at play in the fundamental relationship between analytical properties of Green's function associated with the resolvent on one hand, and spectral properties of the Hamiltonian on the other hand. This relationship takes on its full interest in situations where a direct resolution of Schrödinger equation is not possible. This is the case for so-called disordered systems, such as a quantum particle subjected to a potential depending on one or more random parameters. This is also the case in the context of the many-body problem, where it is useful to study the spectrum of a particle dressed by interactions with its counterparts starting from Green's function reduced to a single body[19].

About perturbative expansions   In general an exact calculation of Green's function is obviously very difficult. It can then be useful to perform a perturbative expansion as described in the first part of this chapter, by carefully selecting the Green's function of reference. Note however that these perturbative expansions may miss essential effects. In the case of δ potential wells studied previously for example, a perturbative expansion of $G_\lambda(x; x')$ in powers of U does not capture the singularity at λ_P. For example this indeed gives, up to first order in U,

$$G_\lambda(x; x') \simeq G_\lambda^{(0)}(x; x') + U\, G_\lambda^{(0)}(x; 0)\, G_\lambda^{(0)}(0; x')$$

so that the localised ground state remains unseen, while this property still holds at any finite order of the perturbative expansion! On the other hand, we note that identity (2.80) can provide more reliable non-perturbative information.

2.2.4   *Scattering by a repulsive potential*

◊ Presentation ◊

Green's function formalism turns out to be also useful for scattering theory in quantum mechanics. Consider a particle submitted to a short-range repulsive potential $V(\mathbf{r})$ generated by a scattering centre. In classical mechanics the aim of scattering is to study deviation of the particle trajectory by this centre, when it comes from infinity with an initial velocity $\mathbf{v}_0$. There is a similar dynamic formulation of scattering in quantum mechanics that we present at the end of this section. It turns out that there is also a stationary formulation, which is to look for the eigenfunctions of Hamiltonian

$$H = -\frac{\hbar^2}{2m}\Delta + V(\mathbf{r}). \tag{2.108}$$

In addition we assume that the scattering potential $V(\mathbf{r})$ is sufficiently weak so that these eigenfunctions are perturbations of plane waves, the eigenfunctions of free Hamiltonian

$$H_0 = -\frac{\hbar^2}{2m}\Delta.$$

---

[19] See also the comment p. 150.

This approach is exposed in a first place.

Let $\psi_{\mathbf{k}}^{(0)}(\mathbf{r})$ be the plane wave

$$\psi_{\mathbf{k}}^{(0)}(\mathbf{r}) = \frac{1}{(2\pi)^{3/2}}\, e^{i\mathbf{k}\cdot\mathbf{r}}\,,$$

eigenfunction of H_0 of energy $E_{\mathbf{k}} = \hbar^2\mathbf{k}^2/(2m)$. Since $V(\mathbf{r})$ is repulsive and decays at infinity, one can show that there is an eigenfunction $\psi_{\mathbf{k}}(\mathbf{r})$ of H with the same eigenenergy $E_{\mathbf{k}} = \hbar^2\mathbf{k}^2/(2m)$. This comes from the fact that we are in the continuous part of the spectrum of the full Hamiltonian. Then $\psi_{\mathbf{k}}$ behaves at infinity as a combination of plane waves with $|\mathbf{k}| = \sqrt{2mE_k}/\hbar$. Writing

$$\psi_{\mathbf{k}}(\mathbf{r}) = \psi_{\mathbf{k}}^{(0)}(\mathbf{r}) + \delta\psi_{\mathbf{k}}(\mathbf{r})\,,$$

we obtain the exact equation

$$(H_0 - E_{\mathbf{k}})\delta\psi_{\mathbf{k}}(\mathbf{r}) = -V(\mathbf{r})\psi_{\mathbf{k}}(\mathbf{r}). \tag{2.109}$$

As already noticed in section 2.2.3, this type of PDE is not readily solvable in terms of Green's functions as the corresponding resolvent $[H_0 - E_{\mathbf{k}}]^{-1}$ is singular! However, as we shall see, the introduction of resolvent $[H_0 + \lambda]^{-1}$ allows to build a perturbative expansion of $\delta\Psi_{\mathbf{k}}(\mathbf{r})$.

◇ **Analysis and solution** ◇

The method relies on the introduction of PDE

$$(H_0 + \lambda)\phi_\lambda(\mathbf{r}) = -V(\mathbf{r})\phi_\lambda(\mathbf{r}), \tag{2.110}$$

where λ is a complex parameter, with $\text{Im}(\lambda) \neq 0$ so that the Green's function $G_\lambda^{(0)}(\mathbf{r};\mathbf{r}')$ associated with resolvent $[H_0 + \lambda]^{-1}$ is well defined. We readily establish the expression of $G_\lambda^{(0)}(\mathbf{r};\mathbf{r}')$ following the one-dimensional calculation on p. 100. We then show how taking the suitable limit where $\lambda \to -E_{\mathbf{k}}$ yields $\delta\psi_{\mathbf{k}}(\mathbf{r})$ from the corresponding solution of PDE (2.110). This generates a perturbative expansion of $\delta\psi_{\mathbf{k}}(\mathbf{r})$ in powers of the potential, for which we explicitly compute the first term at large distances from the scattering centre.

Computation of the free Green's function   Green's function $G_\lambda^{(0)}(\mathbf{r};\mathbf{r}')$ is the solution of a Helmholtz type PDE with complex parameters

$$\left[-\frac{\hbar^2}{2m}\Delta + \lambda\right]G_\lambda^{(0)}(\mathbf{r};\mathbf{r}') = \delta(\mathbf{r} - \mathbf{r}'), \tag{2.111}$$

with homogeneous Dirichlet BC at infinity. The resolution can be performed by Fourier transform, as in the calculation leading to formula (2.48), p. 74. We thus obtain a very similar expression

$$G_\lambda^{(0)}(|\mathbf{r} - \mathbf{r}'|) - \frac{me^{i|\mathbf{r}-\mathbf{r}'|(-2m\lambda/\hbar^2)^{1/2}}}{2\pi\hbar^2|\mathbf{r} - \mathbf{r}'|}. \tag{2.112}$$

Just as in the one-dimensional case, the function $(-Z)^{1/2}$ is defined by the choice of determination

$$(-Z)^{1/2} = \sqrt{|Z|}e^{i(\arg Z + \pi)/2},$$

with a branch cut on the negative real semi-axis and $\arg Z \in\,]-\pi, \pi[$.

Taking the limit The auxiliary field $\phi_\lambda(\mathbf{r})$ solution of PDE (2.110) is simply given by superposition formula

$$\phi_\lambda(\mathbf{r}) = -\int d\mathbf{r}' \, G_\lambda^{(0)}(|\mathbf{r} - \mathbf{r}'|)V(\mathbf{r}')\psi_\mathbf{k}(\mathbf{r}')$$

valid for the infinite system[20]. At this point it is tempting to take the limit $\lambda \to -E_\mathbf{k}$ because the limit of $\phi_\lambda(\mathbf{r})$ should give $\delta\psi_\mathbf{k}(\mathbf{r})$ as PDE (2.110) and (2.109) then become identical. However the point $-E_\mathbf{k}$ is singular because the negative real axis is a cut for $G_\lambda^{(0)}(|\mathbf{r} - \mathbf{r}'|)$. In other words one will not get the same result depending on whether one reaches this point from above or below the cut. Here we take the limit with $\mathrm{Im}(\lambda) < 0$. We will justify this choice later on and briefly comment on the meaning of the alternative choice $\mathrm{Im}(\lambda) > 0$. We thus obtain the integral equation

$$\delta\psi_\mathbf{k}(\mathbf{r}) = -\frac{m}{2\pi\hbar^2} \int d\mathbf{r}' \, \frac{e^{ik|\mathbf{r}-\mathbf{r}'|}}{|\mathbf{r} - \mathbf{r}'|} V(\mathbf{r}')\left[\psi_\mathbf{k}^{(0)}(\mathbf{r}') + \delta\psi_\mathbf{k}(\mathbf{r}')\right]. \qquad (2.113)$$

Born approximation We generate the perturbative expansion of $\delta\psi_\mathbf{k}(\mathbf{r})$ in powers of the potential by iterating integral equation (2.113). Within Born approximation we keep only the first term and get

$$\delta\psi_\mathbf{k}^B(\mathbf{r}) = -\frac{m}{(2\pi)^{5/2}\hbar^2} \int d\mathbf{r}' \, \frac{e^{ik|\mathbf{r}-\mathbf{r}'|}}{|\mathbf{r} - \mathbf{r}'|} V(\mathbf{r}')e^{i\mathbf{k}\cdot\mathbf{r}'}. \qquad (2.114)$$

It is interesting to study the structure of this deviation at large distance r from the scattering centre. Dominant contributions to integral (2.114) come from finite distances $|\mathbf{r}'|$ of the same order as the range of potential V, which are very small compared with r. We can then expand the phase $k|\mathbf{r} - \mathbf{r}'| = kr - k\mathbf{r} \cdot \mathbf{r}'/r + O(1/r)$, leading to the asymptotic formula

$$\psi_\mathbf{k}^B(\mathbf{r}) \simeq \frac{1}{(2\pi)^{3/2}}\left[e^{i\mathbf{k}\cdot\mathbf{r}} + f_\mathbf{k}(\hat{\mathbf{r}})\frac{e^{ikr}}{r}\right] \quad \text{when } r \to \infty, \qquad (2.115)$$

with

$$f_\mathbf{k}(\hat{\mathbf{r}}) = -\frac{m}{2\pi\hbar^2} \int d\mathbf{r}' \, \exp\left[i(\mathbf{k} - k\hat{\mathbf{r}}) \cdot \mathbf{r}'\right]V(\mathbf{r}'), \qquad (2.116)$$

$\hat{\mathbf{r}} = \mathbf{r}/r$ being the unitary vector along $\mathbf{r}$. Based on these formulas, we see that deviation from the plane wave is an anisotropic spherical wave whose amplitude decays as $1/r$ at infinity. This structure is that of the exact wave function at large distances. Anisotropic amplitude factor $f_\mathbf{k}(\hat{\mathbf{r}})$ obtained within the Born approximation is in general not accurate due to multiple scattering.

◊ Interpretation ◊

It is instructive to show that a dynamical approach to scattering, more in line with experimental situations, indeed gives results obtained by the purely stationary method. We briefly summarise this approach and we finally illustrate it with an explicit calculation for Yukawa potential.

---

[20]Strictly speaking, we should work in a large domain, and then send the boundaries to infinity. Again, we free ourselves from this step by working directly on the infinite system, without worrying about BC. Here this procedure is ultimately legitimate.

Perturbative dynamical approach   In order to mimic the dynamic of the scattering process, it is convenient to imagine that the particle is in an initial state of wave function $\psi_{\mathbf{k}}^{(0)}(\mathbf{r})$ at $t_0 = -\infty$. The potential $V(\mathbf{r})$ is then triggered by adiabatic connection at $t_0 = -\infty$. In other words the particle is submitted to a time-dependent potential $W(\mathbf{r}, t) = e^{\epsilon t}V(\mathbf{r})$. Wave function evolution can then be obtained by perturbative expansions presented in Chapter 3. Their basic ingredient is the causal free Green's function $G^{+(0)}(|\mathbf{r} - \mathbf{r}'|; \tau)$. We invite the reader to apply the expansion (3.73), p. 154, keeping only the first order term in W. One can then explicitly check by taking the limit $\epsilon \to 0^+$ that

$$\lim_{\epsilon \to 0^+} \phi(\mathbf{r}, t)e^{iE_{\mathbf{k}}(t-t_0)/\hbar} = \psi_{\mathbf{k}}^{B}(\mathbf{r}),$$

using the fundamental identity

$$\lim_{\epsilon \to 0^+} \int_0^{+\infty} \mathrm{d}\tau\, G^{+(0)}(|\mathbf{r} - \mathbf{r}'|; \tau)\, e^{-i(-E_{\mathbf{k}} - i\epsilon\hbar)\tau/\hbar} = -i\hbar \lim_{\epsilon \to 0^+} G_{-E_{\mathbf{k}} - i\epsilon\hbar}^{(0)}(|\mathbf{r} - \mathbf{r}'|).$$

So the particle after diffusion by potential $V(\mathbf{r})$ is in the eigenstate corresponding to the eigenenergy $E_{\mathbf{k}}$ of Hamiltonian H. Furthermore, if the incoming wave is a plane wave propagating along $\mathbf{k}$, the outgoing wave has an outgoing scattered component of spherical wave type. This completes the justification for the choice of sign for $\mathrm{Im}(\lambda)$ in the stationary method: The other choice corresponds to an incoming spherical wave which is not allowed by causality.

Application to Yukawa potential   The repulsive Yukawa potential is $V(\mathbf{r}) = Ue^{-\alpha r}/r$. Anisotropic amplitude $f_{\mathbf{k}}(\hat{\mathbf{r}})$ given by formula (2.116) is simply proportional to the Fourier transform of potential $V(\mathbf{r})$. Here the calculation is obviously the same as for Green's function of Helmholtz operator, leading to

$$\boxed{f_{\mathbf{k}}(\hat{\mathbf{r}}) = \frac{-2mU}{\hbar^2}\frac{1}{4k^2\sin^2(\theta/2) + \alpha^2},}$$

where θ is the angle between $\mathbf{k}$ and $\mathbf{r}$. The differential scattering cross section is easily obtained from this expression. Note that in the limit $\alpha \to 0^+$, we find the exact formula of Rutherford cross section for Coulomb potential. The latter can be obtained by looking for stationary solutions of Schrödinger equation for positive energies. It moreover coincides with its classical value.

2.2.5   *Simple model for the wind blowing on a wall*

◊ **Presentation** ◊

This example is a practical implementation of the method based on two-dimensional conformal transformations, and exposed pp. 78-82. Imagine the wind blowing on an empty plain, with a wall as the only possible shelter. We would like to know which side of the wall is the best shelter. We use a simplified yet representative model of the situation to answer this question.

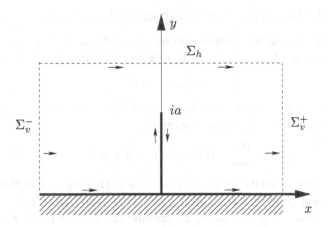

Fig. 2.15   Problem geometry: The wall of height a is at $x = 0$, Σ_h and $\Sigma_v^{\pm}$ are fictitious walls at infinity.

Let us consider the following two-dimensional problem: an incompressible fluid flows in an irrotational way in a domain $\mathcal{D}$ formed by the half-plane $\mathbb{C}^+$, but a vertical wall of height a is located at $x = 0$. At infinity the fluid velocity is given by $\mathbf{u} = u_0\, \mathbf{e}_x$. As shown at the beginning of section 2.2.2, the velocity derives from a potential $\phi(\mathbf{r})$, $\mathbf{u}(\mathbf{r}) = \boldsymbol{\nabla}\phi(\mathbf{r})$, solution of Laplace equation

$$\boxed{\Delta\phi = 0}$$

in $\mathcal{D}$. In order to specify the boundary conditions that ϕ must satisfy, we introduce three fictitious walls at infinity: Σ_h is horizontal and the two walls $\Sigma_v^{\pm}$ are vertical (see Figure 2.15). Boundary conditions are then of Neumann type,

$$\mathbf{n}\cdot\boldsymbol{\nabla}\phi = 0 \text{ on the ground, the wall and } \Sigma_h \; ; \; \mathbf{n}\cdot\boldsymbol{\nabla}\phi = \pm u_0 \text{ on } \Sigma_v^{\pm}.$$

◊ Analysis and solution ◊

As shown on p. 81, there exists an analytic function $\mathcal{A}$ of variable $z = x + iy$,

$$\boxed{\mathcal{A} = \phi + i\psi,}$$

whose real part is ϕ. First we translate the boundary conditions in terms of $\mathcal{A}$. We will then perform a conformal transformation allowing to reformulate the problem on a simpler domain to determine the velocity field on the original domain $\mathcal{D}$.

Rewriting boundary conditions   The interpretation of boundary conditions is easier here than in the general case p. 81. Indeed, since

$$u_x = \frac{\partial\phi}{\partial x} = \frac{\partial\psi}{\partial y} \quad \text{and} \quad u_y = \frac{\partial\phi}{\partial y} = -\frac{\partial\psi}{\partial x}, \tag{2.117}$$

by Cauchy's relations, the derivative of $\mathcal{A}$ with respect to z is simply related to the velocity field:

$$\boxed{\mathcal{A}'(z) = u_x - iu_y.} \tag{2.118}$$

Boundary conditions can therefore be written on $\mathcal{A}$ as follows.

Let us show that on the ground and on each side of the wall, i.e. respectively for $z = x$ and $z = 0^{\pm} + iy$ with $0 \leq y \leq a$, it is sufficient to impose

$$\text{Im}[\mathcal{A}(z)] = \alpha \qquad (2.119)$$

where α is a constant. Indeed, since $\text{Im}\,\mathcal{A} = \psi$ and x varies with $y = 0$ fixed on the ground, condition (2.119) therefore implies

$$\frac{\partial \psi}{\partial x} = 0 \text{ for } y = 0 \text{ and } x \neq 0.$$

By virtue of Cauchy's relations (2.117), we recover that $u_y = 0$ on the ground. On the wall, it is now y which varies at $x = 0^{\pm}$ fixed: condition (2.119) now implies $(\partial \psi / \partial y) = 0$, or $u_x = 0$ by Cauchy's relations (2.117). It is clear here that the constant α appearing in condition (2.119) can be set to zero[21]. In summary we then write the BC both on the ground and the wall as

$$\boxed{\text{Im}[\mathcal{A}(z)] = 0 \qquad \text{for} \quad z = x \quad \text{and for} \quad z = 0^{\pm} + iy \quad \text{with} \quad 0 \leq y \leq a.}$$

These BC are completed with the conditions on the fictitious vertical and horizontal walls, respectively $\Sigma_v^{\pm}$ and Σ_h, obviously leading to

$$\boxed{\text{Re}[\mathcal{A}'(z)] = u_0 \quad \text{when} \quad x \to \pm\infty \quad ; \quad \text{Im}[\mathcal{A}'(z)] = 0 \quad \text{when} \quad y \to +\infty.}$$

Conformal transformations We now try to find a conformal transformation whose image of the domain $\mathcal{D}$ is the upper half-plane without the wall. To do so, we proceed by stages with three transformations, represented in Figure 2.16. In the first transformation,

$$z \mapsto w = z^2,$$

the ground and the wall become the real half-axis $[-a^2, +\infty[$. Then by a simple translation,

$$w \mapsto t = w + a^2,$$

the image of the previous half-axis becomes $\mathbb{R}^+$. Finally the transformation

$$t \mapsto \tilde{z} = t^{1/2}$$

leads to a final domain which is the upper half-plane without wall! The composition of all these transformations provides the function $\mathcal{F}(z)$ defined by

$$\boxed{z \mapsto \tilde{z} = \mathcal{F}(z) = \left(z^2 + a^2\right)^{1/2},}$$

with branch points at $\pm ia$ and a branch cut reducing to the purely imaginary segment $[-ia, ia]$. In addition the determination chosen is such that[✠]

---

[21]This is analogous to the point presented in the general case p. 82. This possibility arises from the freedom in the choice of $\mathcal{A}$, to which a constant can be added without changing the velocity field.

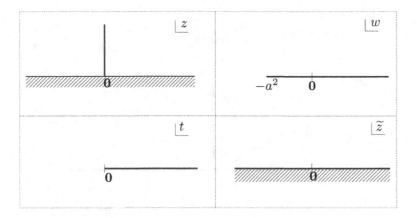

Fig. 2.16 Visualisation of the successive conformal transformations $z \to w \to t \to \widetilde{z}$.

$$\mathcal{F}(x) = \sqrt{x^2 + a^2} \text{ if } x > 0$$

and

$$\mathcal{F}(x) = -\sqrt{x^2 + a^2} \text{ if } x < 0.$$

Note that part of the space under ground and inside (!) the wall of zero thickness has its image by $\mathcal{F}(z)$ under the new horizontal ground. The empty part where the wind blows remains above the ground as expected.

✠ **Comment:** The change of sign of $\mathcal{F}(x)$ is due to the presence of the cut $[-ia, ia]$ which cannot be crossed. To go from $x = 0^+$ to $x = 0^-$ we must get around this cut, passing above the branching point ia for example. The argument of complex number $Z = z^2 + a^2$ rotates by 2π along such a path so that the sign of factor $e^{i \arg Z/2}$, defining the determination, indeed changes.

Calculation of velocity field The problem with the half-plane is elementary. The analytic function

$$\widetilde{\mathcal{A}}(\widetilde{z}) = u_0 \, \widetilde{z},$$

obviously satisfies the boundary conditions $\mathrm{Im}\big[\widetilde{\mathcal{A}}(\widetilde{z})\big] = 0$ on the real axis, and $(\mathrm{d}\widetilde{\mathcal{A}}/d\widetilde{z}) = u_0$ on fictitious walls. We thus obtain the analytic function $\mathcal{A}(z)$ on the original domain $\mathcal{D}$,

$$\mathcal{A}(z) = \widetilde{\mathcal{A}}\big[\mathcal{F}(z)\big] = u_0 \, \big(z^2 + a^2\big)^{1/2},$$

as well as its derivative. The velocity field at each point is obtained using

$$\mathcal{A}'(z) = u_0 \, \frac{z}{(z^2 + a^2)^{1/2}}$$

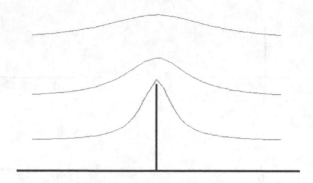

Fig. 2.17   Velocity field lines.

and equation (2.118). We can now analyse the form of the velocity field on various surfaces, with the following results:

$$u_x = u_0 \, \frac{|x|}{\sqrt{x^2 + a^2}}; \qquad\qquad u_y = 0$$

on the ground,

$$u_x = 0; \qquad\qquad u_y = \pm u_0 \, \frac{y}{\sqrt{a^2 - y^2}}$$

respectively on the left and right sides of the wall and

$$u_x = u_0 \, \frac{y}{\sqrt{y^2 - a^2}}; \qquad\qquad u_y = 0$$

on the vertical axis $x = 0$ with $y > a$. The velocity field lines are plotted in Figure 2.17.

◊ **Interpretation** ◊

The velocity field profile shows that it is useless to hide behind the wall to get protection from wind blowing as hard before and after the wall. Another important observation concerns the velocity field singularity at the tip of the wall. It is a peak effect that gives rise to very high speeds. In fact, for a wall of finite thickness the flow ceases to be laminar and becomes turbulent for large enough velocities u_0.

2.3 Exercises

■ **Exercise 2.1. Green's function G_∞ for the Laplacian in 3d**

The aim of this exercise is to determine, in three dimensions, the Green function G_∞ of the Laplacian vanishing at infinity by another method than the one described on p. 68. The goal is therefore to find the function $G_\infty(\mathbf{r})$ solution of

$$-\Delta G_\infty(\mathbf{r}) = \delta(\mathbf{r}) \tag{2.120}$$

with $G_\infty(\mathbf{r}) \to 0$ when $r \to \infty$.

1. We recall that G_∞ depends only on $r = |\mathbf{r}|$. Infer the differential equation satisfied by $G_\infty(r)$ for $r \neq 0$.

2. What is the most general solution of this differential equation that is consistent with the boundary condition at infinity?

3. The above reasoning allows to find the function $G_\infty(r)$ up to a prefactor. How to fix this prefactor? Find then Green's function $G_\infty(r) = 1/(4\pi r)$.

Solution p. 302.

■ **Exercise 2.2. Green's function G_∞ for the Laplacian in dimension $d \geq 3$**

Apply the method of the previous exercise to determine in dimension $d \geq 3$ the Green's function $G_\infty(r)$ of the Laplacian vanishing at infinity. Give the result in terms of $r = |\mathbf{r}|$ and the total solid angle Ω_d in d dimensions. To do so, use the fact that the Laplacian of the function $G_\infty(r)$, depending only on $r = |\mathbf{r}|$, reads:

$$\Delta G_\infty = \frac{d-1}{r}\frac{\mathrm{d}G_\infty}{\mathrm{d}r} + \frac{\mathrm{d}^2 G_\infty}{\mathrm{d}r^2} = \frac{1}{r^{d-1}}\frac{\mathrm{d}}{\mathrm{d}r}\left(r^{d-1}\frac{\mathrm{d}G_\infty}{\mathrm{d}r}\right). \tag{2.121}$$

Solution p. 303.

■ **Exercise 2.3. Green's functions for the Laplacian in 1d and 2d**

1. Explicitly check that the function

$$G(x - x') = -\frac{|x - x'|}{2}$$

is a Green's function of the Laplacian in one dimension. This can be done by applying a test function (vanishing at infinity) to ΔG.

2. Same question for the two-dimensional case and the function

$$G(\mathbf{r} - \mathbf{r}') = -\frac{\ln(|\mathbf{r} - \mathbf{r}'|)}{2\pi}.$$

It will be worth using expression (2.121) mentioned in exercise 2.2.

Solution p. 303.

■ Exercise 2.4. Symmetry of Laplacian Green's functions with homogeneous Dirichlet BC

Show that Green's functions of the Laplacian with homogeneous Dirichlet BC are symmetric.

Solution p. 304.

■ Exercise 2.5. Special Neumann Green's function of the Laplacian

Let $G_{\bar{N}}$ be a Laplacian special Neumann Green's function on a domain $\mathcal{D}$ with boundary $\partial\mathcal{D}$. It satisfies the BC

$$\mathbf{n} \cdot \boldsymbol{\nabla}_{\mathbf{r}} G_{\bar{N}}(\mathbf{r}; \mathbf{r}') = -\frac{1}{s}, \tag{2.122}$$

$$G_{\bar{N}}(\mathbf{r}_0; \mathbf{r}') = c(\mathbf{r}') \qquad \forall \mathbf{r}' \in \mathcal{D}. \tag{2.123}$$

1. Using the second Green's formlula show that

$$G_{\bar{N}}(\mathbf{r}; \mathbf{r}') - G_{\bar{N}}(\mathbf{r}'; \mathbf{r}) = F(\mathbf{r}) - F(\mathbf{r}')$$

and give the expression of F. Conclude that function

$$\tilde{G}_{\bar{N}}(\mathbf{r}; \mathbf{r}') = G_{\bar{N}}(\mathbf{r}; \mathbf{r}') + F(\mathbf{r}')$$

is symmetric special Neumann Green's function.

2. $\phi(\mathbf{r})$ is the solution of Laplace equation with Neumann BC on $\partial\mathcal{D}$. Check that expressions obtained from equation (2.39), p. 71 lead to the same result for $\tilde{G}_{\bar{N}}$ on the one hand and for $G_{\bar{N}}$ on the other hand.

Solution p. 304.

■ Exercise 2.6. Sum rules and resolvent

Let G_λ be the Green's function satisfying equation

$$(\mathcal{O} + \lambda)\, G_\lambda(x; x') = \delta(x - x') \tag{2.124}$$

and homogeneous Dirichlet conditions in $x = 0$ and $x = 1$. Suppose that the operator $\mathcal{O}$ has a complete and orthonormal basis of eigenfunctions $\psi_n(x)$ of eigenvalues λ_n satisfying the same homogeneous Dirichlet conditions.

1. Using the spectral representation of G_λ, write

$$\int_0^1 \prod_{i=1}^m dy_i\;\; G_\lambda(y_1; y_2)G_\lambda(y_2; y_3) \cdots G_\lambda(y_{m-1}; y_m)G_\lambda(y_m; y_1)$$

in terms of λ, m and λ_n.

2. Take for $\mathcal{O}$ the one-dimensional Laplacian operator, $\mathcal{O} = -(\partial^2/\partial x^2)$. Show that $\lambda = \omega^2\pi^2$ with ω real.

$$G_\lambda(x; x') = -\frac{\operatorname{sh}\omega\pi x_<\; \operatorname{sh}\omega\pi(x_> - 1)}{\omega\pi \qquad \omega\pi}$$

with $x_< = \inf(x, x')$ and $x_> = \sup(x, x')$.

3. The set of functions $\psi_n(x) = \sqrt{2}\sin n\pi x$ for $n \in \mathbb{N}^*$ constitutes a complete and orthonormal basis of eigenfunctions of the operator $\mathcal{O}$, satisfying the homogeneous Dirichlet conditions at $x = 0$ and $x = 1$. Show then that

$$\sum_{n=1}^\infty \frac{1}{n^2 + \omega^2} = \frac{1}{2\omega^2}\left(\omega\pi \coth(\omega\pi) - 1\right).$$

4. Give the explicit expression for $G_0(x; x')$ and conclude that

$$\sum_{n=1}^\infty \frac{1}{n^2} = \frac{\pi^2}{6}.$$

Solution p. 304.

■ Exercise 2.7. Conductive plane

In this exercise we come back to the electrostatic configuration discussed in example 2.2.1, p. 90. Consider a conducting plane located in $x = 0$ and maintained at zero potential. We propose to compute the Green's function G_{DH} of Laplace operator with homogeneous Dirichlet BC without using the method of image charges.

1. Let $\widehat{G}_{\text{DH}}(x; x'; \mathbf{k})$ be the Fourier transform along y and z of G_{DH}. What are the differential equation and the boundary conditions satisfied by $\widehat{G}_{\text{DH}}$?

2. Compute $\widehat{G}_{\mathrm{DH}}$ using for instance the method of variation of constants.

3. Compute then the inverse Fourier transform using polar coordinates and the residue theorem.

Solution p. 305.

■ Exercise 2.8. Green's functions for Laplace operator in spherical coordinates

We propose to determine the expression, in spherical coordinates, of the Green's function G_∞ of the Laplace operator. We thus decompose a Green's function G of the Laplacian in spherical harmonics:

$$G(\mathbf{r};\mathbf{r}') = \sum_{l=0}^{\infty} \sum_{m=-l}^{l} g_{lm}(r;r') Y_{lm}^*(\theta',\varphi') Y_{lm}(\theta,\varphi)$$

where Y_{lm} are spherical harmonics whose properties are summarised in Appendix G.

1. Determine the PDE satisfied by $g_{lm}(r;r')$ using result (G.3), p. 288 to rewrite $\delta(\mathbf{r}-\mathbf{r}')$ in terms of spherical harmonics. What is the value of the solutions of the homogeneous equation associated with this PDE?

2. We first restrict ourselves to the volume between concentric spheres of radii a and $b > a$, and look for the Green's function $G_{a,b}$ satisfying homogeneous Dirichlet BC. Using the result (C.6) from variation of constants, p. 273, show that

$$G_{a,b}(\mathbf{r};\mathbf{r}') = \sum_{l=0}^{\infty} \sum_{m=-l}^{l} \frac{Y_{lm}^*(\theta',\varphi') Y_{lm}(\theta,\varphi)}{(2l+1)\left[1-(\frac{a}{b})^{2l+1}\right]} \left(r_<^l - \frac{a^{2l+1}}{r_<^{l+1}}\right) \left(\frac{1}{r_>^{l+1}} - \frac{r_>^l}{b^{2l+1}}\right) \tag{2.125}$$

with

$$r_< = \min\{r,r'\} \quad and \quad r_> = \max\{r,r'\}.$$

3. Conclude that the Green's function G_b for the sphere of radius b and homogeneous Dirichlet BC reads

$$G_b(\mathbf{r};\mathbf{r}') = \sum_{l=0}^{\infty} \sum_{m=-l}^{l} \left(\frac{r_<^l}{r_>^{l+1}} - \frac{(rr')^l}{b^{2l+1}}\right) \frac{Y_{lm}^*(\theta',\varphi') Y_{lm}(\theta,\varphi)}{(2l+1)}. \tag{2.126}$$

4. Conclude also that the Green's function G_a for the domain $r > a$ and homogeneous dirichlet BC is

$$G_a(\mathbf{r};\mathbf{r}') = \sum_{l=0}^{\infty} \sum_{m=-l}^{l} \left(\frac{r_<^l}{r_>^{l+1}} - \frac{a^{2l+1}}{(rr')^{l+1}}\right) \frac{Y_{lm}^*(\theta',\varphi') Y_{lm}(\theta,\varphi)}{(2l+1)}. \tag{2.127}$$

5. Show that

$$\boxed{\frac{1}{4\pi|\mathbf{r}-\mathbf{r}'|} = \sum_{l=0}^{\infty}\sum_{m=-l}^{l}\frac{r_<^l}{(2l+1)r_>^{l+1}}Y_{lm}^*(\theta',\varphi')Y_{lm}(\theta,\varphi)} \qquad (2.128)$$

and then

$$\frac{1}{|\mathbf{r}-\mathbf{r}'|} = \sum_{l=0}^{\infty}\frac{r_<^l}{r_>^{l+1}}P_l(\cos\tilde{\theta}) \qquad (2.129)$$

where $\tilde{\theta}$ is the angle between $\mathbf{r}$ and $\mathbf{r}'$.

Solution p. 306.

■ **Exercise 2.9. Point charge in a conducting sphere**

This exercise is an extension of exercise 2.8. It proposes indeed to now start with the expression (2.128) for the Green's function $G_\infty(\mathbf{r};\mathbf{r}')$ of Laplace operator and to use the method of image charges to determine the Green's function G_a of Laplace operator in the sphere of radius a with homogeneous Dirichlet BC.

1. Give the Green's function G_a by placing a fictitious charge in the same radial direction as $\mathbf{r}'$ but for a radius larger than a.

2. Obtain in the same way the Green's function for homogeneous Dirichlet boundary conditions in the domain formed by the half-sphere $r < a$, $x > 0$.

3. Infer the Green's function of the infinite half-volume $x > 0$ with homogeneous Dirichlet boundary conditions, by considering the suitable limit.

Solution p. 307.

■ Exercise 2.10. Point charge and dielectric sphere

We seek to determine the electric potential $\phi(\mathbf{r})$ created by a point charge q located in $\mathbf{r}_0$ in the presence of a dielectric sphere of radius R, with $R < r_0$ and whose centre is located in $r = 0$ (see Figure 2.18). The z-axis is chosen so that

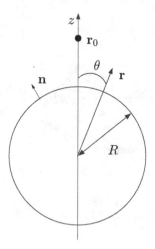

Fig. 2.18   Dielectric sphere of radius R and point charge in $\mathbf{r}_0$.

it passes through the position of the point charge and the sphere centre. We thus have a problem with azimuthal symmetry. The equation[22] satisfied by $\phi(\mathbf{r})$ is $\Delta\phi = -q\delta(\mathbf{r} - \mathbf{r}_0)$. The boundary conditions are chosen such that $\phi(\mathbf{r}) \to 0$ when $r \to \infty$. The matching conditions at the surface of the sphere are given by

$$\phi(R^+, \theta) = \phi(R^-, \theta) \qquad \text{and} \qquad \mathbf{n} \cdot \boldsymbol{\nabla}\phi(R^+, \theta) = \epsilon\, \mathbf{n} \cdot \boldsymbol{\nabla}\phi(R^-, \theta)$$

where $\mathbf{n}$ is the vector normal to the sphere, ϵ is the relative permittivity of the sphere and $R^\pm$ is a shorthand notation to denote the limit $r \to R^\pm$.

1. We write $\phi = \phi_0 + \psi$ where ϕ_0 is the potential in the absence of sphere. Specify the differential equation and the boundary conditions satisfied by ψ. Give the general form of the functions satisfying these conditions in spherical coordinates, inside and outside of the sphere.

2. Complete the determination of ϕ using the result (2.129) of exercise 2.8 and matching conditions in $r = R$.

Solution p. 308.

---

[22]By setting $\epsilon_0 = 1$.

■ **Exercise 2.11. Green's function G_∞ for Laplace operator in cylindrical coordinates**

The goal of this exercise is to determine the expression of G_∞ in cylindrical coordinates.

1. Show that

$$\delta(\mathbf{r} - \mathbf{r}') = \frac{1}{\rho}\delta(\rho - \rho')\frac{1}{2\pi^2} \sum_{m=-\infty}^{+\infty} e^{im(\varphi-\varphi')} \int_0^{+\infty} dk \cos k(z - z') \qquad (2.130)$$

where (ρ, z, φ) and (ρ', z', φ') are respectively the cylindrical coordinates of $\mathbf{r}$ and $\mathbf{r}'$. Note that for angular variables φ and φ':

$$2\pi\delta(\varphi - \varphi') = \sum_{m=-\infty}^{\infty} e^{im(\varphi-\varphi')}.$$

2. Let $G_\infty(\mathbf{r}; \mathbf{r}')$ be the Green's function for the Laplacian vanishing at infinity, and set

$$G_\infty(\mathbf{r}; \mathbf{r}') = \frac{1}{2\pi^2} \sum_{m=-\infty}^{+\infty} e^{im(\varphi-\varphi')} \int_0^{+\infty} dk g_{mk}(\rho; \rho') \cos k(z - z').$$

What is the PDE satisfied by g_{mk}? Solutions of the homogeneous PDE associated with this PDE are the modified Bessel functions[23]. Thus the modified Bessel functions $I_m(k\rho)$ and $K_m(k\rho)$ are two solutions with Wronskian $-1/\rho$. Moreover $I_m(x)$ is finite if $x \to 0$ and $K_m(x) \to 0$ when $x \to +\infty$. By first studying the case $\rho < \rho'$, show that

$$\boxed{\frac{1}{4\pi}\frac{1}{|\mathbf{r} - \mathbf{r}'|} = \frac{1}{2\pi^2} \sum_{m=-\infty}^{+\infty} e^{im(\varphi-\varphi')} \int_0^{+\infty} dk I_m(k\rho_<)K_m(k\rho_>) \cos k(z - z')}$$

with $\rho_< = \inf(\rho, \rho')$ and $\rho_> = \sup(\rho, \rho')$.

Solution p. 309.

■ **Exercise 2.12. Oseen's tensor**

The behaviour of a fluid at very low Reynolds number in the presence of a vesicle can be modelled by equations:

$$\eta\, \Delta\mathbf{v} - \boldsymbol{\nabla}p = \mathbf{f}$$

and

$$\boldsymbol{\nabla}\cdot\mathbf{v} = 0,$$

---

[23]The reader may find definitions and other properties of Bessel functions in the literature [26] or [4].

where $\mathbf{v}$ and p stand respectively for the velocity and pressure fields, and $\mathbf{f}$ for a source (the vesicle).

1. We work in a domain $\mathcal{D}$ of $\mathbb{R}^n$, of boundary $\partial\mathcal{D}$. What conditions is it possible to impose on $\mathbf{v}$ and p so that they are uniquely determined in the whole domain $\mathcal{D}$?

2. Assume $\mathbf{f}$ localised, and look for solutions for $\mathbf{v}$ and p vanishing at infinity. Give p in terms of $\mathbf{f}$. How is p modified if we change $\mathbf{f}$ by $\mathbf{f} + \mathbf{g}$ with $\boldsymbol{\nabla} \cdot \mathbf{g} = 0$?

3. Finally give $\mathbf{v}$ as a function of $\mathbf{f}$. How does this function change if $\mathbf{f}$ becomes $\mathbf{f} + \boldsymbol{\nabla}h$ with h an arbitrary function?

Solution p. 309.

■ **Exercise 2.13. Green's function and elasticity theory**

In elasticity theory the local deformation vector $\mathbf{u}$ satisfies

$$\Delta\mathbf{u} + \frac{1}{1 - 2\sigma}\boldsymbol{\nabla}\left(\boldsymbol{\nabla} \cdot \mathbf{u}\right) = \mathbf{F}(\mathbf{r}) \tag{2.131}$$

where σ is Poisson's ratio relating the rate of transverse contraction to the rate of linear expansion. $\mathbf{F}$ is a vector function given by the constraints imposed on the solid and local temperature variations. We propose to calculate the vectorial Green's function $\mathbf{G}^i(\mathbf{r} - \mathbf{r}')$ with BC at infinity, satisfying the equation:

$$\Delta\mathbf{G}^i(\mathbf{r}) + \frac{1}{1 - 2\sigma}\boldsymbol{\nabla}\left(\boldsymbol{\nabla} \cdot \mathbf{G}^i(\mathbf{r})\right) = -\mathbf{e}_i\delta(\mathbf{r}) \tag{2.132}$$

where $\mathbf{e}_i$ is the unitary vector along direction i. The components of vector $\mathbf{G}^i$ are denoted by G^{ij}.

1. Let us write $\mathbf{G}^i(\mathbf{r})$ as $\mathbf{G}^i(\mathbf{r}) = \mathbf{g}_0^i(\mathbf{r}) + \mathbf{g}_1^i(\mathbf{r})$, with $\mathbf{g}_0^i(\mathbf{r})$ and $\mathbf{g}_1^i(\mathbf{r})$ vanishing when $r \to \infty$ and where $\mathbf{g}_0^i(\mathbf{r})$ is solution of:

$$\Delta\mathbf{g}_0^i(\mathbf{r}) = -\mathbf{e}_i\delta(\mathbf{r}). \tag{2.133}$$

Give the expression for $\mathbf{g}_0^i(\mathbf{r})$ and the equation satisfied by $\mathbf{g}_1^i(\mathbf{r})$.

2. Conclude that $\Delta\left(\boldsymbol{\nabla} \wedge \mathbf{g}_1^i(\mathbf{r})\right) = 0$. We recall that a harmonic function in the whole space and vanishing asymptotically at infinity is null everywhere. Conclude that $\mathbf{g}_1^i(\mathbf{r})$ can be written as

$$\mathbf{g}_1^i(\mathbf{r}) = \boldsymbol{\nabla}\phi^i(\mathbf{r})$$

where ϕ^i is a scalar function.

3. Write down the equation satisfied by ϕ^i and show that the solution is:

$$\phi^i(\mathbf{r}) = -\frac{1}{16\pi(1 - \sigma)}\frac{\partial}{\partial x_i}r. \tag{2.134}$$

4. Give then the components of Green's tensor $G^{ij}(\mathbf{r})$.

Solution p. 310.

■ **Exercise 2.14. Discrete Laplacian and resistors network**

Consider an infinite d-dimensional network consisting of all points $\mathbf{r}$ whose positions are specified by d integers $l_1, \cdots, l_d$ i.e.

$$\mathbf{r} = \sum_{i=1}^{d} l_i \mathbf{e}_i, \qquad \text{with} \qquad \mathbf{e}_i \cdot \mathbf{e}_j = \delta_{ij}.$$

Each link between neighbouring sites corresponds to a conducting wire of electrical resistance R (see Figure 2.19). The goal of this exercise is to calculate the electrical

Fig. 2.19 Resistor network in the two-dimensional case: each wire between two points • has resistance R.

resistance between two arbitrary points on the network.

1. Let $V(\mathbf{r})$ be the electric potential on network site $\mathbf{r}$. Show that Ohm's and Kirchhoff's laws imply that $\Delta V(\mathbf{r})$, defined by

$$\Delta V(\mathbf{r}) = \sum_{i=1}^{d} \left[V(\mathbf{r} + \mathbf{e}_i) + V(\mathbf{r} - \mathbf{e}_i) - 2V(\mathbf{r}) \right],$$

vanishes. Operator Δ corresponds to the discrete Laplacian.

2. To measure the resistance $R(\mathbf{r}_0)$ between points 0 and $\mathbf{r}_0$ on the network, we connect an ohmmetre which injects a current I on the site $\mathbf{r}_0$ and extract the same quantity at the site $\mathbf{0}$. Give the equation satisfied by $V(\mathbf{r})$.

3. Give the expression for $R(\mathbf{r}_0)$ in terms of R and the Green's function $G(\mathbf{r} - \mathbf{r}')$ for the discrete Laplacian operator, solution of

$$\Delta_{\mathbf{r}} G(\mathbf{r} - \mathbf{r}') = -\delta_{\mathbf{r}, \mathbf{r}'}.$$

4. Give the expression of $R(\mathbf{r}_0)$ in terms of the network Fourier modes

$$\frac{1}{(2\pi)^{d/2}} e^{i\mathbf{k} \cdot \mathbf{r}}, \qquad -\pi \le k_i \le \pi,$$

with

$$\int_{-\pi}^{\pi} \frac{d^d k}{(2\pi)^d} e^{i\mathbf{k}\cdot(\mathbf{r}-\mathbf{r}')} = \delta_{\mathbf{r},\mathbf{r}'}.$$

Compute then $R(\mathbf{r}_0)$ in the one-dimensional case.

5. Give the dominant term in the asymptotic expansion of $R(\mathbf{r}_0)$ when $r_0 \gg 1$ for $d = 2$. Show also for $d = 2$ that the resistance between two neighbouring sites is $R/2$.

Solution p. 311.

■ **Exercise 2.15. Method of image charges for a bidimensional problem**

Suppose we want to calculate the electric potential in a two-dimensional geometry defined by the half-plane $x > 0$. More specifically we want to compute the electric potential which satisfies

$$\Delta\Phi(x,y) = 0, \qquad \forall x > 0,$$

and the condition imposed by the conductor in $x = 0$, $\Phi(0,y) = f(y)$.

1. Give an integral expression for potential Φ as a function of $f(y)$ and the appropriate Green's function that should be computed.

2. Explicitly show that the final expression for Φ meets imposed boundary conditions.

3. Show using arguments based on analyticity that $\Delta\Phi(x,y) = 0$ for all x strictly positive.

Solution p. 312.

■ **Exercise 2.16. Semi-cylindrical warehouse exposed to wind**

Consider the air as an incompressible perfect fluid of mass density ρ. We want to study the steady and irrotational flow over a warehouse of radius R, as shown in Figure 2.20. The warehouse length L along Oz is large enough so that the problem is purely two-dimensional in the plane (xOy). Note $\mathbf{u}$ the velocity field and ϕ the corresponding potential defined as $\mathbf{u} = \boldsymbol{\nabla}\phi$.

1. Give the boundary conditions satisfied by $\mathbf{u}$.

2. Find the conformal representation mapping the area accessible to the flow onto the upper half complex plane.

3. Deduce then the potential $\phi(r,\theta)$ and the velocity field $\mathbf{u}$.

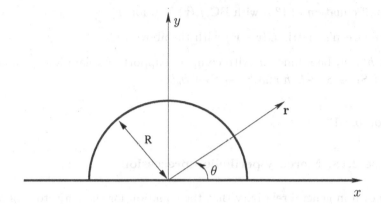

Fig. 2.20 Semi-cylindrical warehouse.

4. Obtain the same result by decomposing $\phi(r, \theta)$ on a suitable basis of functions.

5. Assume that the warehouse has a small aperture (e.g. an open window). Calculate the force exerted by the wind on the roof assuming that the velocity field inside the warehouse is identically zero. What are the nature and the consequences of this force?

Solution p. 313.

■ **Exercise 2.17. Dirac operator**

We propose to study the following differential equation on $\mathbb{R}^2$:

$$\begin{pmatrix} \partial_x f_1(\mathbf{r}) + \partial_y f_2(\mathbf{r}) \\ \partial_y f_1(\mathbf{r}) - \partial_x f_2(\mathbf{r}) \end{pmatrix} = \begin{pmatrix} \partial_x & \partial_y \\ \partial_y & -\partial_x \end{pmatrix} \begin{pmatrix} f_1(\mathbf{r}) \\ f_2(\mathbf{r}) \end{pmatrix} = \begin{pmatrix} S_1(\mathbf{r}) \\ S_2(\mathbf{r}) \end{pmatrix}. \qquad (2.135)$$

Suppose $S_i(\mathbf{r})$ with compact support (localised) and introduce Green's matrix,

$$G = \begin{pmatrix} G_{11} & G_{12} \\ G_{21} & G_{22} \end{pmatrix},$$

satisfying

$$\begin{pmatrix} \partial_x & \partial_y \\ \partial_y & -\partial_x \end{pmatrix} G(\mathbf{r} - \mathbf{r}') = \delta(x - x')\delta(y - y')\mathcal{I} = \delta(\mathbf{r} - \mathbf{r}')\mathcal{I}$$

where $\mathcal{I}$ is the 2×2 identity matrix, with BC $G_{ij} \to 0$ for $|\mathbf{r}| \to \infty$.

1. Show that

$$f_i(\mathbf{r}) = \int_{\mathbb{R}^2} \mathrm{d}x' \, \mathrm{d}y' \sum_j \left[G_{ij}(\mathbf{r} - \mathbf{r}') \, S_j(\mathbf{r}') \right] \quad ; \quad i = 1, 2$$

is solution of equation (2.135) with BC $f_i(\mathbf{r}) \to 0$ for $|\mathbf{r}| \to \infty$.

2. Give Green's matrix $G(\mathbf{r} - \mathbf{r}')$ with the above BC.

3. Let $h(x, y)$ be a function with compact support. Explain how f_1 and f_2 are modified if $S_1 \to S_1 + \partial_x h$ and $S_2 \to S_2 + \partial_y h$.

Solution p. 313.

■ **Exercise 2.18. Mercury perihelion precession**

It is shown in general relativity that the equation for the trajectory of a planet (assumed as a point mass) around the Sun can be written as

$$\ddot{u} + u = \frac{MG}{L^2} + \frac{3MG}{c^2} u^2 \tag{2.136}$$

where $(1/u, \phi)$ are polar coordinates in the plane of the trajectory, $\dot{u} = (\mathrm{d}u/\mathrm{d}\phi)$, M the mass of the sun, G Newton's constant, c the speed of light and L, the orbital angular momentum of the planet divided by its mass[24]. The term proportional to u^2 is interpreted as the correction from general relativity to the classical Newton's law. In the following, we note

$$\alpha = \frac{MG}{L^2} \qquad \text{and} \qquad \beta = \frac{3MG}{c^2}.$$

Since $(\beta u^2/\alpha) = 3(L^2 u^2/c^2)$, this ratio is of order v^2/c^2 where v is the planet speed: it is therefore very small. We have for instance for Mercury

$$\frac{MG}{c^2} = 1.475 \, \text{km} \qquad \text{and} \qquad \frac{L^2}{MG} = 55.3 \, 10^6 \, \text{km}.$$

1. Let $G(\phi - \phi')$ be the Green's function solution of

$$\ddot{G} + G = \delta(\phi - \phi')$$

with $G(\phi - \phi') = 0$ for $\phi < \phi'$. Give an expression for G.

2. Show that the solution of equation (2.136) satisfies the integral equation

$$u(\phi) = h(\phi) + \int_0^\phi \mathrm{d}\phi' \, G(\phi - \phi') \left[\alpha + \beta u^2(\phi') \right], \tag{2.137}$$

with $h(\phi)$ a solution of the homogeneous equation.

3. In the case where $\beta = 0$ give the function $h(\phi)$ corresponding to the solution $u(\phi) = \alpha(1 + e \cos \phi)$. We recall that the semi-major axis of the ellipse is

$$a = \frac{L^2}{MG} \frac{1}{1 - e^2}.$$

---

[24]Let us briefly remind that this equation is obtained from the equation of time-like geodesics in the Schwarzschild metric associated with the Sun (see e.g. [94]).

4. We come back to the case where $\beta \neq 0$ keeping the same BC as before. Show using a perturbative expansion that

$$u(\phi) \simeq \alpha\big[(1 + e\cos\phi) + \beta\alpha\left\{A + B\cos\phi + C\cos 2\phi + D\phi\sin\phi\right\}\big].$$

Give expressions for A, B, C and D. Discuss the validity of this expansion.

5. The perihelion is the point of the trajectory farthest from the Sun. When $\beta = 0$ the trajectory is periodic and the difference $\Delta\phi$ between two perihelion passages is then 2π. When $\beta \neq 0$ the trajectory is no longer periodic and we then have $\Delta\phi = 2\pi + \epsilon$. Show that

$$\epsilon \simeq \frac{2\pi D\beta\alpha}{e}.$$

6. Conclude with a numerical estimation for the planet Mercury: give the perihelion advance in seconds of arc per century knowing that Mercury revolution period is 87.97 days.

Solution p. 314.

■ **Exercise 2.19. Harmonic oscillator in the presence of an impurity**

Let us consider a quantum particle of mass m submitted to a harmonic potential $U(x) = \frac{1}{2}m\omega^2 x^2$. The corresponding Hamiltonian $\mathcal{H}_0$ reads

$$\mathcal{H}_0 = -\frac{\hbar^2}{2m}\frac{\mathrm{d}^2}{\mathrm{d}x^2} + \frac{1}{2}m\omega^2 x^2.$$

The spectrum of $\mathcal{H}_0$ is completely discrete. Energy levels are written as $E_n = (n + \frac{1}{2})\hbar\omega$ with $n \in \mathbb{N}$ and the corresponding normalised wave functions are given by

$$\psi_n(x) = \frac{1}{\pi^{1/4}2^{n/2}(n!)^{1/2}l^{1/2}}e^{-\frac{x^2}{2l^2}}H_n\left(\frac{x}{l}\right)$$

with $l = \sqrt{\hbar/m\omega}$ and where $H_n(\xi)$ the Hermite polynomial of degree n. In the following it will only be necessary to know that, for $p \in \mathbb{N}$,

$$H_{2p}(0) = (-1)^p 2^p (2p-1)!!, \qquad H_{2p+1}(0) = 0.$$

1. Let us define the thermal Green's function as

$$G_0(x_a, x_b; \beta) = \langle x_a|e^{-\beta\mathcal{H}_0}|x_b\rangle$$

with $\beta = 1/(kT)$. Write the Laplace transform

$$\hat{G}_0(x_a, x_b; z) = \int_0^{+\infty} \mathrm{d}\beta\, e^{-\beta z} G_0(x_a, x_b; \beta)$$

in terms of wave functions ψ_n and energy levels E_n. Compute $\hat{G}_0(0, 0; z)$ and $G_0(0, 0; \beta)$ by summing the series defining $G_0(0, 0; \beta)$ using

$$\frac{d^p}{d\xi^p}(1 - \xi)^{-\frac{1}{2}}|_{\xi=0} = (p - \frac{1}{2})(p - \frac{3}{2}) \cdots \frac{1}{2}.$$

2. Express the density of states $\rho_0(E)$ in terms of a series, starting from the general formula

$$\rho_0(E) = -\frac{1}{\pi} \lim_{\epsilon \to 0^+} \int_{-\infty}^{\infty} dx \hat{G}_0(x, x; -E + i\epsilon).$$

3. Introduce then an impurity at the origin creating a potential of form $Vl\delta(x)$. The Hamiltonian $\mathcal{H}$ of the particle in the presence of the impurity reads

$$\mathcal{H} = \mathcal{H}_0 + Vl\delta(x).$$

Let G and $\hat{G}$ be the corresponding Green's functions. Show that $\hat{G}$ satisfies

$$\hat{G}(x_a, x_b; z) = \hat{G}_0(x_a, x_b; z) - Vl\hat{G}(x_a, 0; z)\hat{G}_0(0, x_b; z). \qquad (2.138)$$

Give then an expression for $\hat{G}$ as function of $\hat{G}_0$, V and l.

4. Show that equation (2.138) implies that the Green's function still has poles in $z = -(2p + 1 + \frac{1}{2})\hbar\omega$ but no longer in $z = -(2p + \frac{1}{2})\hbar\omega$. Conclude that the new density of state takes the form

$$\rho(E) = \sum_{n=0}^{+\infty} \delta(E - E_n(V)),$$

where $E_n(V)$ are the new eigenenergies. Give the implicit equation allowing to determine $E_{2p}(V)$ by inverting a series.

5. From now on we focus only on the fundamental level $E_0(V)$. Show the following asymptotic behaviours:

$$E_0(V) \simeq -\frac{V^2}{2\hbar\omega} \qquad\qquad \text{when } V \to -\infty,$$

$$E_0(V) = \frac{1}{2}\hbar\omega + \frac{V}{\sqrt{\pi}} + O(V^2) \qquad\qquad \text{when } V \to 0,$$

$$E_0(V) = E_0^\infty - \frac{\sqrt{\pi}}{S(E_0^\infty)V} + O(\frac{1}{V^2}) \qquad\qquad \text{when } V \to +\infty$$

with $\frac{1}{2}\hbar\omega < E_0^\infty < \frac{5}{2}\hbar\omega$. It is useful to write the implicit equation defining E_0^∞ and $S(E_0^\infty)$ as a series expansion.

Solution p. 315.

Chapter 3

Dynamical Green's functions

This chapter is dedicated to Green's functions in time-dependent situations. The motivation, the interest and the definition of Green's functions studied here are similar to those of the previous chapter. The reader is referred to the introduction of the latter for a general discussion of the interest of Green's functions. In contrast, we focus here on a new feature linked to time dependence. Indeed, for each studied physical situation the corresponding Green's functions give, at point $\mathbf{r}$ of space and time t, the field value created by a point source localised both in space and time, i.e. at a point $\mathbf{r}_0$ and a time t_0. In electromagnetism for example, this particular primary source is a flash of light. The essential difference with the stationary case lies in the field propagation between the space-time points $(\mathbf{r}_0, t_0)$ and $(\mathbf{r}, t)$. This first observation immediately leads to another one with major consequences: the propagation between these two points in space-time must satisfy the causality principle. But as we have seen in Chapter 1 within the framework of linear response theory, causality leads to some analyticity properties. So we shall find here on the one hand the general properties of the static Green's functions as presented in Chapter 2, and on the other hand the properties of response functions described in Chapter 1.

In the first part of this chapter, we start by presenting the general problem of boundary conditions for some linear PDEs involving both space and time variables. Causality principle leads us to formulate the temporal part of these conditions in terms of initial conditions. This principle also gives a particular interest to the so-called causal Green's functions, naturally interpreted as characteristic response functions of the field under consideration. We then explain how the particular form of most operators under consideration allows to separate the spatial and temporal dependencies. Causal homogeneous Green's functions of these operators then admit simple spectral representations. In addition a Laplace transform brings us back to a purely static PDE whose properties have already been studied in Chapter 2.

We then study separately various operators. As in Chapter 2, we address issues concerning boundary conditions and uniqueness of solutions, reciprocal relations as well as expression of the general solution for any source in terms of Green's functions. The first operator studied is associated with the diffusion equation, which governs the evolution of many different physical quantities. We then move to Schrödinger equation, a fundamental equation of quantum mechanics. Bloch equation, appearing in statistical mechanics, is then briefly presented. We end with d'Alembert operator, which is a fundamental operator describing wave propagation, in particular (but not exclusively) in electromagnetism. Let us immediately notice that d'Alembert equation leads to very different propagation properties from those governed by diffusion processes. As we shall see this difference appears in the structure of the corresponding Green's functions.

The analysis of previous PDEs sheds light on the dual nature of dynamical Green's functions. As in d'Alembert equation, they appear when sources are present as response functions of the field to these sources. In the absence of sources as in Schrödinger equation, Green's functions describe the intrinsic propagation of an elementary entity associated with the field of interest. Beyond this duality of interpretation, Fourier transforms of causal Green's functions have similar analytical properties to those of susceptibilities. Their singularities are moreover induced by eigenmodes of the unperturbed system without external sources. Their positions in the complex plane depend crucially on the dissipative nature of the evolution.

Applications and examples discussed in the second part of this chapter are as follows. The first example is one-dimensional diffusion on a segment that simply illustrates the properties described in the first part. We then treat Fraunhofer diffraction as a fundamental application of d'Alembertian Green's functions. We discuss carefully the precise role and the choice of boundary conditions on this example. The next example deals with the transmission of sound waves by an object moving in a fluid. Another interesting application of the d'Alembertian Green's functions concerns the analysis of propagation of shock waves emitted by a supersonic object. We then study Cattaneo equation, presenting both a propagating and a diffusive character. We finally determine the polarisability of the hydrogen atom by perturbative methods.

3.1   General properties

3.1.1   *Green's functions and causality*

Consider a system contained in a domain $\mathcal{D}$ which is part of a d-dimensional space. Unlike the situation studied in the previous chapter, the source density $\rho(\mathbf{r}, t)$ now depends on time t. These sources generate a field $\phi(\mathbf{r}, t)$, which is a particular solution of the PDE

$$\mathcal{O}\phi(\mathbf{r}, t) = \rho(\mathbf{r}, t), \tag{3.1}$$

where $\mathcal{O}$ is a linear operator depending on spatial and temporal coordinates.

The field ϕ shall of course be determined by the PDE and boundary conditions specific to the problem under consideration. Mathematically speaking, everything happens as if we had simply considered a PDE in a space of $d + 1$ dimension. However the temporal dimension actually plays a special role due to the existence of the arrow of time. According to the principle of causality, evolution of a physical system is partly determined by the initial conditions at some time t_0 taken as origin. For a system in classical mechanics, the trajectory of a particle subjected to a given force depends only on the initial positions and velocities. Here the temporal part of the boundary conditions thus consists in setting values for the field and some of its partial temporal derivatives at the time origin t_0, and anywhere on the domain $\mathcal{D}$. These initial conditions are denoted $\text{IC}(t_0)$.

In addition to the initial conditions we must of course add spatial boundary conditions. It typically consists in imposing ϕ and/or its spatial partial derivatives on the boundary $\partial \mathcal{D}$ of the domain, as in the static case. Note that these values set at the boundary may depend on time as a result of the action of an observer outside the system. In electromagnetism an AC generator can impose for instance a given time-dependent voltage between the plates of a capacitor containing a dielectric medium. Boundary conditions are denoted $\text{BC}(\phi|\partial \mathcal{D})$. Ultimately the field of interest is the unique solution of the system

$$\boxed{\begin{aligned} \mathcal{O}\phi(\mathbf{r}, t) &= \rho(\mathbf{r}, t) \\ \text{BC}(\phi|\partial \mathcal{D}) \quad &\text{and} \quad \text{IC}(\phi|t_0). \end{aligned}}$$ (3.2)

◊ Definition of Green's functions ◊

Viewing PDE (3.1) in a space of dimension $d+1$ leads to a natural generalisation of the stationary case definition (2.3). Here a dynamical Green's function is the field created by a flash-type point source with appropriate boundary conditions, i.e.

$$\boxed{\begin{aligned} \mathcal{O}_{\mathbf{r},t}G(\mathbf{r}, t; \mathbf{r}', t') &= \delta(\mathbf{r} - \mathbf{r}')\delta(t - t'), \\ \text{BC}(G). \end{aligned}}$$ (3.3)

As in the static case, boundary conditions on G can be chosen arbitrarily regardless of those imposed on ϕ. However, in determining the physical field ϕ it seems very reasonable to expect that causal Green's functions satisfying the principle of causality, denoted G^+, will play a particularly interesting role. Since the flash source is switched off for all time $t < t'$, a causal Green's function necessarily vanishes identically prior to lighting this flash. So the condition

$$\boxed{\textbf{Causal Green's function:} \quad G^+(\mathbf{r}, t; \mathbf{r}', t') = 0 \text{ for } t < t',}$$ (3.4)

is expected to completely define the temporal part of the boundary conditions $\text{CL}(G)$. In other words, there is no need then to introduce additional initial conditions[1] at $t = t'$. Of course we must also impose boundary conditions on the domain surface to uniquely determine a causal Green's function.

---

[1] For operators considered in section 3.1.2 we will show that causality condition implies that G^+ and possibly some of its time derivatives vanish at $t = t'$.

The flash source being active only at time t', a causal Green's function appears as an elementary field emitted at this time t' at a point $\mathbf{r}'$, and observed at a later time $t > t'$ at a point $\mathbf{r}$. $G^+(\mathbf{r}, t; \mathbf{r}', t')$ thus takes into account the signal propagation[*] between the points $\mathbf{r}'$ and $\mathbf{r}$ for a duration $t - t'$.

This propagative aspect highlights the major interest of causal Green's functions in characterising and understanding the system under consideration.

Note that it is not forbidden to introduce temporal conditions which do not comply with the principle of causality.

> [*] **Comment:** Using the term propagation for a generic operator $\mathcal{O}$ at this level implies that it is possible for the propagation speed to be infinite. This is allowed in non-relativistic physics, but is excluded in special relativity. This slight difference appears in the comparison of the causal Green's functions associated with the diffusion equation, the d'Alembert operator, and the Cattaneo equation (sections 3.1.3, 3.1.6 and 3.2.5 respectively).

If the corresponding acausal Green's functions have no physical interpretation, they can nevertheless be useful in solving some problems. In fact, all dynamical Green's functions are potentially interesting because the total field

$$\int_{-\infty}^{\infty} dt' \int_{\mathcal{D}} d\mathbf{r}' \; G(\mathbf{r}, t; \mathbf{r}', t') \, \rho(\mathbf{r}', t') \tag{3.5}$$

is always a particular solution of the original PDE (3.1). One can check this property by applying operator $\mathcal{O}_{\mathbf{r},t}$ to this integral expression, and then using the linearity of this operator.

◊ Causal Green's functions and linear response theory ◊

We shall now rephrase the problem within the general context of linear response theory described in Chapter 1. The observable of interest is here the field ϕ, and sources play the role of external forcing. A given causal Green's function then appears as a particular response function associated with this observable: the flash source is indeed exactly analogous to an impulse excitation. Under additional assumptions on the nature of boundary conditions and the system invariance by translation in time, this analogy can be pursued and clarified as follows.

Let us consider a time-translation invariant operator $\mathcal{O}$. To illustrate this analogy, we further assume that the field satisfies Dirichlet boundary conditions BC on $(\phi|\partial\mathcal{D})$, with a static boundary function,

$$\phi(\mathbf{r}, t) = D(\mathbf{r}) \qquad \text{for } \mathbf{r} \in \partial\mathcal{D} \qquad \forall t. \tag{3.6}$$

Suppose that no source is present at the initial time $t_0 = -\infty$. In addition we assume that the initial field $\phi_0(\mathbf{r})$ is a static solution of PDE (3.1) without sources

$$\mathcal{O}\phi_0(\mathbf{r}) = 0$$

and satisfying boundary conditions (3.6). This static field ϕ_0 clearly represents the value of the observable in the unperturbed stationary state and thus plays the role of the quantity A_0 introduced on p. 3. We then connect adiabatically the source $\rho(\mathbf{r}, t)$ from $t_0 = -\infty$, i.e. in such a way that $\rho(\mathbf{r}, t)$ tends exponentially fast to 0 when $t \to -\infty$. This source acts as a forcing and induces a modification in the field ϕ, solution of the PDE (3.1) with boundary conditions (3.6) and initial condition[2]

$$\lim_{t \to -\infty} \phi(\mathbf{r}, t) = \phi_0(\mathbf{r}). \tag{3.7}$$

We then introduce the causal homogeneous Dirichlet Green's function G_{HD}^+ defined by the boundary conditions

$$G_{\mathrm{HD}}^+(\mathbf{r}, t; \mathbf{r}', t') = 0 \qquad \text{for } \mathbf{r} \in \partial\mathcal{D}. \tag{3.8}$$

Since the operator $\mathcal{O}$, the causality condition (3.4) and the BC (3.8) are invariant by translation in time, G_{HD}^+ depends only on the difference $t - t'$. It is then clear that

$$\phi(\mathbf{r}, t) = \phi_0(\mathbf{r}) + \int_{-\infty}^{t} dt' \int_{\mathcal{D}} d\mathbf{r}' \, G_{\mathrm{HD}}^+(\mathbf{r}; \mathbf{r}'; t - t') \, \rho(\mathbf{r}', t') \tag{3.9}$$

is solution of the PDE (3.1) with BC (3.6) and initial condition (3.7). This formula is a particular form of the general response (1.30), p. 17, for a system submitted to an inhomogeneous perturbation. The causal homogeneous Dirichlet Green's function therefore plays the role of the response function for the quantity ϕ. Note that formula (3.9) is exact because the internal dynamics of the considered system is linear, while general expression (1.30) is the first term of a perturbative expansion in powers of the forcing.

Seen as a response function G_{HD}^+ must have the properties described in Chapter 1. In particular, if the domain $\mathcal{D}$ is the entire space, and if the operator $\mathcal{O}$ is invariant under spatial translation, the Fourier transform of G_{HD}^+ becomes a response function for the mode $\mathbf{k}$. Consequently, its Laplace transform with respect to time must satisfy generic analyticity properties similar to those seen in Chapter 1, including Kramers-Kronig relations.

---

[2]This is a sufficient boundary condition to ensure uniqueness. Note that all temporal partial derivatives of ϕ vanish at $t_0 = -\infty$, which is compatible with an exponentially fast decay of the source when $t \to -\infty$.

3.1.2 *Separable operators*

In several physical situations the operator $\mathcal{O}$ takes the form

$$\mathcal{O} = \mathcal{O}_{\mathbf{r}} + \mathcal{O}_t,$$

$$\mathcal{O}_t = a_p \frac{\partial^p}{\partial t^p} + a_{p-1} \frac{\partial^{p-1}}{\partial t^{p-1}} + \cdots + a_1 \frac{\partial}{\partial t} = P_p(\tfrac{\partial}{\partial t}),$$

(3.10)

where $\mathcal{O}_{\mathbf{r}}$ and $\mathcal{O}_t$ are linear operators involving respectively only position $\mathbf{r}$ and time t. Furthermore, the temporal part of $\mathcal{O}_t$ is a linear combination of partial derivatives with respect to time, with constant coefficients. In other words, $\mathcal{O}_t$ is invariant under any time translation. In the remaining of this chapter we shall only consider operators with structure (3.10), as it allows to go deeper into the general discussion. In a first step, we show that causal homogeneous Green's functions of $\mathcal{O}$ are simply expressed in terms of the spectrum of the (static) spatial part $\mathcal{O}_{\mathbf{r}}$ of operator $\mathcal{O}$. We then show how the temporal part of PDE (3.2) can be integrated by Laplace transform.

◊ Spectral representation ◊

Introduction of the spectrum of the static operator Consider a causal homogeneous Green's function G_{H}^+ satisfying homogeneous boundary conditions of Dirichlet or Neumann type for instance. Suppose then that with these homogeneous boundary conditions, the static operator $\mathcal{O}_{\mathbf{r}}$ has an orthonormal basis of eigenfunctions $\psi_n(\mathbf{r})$ associated with eigenvalues λ_n. It means on the one hand that each $\psi_n(\mathbf{r})$ satisfies the static EDP

$$\mathcal{O}_{\mathbf{r}}\psi_n(\mathbf{r}) = \lambda_n \psi_n(\mathbf{r}),$$

and on the other hand that the set of $\psi_n(\mathbf{r})$ satisfies the completeness relation

$$\delta(\mathbf{r} - \mathbf{r}') = \sum_n \psi_n(\mathbf{r})\psi_n^*(\mathbf{r}').$$

Let us fix two times t and t'. Just as in the static case $G_{\mathrm{H}}^+(\mathbf{r};\mathbf{r}';t-t')$ can then be decomposed as (see Appendix D, p. 277)

$$G_{\mathrm{H}}^+(\mathbf{r};\mathbf{r}';t-t') = \sum_{m,n} g_{mn}(t-t')\psi_m(\mathbf{r})\psi_n^*(\mathbf{r}').$$

(3.11)

The spatial dependence of $G_{\mathrm{H}}^+(\mathbf{r};\mathbf{r}';t-t')$ is controlled by the eigenfunctions ψ_n of the static operator $\mathcal{O}_{\mathbf{r}}$. The time dependence is entirely embedded in the functions g_{mn} that we are going to determine.

Reduced temporal Green's functions The double expansion analogous to (3.11) for the Dirac distribution $\delta(\mathbf{r} - \mathbf{r}')$ is simply given by completeness relation (D.1). Injecting this expansion into the PDE satisfied by G_H^+, and using the equation for the eigenvalues of $\mathcal{O}_\mathbf{r}$ then leads to

$$\sum_{m,n} (\mathcal{O}_t g_{mn} + \lambda_m g_{mn} - \delta_{mn}\delta(t - t')) \; \psi_m(\mathbf{r})\psi_n^*(\mathbf{r}') = 0,$$

where δ_{mn} is the Kronecker symbol, $\delta_{mn} = 0$ for $m \neq n$ and $\delta_{mm} = 1$. By uniqueness of the representation (3.11), we find the ordinary differential equation for $m \neq n$

$$\mathcal{O}_t g_{mn}(t - t') + \lambda_m g_{mn}(t - t') = 0.$$

The solution satisfying the causality condition $g_{mn}(t - t') = 0$ for $t < t'$ actually vanishes at all times. We then find, after renaming the function g_{nn} as g_n,

$$(\mathcal{O}_t + \lambda_n) \, g_n(t - t') = \delta(t - t'), \tag{3.12}$$

with $g_n(t - t') = 0$ for $t < t'$. This ordinary differential equation shows that g_n can be seen as a purely temporal reduced Green's function. In the context of linear response, g_n is also interpreted as a response function associated with the operator $\mathcal{O}_t + \lambda_n$.

Remember that the temporal operator $\mathcal{O}_t$ is supposed to be a polynomial of degree p in the partial derivative $(\partial/\partial t)$ (see equation (3.10)). There is then a systematic method for determining g_n in terms of a particular solution Z_n of the homogeneous version of differential equation (3.12),

$$(\mathcal{O}_t + \lambda_n) \, Z_n(t - t') = 0. \tag{3.13}$$

This method, recalled in Appendix C p. 271, leads to

$$g_n(t - t') = \theta(t - t')Z_n(t - t')$$

where θ is the Heaviside function and Z_n, the particular solution of the homogeneous equation with initial conditions for $p \geq 3$

$$Z_n(0) = 0, \quad Z_n^{(1)}(0) = 0, \quad \ldots \quad Z_n^{(p-2)}(0) = 0, \quad Z_n^{(p-1)}(0) = 1/a_p, \tag{3.14}$$

where we introduced the notation $Z_n^{(j)} = (\partial^j Z_n/\partial t^j)$. For $p = 1$ and $p = 2$, these conditions are respectively $Z_n^{(0)} = 1/a_1$ and $Z_n^{(0)} = 0$, $Z_n^{(1)} = 1/a_2$. It completes the determination of the functions g_{mn} in double expansion (3.11), which becomes

$$\boxed{G_H^+(\mathbf{r}; \mathbf{r}'; t - t') = \theta(t - t') \sum_n Z_n(t - t')\psi_n(\mathbf{r})\psi_n^*(\mathbf{r}').} \tag{3.15}$$

This spectral representation is the analogue of the one obtained in Chapter 2, p. 60, for static homogeneous Green's functions.

Reinterpretation of the causality condition It is easy to show from representation (3.15) that G_{H}^{+} and its first $p-2$ partial derivatives with respect to time all tend to zero as $t-t' \to 0^{+}$,

$$
\lim_{t-t' \to 0^{+}} G_{\mathrm{H}}^{+}(\mathbf{r}; \mathbf{r}'; t-t') = 0,
$$

$$
\lim_{t-t' \to 0^{+}} \frac{\partial^{j}}{\partial t^{j}} G_{\mathrm{H}}^{+}(\mathbf{r}; \mathbf{r}'; t-t') = 0, \qquad 1 \le j \le p-2,
$$

(3.16)

thus ensuring continuity of the corresponding functions in $t = t'$. A method identical to that of p. 271 indeed leads to

$$
\frac{\partial^{j}}{\partial t^{j}} G_{\mathrm{H}}^{+}(\mathbf{r}; \mathbf{r}'; t-t') = \theta(t-t') \sum_{n} Z_{n}^{(j)}(t-t') \psi_{n}(\mathbf{r}) \psi_{n}^{*}(\mathbf{r}')
$$

for $j \le p-2$. On the other hand the derivative of order $(p-1)$ is discontinuous at $t = t'$, according to the following asymptotic behaviour

$$
\lim_{t-t' \to 0^{+}} \frac{\partial^{p-1}}{\partial t^{p-1}} G_{\mathrm{H}}^{+}(\mathbf{r}; \mathbf{r}'; t-t') = \frac{1}{a_{p}} \delta(\mathbf{r} - \mathbf{r}'),
$$

(3.17)

coming from the condition $Z_{n}^{(p-1)}(0) = 1/a_{p}$ combined with the completeness relation. This discontinuity is of course induced by the ignition of the flash at $t = t'$. Note that the causality condition (3.4) is ultimately equivalent to imposing p initial conditions sufficient to guarantee the uniqueness of the causal Green's function under consideration.

The spectral representation (3.15) is method for computing causal Green's functions G_{H}^{+}. Let us step back a moment and address the issue of determining *a priori* these causal Green's functions. Two other methods come to mind.

Time translation invariance of both PDE (3.3) and causality condition (3.4) suggests to perform a Fourier transform with respect to time variable $t - t'$. The Fourier transform could however be ill defined depending on whether or not the underlying physics is dissipative. Indeed we noted on p. 129 that Green's functions G_{H}^{+} can be interpreted as response functions. We also saw in Chapter 1 p. 8 that in this case, the Fourier transform

$$
\int_{-\infty}^{+\infty} d\tau \, e^{i\omega \tau} G_{\mathrm{H}}^{+}(\mathbf{r}; \mathbf{r}'; \tau)
$$

can be singular because of undamped eigenmodes. So $G_{\mathrm{H}}^{+}(\mathbf{r}; \mathbf{r}'; \omega)$ must in fact be defined as a limit process,

$$
\widehat{G}_{\mathrm{H}}^{+}(\mathbf{r}; \mathbf{r}'; \omega) = \lim_{\epsilon \to 0^{+}} \int_{-\infty}^{+\infty} d\tau \, e^{iz\tau} G_{\mathrm{H}}^{+}(\mathbf{r}; \mathbf{r}'; \tau),
$$

(3.18)

with $z = \omega + i\epsilon$, the positive imaginary part of z ensuring the convergence of the integral when $\tau \to +\infty$. Note that this positive imaginary part does not raise any problem when $\tau \to -\infty$ as causality of $G_{HD}^+(\mathbf{r}; \mathbf{r}'; \tau)$ limits the integration domain to $[0, +\infty[$.

The other method is suggested by the following remark. For $t > t'$, G_H^+ is solution of the homogeneous version of PDE (3.1), i.e.

$$\mathcal{O}_{\mathbf{r},t} G_H^+(\mathbf{r}; \mathbf{r}'; t - t') = 0$$

with initial conditions (3.16) and (3.17). The determination of the causal Green's function thus appears like solving a PDE with initial conditions, just as for the field $\phi(\mathbf{r}, t)$. However a standard method for solving a PDE with initial conditions is to perform a Laplace transform. We implement this method in the following, before establishing the correspondence between spectral representation, Laplace transform and Fourier transform.

$$\Diamond \textbf{ Laplace transform } \Diamond$$

As mentioned above the temporal part of the general PDE (3.1) is integrated easily by the usual method of Laplace transform, thanks to the very simple structure (3.10) of operator $\mathcal{O}$. Thus this PDE becomes purely spatial in Laplace space, with an additional source term automatically including the initial conditions $IC(\phi|t_0)$. We are thus brought back to the problem of a purely static operator described in Chapter 2.

Static PDE in Laplace space Definition and properties of the Laplace transform are given in Appendix B. Let us recall here for convenience that the Laplace transform of any function $f(\tau)$, defined for $\tau \geq 0$, is the function $\mathcal{L}[f](s)$ of parameter[3] s,

$$\mathcal{L}[f](s) = \int_0^{+\infty} d\tau \, e^{-s\tau} \, f(\tau),$$

also noted $\widetilde{f}(s)$.

Let us take the Laplace transform side by side of the PDE (3.1),

$$\mathcal{O}\phi(\mathbf{r}, t) = \rho(\mathbf{r}, t),$$

with respect to variable $\tau = t - t_0$. The right-hand side gives of course the Laplace transform of the source term, $\widetilde{\rho}(\mathbf{r}, s)$. By linearity of this transformation, the left-hand side gives the sum of the transforms of each term, i.e.

$$\mathcal{L}[\mathcal{O}\phi](\mathbf{r}, s) = \mathcal{L}[\mathcal{O}_{\mathbf{r}}\phi](\mathbf{r}, s) + \mathcal{L}[\mathcal{O}_t\phi](\mathbf{r}, s).$$

We then obtain, again by virtue of the linearity of spatial operator $\mathcal{O}_{\mathbf{r}}$,

$$\mathcal{L}[\mathcal{O}_{\mathbf{r}}\phi](\mathbf{r}, s) = \mathcal{O}_{\mathbf{r}}\widetilde{\phi}(\mathbf{r}, s).$$

We then focus on the calculation of $\mathcal{L}[\mathcal{O}_t\phi](\mathbf{r}, s)$ and remember that $\mathcal{O}_t = P_p(\frac{\partial}{\partial t})$ where P_p is a polynomial of degree p. Laplace transform of $\frac{\partial^j}{\partial t^j}\phi(\mathbf{r}, t)$ takes the form $s^j \widetilde{\phi}(\mathbf{r}, s)$,

---

[3]See Appendix B for the domain of definition of $\widetilde{f}(s)$.

plus a polynomial in s of degree $j - 1$ whose coefficients are partial time derivatives of $\phi(\mathbf{r}, t)$ of order l with $0 \le l \le j-1$, taken at the initial time t_0. These coefficients are then fully determined by $IC(\phi|t_0)$. It then follows that the Laplace transform of the temporal part of the PDE reads

---

> ✠ **Comment:**   Let us consider an explicit example,
> $$\mathcal{O}_t = a_2 \frac{\partial^2}{\partial t^2} + a_1 \frac{\partial}{\partial t},$$
> and note $\dot\phi = (\partial\phi/\partial t)$ to simplify the notations. Property (B.1), p. 268 then leads to
> $$\mathcal{L}[\dot\phi](\mathbf{r}, s) = s\widetilde\phi(\mathbf{r}, s) - \phi(\mathbf{r}, t_0),$$
> $$\mathcal{L}[\ddot\phi](\mathbf{r}, s) = s^2\widetilde\phi(\mathbf{r}, s) - s\phi(\mathbf{r}, t_0) - \dot\phi(\mathbf{r}, t_0).$$
> Thus the Laplace transform
> $$\mathcal{L}[\mathcal{O}_t\phi](\mathbf{r}, s)$$
> is indeed of form (3.19) with
> $$P_2(s) = a_2 s^2 + a_1 s,$$
> $$I_1(\mathbf{r}, s) = -a_2 s\phi(\mathbf{r}, t_0) - a_2\dot\phi(\mathbf{r}, t_0) - a_1\phi(\mathbf{r}, t_0)$$

$$\mathcal{L}[\mathcal{O}_t\phi](\mathbf{r}, s) = P_p(s)\widetilde\phi(\mathbf{r}, s) + I_{p-1}(\mathbf{r}, s), \tag{3.19}$$

where $I_{p-1}(\mathbf{r}, s)$ is a polynomial in s of degree $p - 1$. Coefficients of $P_p(s)$ are the constants of the polynomial in $\partial/\partial t$ defining $\mathcal{O}_t$. Coefficients of $I_{p-1}(\mathbf{r}, s)$ are entirely determined by the $IC(\phi|t_0)$, and they depend on $\mathbf{r}$.

One then finds in Laplace space

$$\boxed{\begin{array}{c} [P_p(s) + \mathcal{O}_\mathbf{r}]\,\widetilde\phi(\mathbf{r}, s) = \widetilde\rho(\mathbf{r}, s) - I_{p-1}(\mathbf{r}, s), \\[2mm] BC(\widetilde\phi|\partial\mathcal{D}) \end{array}} \tag{3.20}$$

where the BC on $(\widetilde\phi|\partial\mathcal{D})$ are simply derived from those on ϕ by Laplace transform. In conclusion, for each given value of s, $\widetilde\phi(\mathbf{r}, s)$ is solution of a static PDE of the form (2.73), p. 87 discussed in Chapter 2, i.e. associated with operator $\lambda + \mathcal{O}_\mathbf{r}$ for $\lambda = P_p(s)$. Here the source involves, in addition to the obvious contribution $\widetilde\rho(\mathbf{r}, s)$, the term $-I_{p-1}(\mathbf{r}, s)$ which automatically takes into account initial conditions on the initial field $\phi(\mathbf{r}, t)$ at $t = t_0$.

We should highlight that there is a different PDE (3.20) for each IC $(\phi|t_0)$ since the polynomial I_{p-1} depends on the initial conditions[4]. Moreover, the question of which boundary conditions $BC(\phi|\partial\mathcal{D})$ ensure uniqueness of $\phi(\mathbf{r}, t)$ simply boils down to the study of boundary conditions for the static operator $\lambda + \mathcal{O}_\mathbf{r}$! This problem has already been addressed in Chapter 2 for simple and usual forms of $\mathcal{O}_\mathbf{r}$ occurring later in this chapter.

More generally PDE (3.20) for $\widetilde\phi(\mathbf{r}, s)$ is particularly useful in solving the original problem. Thus the corresponding expressions for the general solution in terms of

---

[4]Then a heavier but more correct notation would have been $I_{p-1}(\mathbf{r}; s; C.I.(\phi|t_0))$.

Green's functions will easily lead to similar expressions for $\phi(\mathbf{r}, t)$ by inverse Laplace transform. These Green's functions are the matrix elements of the resolvent of $[\lambda + \mathcal{O}_\mathbf{r}]^{-1}$. We show in the following section, that they indeed coincide with Laplace transforms of the causal Green's functions.

Causal Green's functions in Laplace space Let us introduce now the Laplace transform $\widetilde{G}_\mathrm{H}^+$ of the homogeneous causal Green's function G_H^+,

$$\widetilde{G}_\mathrm{H}^+(\mathbf{r}; \mathbf{r}'; s) = \int_0^{+\infty} d\tau\ e^{-s\tau} G_\mathrm{H}^+(\mathbf{r}; \mathbf{r}'; \tau),$$

with $\tau = t - t'$. As mentioned earlier, $G_\mathrm{H}^+(\mathbf{r}; \mathbf{r}'; t - t')$ is solution of the homogeneous version of the PDE (3.1), for $t > t'$. The effective source term then reduces to $-I_{p-1}$ in the PDE of type (3.20) satisfied by $\widetilde{G}_\mathrm{H}^+$. Since G_H^+ and its first $(p-2)$ partial derivatives vanish at time $t = t'$, the only contribution of initial conditions to the polynomial I_{p-1} comes from $\partial^{p-1} G_\mathrm{H}^+ / \partial t^{p-1}(\mathbf{r}; \mathbf{r}'; 0^+)$, more precisely[5]

$$I_{p-1}(\mathbf{r}; \mathbf{r}'; s) = -a_p \frac{\partial^{p-1} G_\mathrm{H}^+}{\partial t^{p-1}}(\mathbf{r}; \mathbf{r}'; 0^+).$$

Using the limit expression (3.17) we find here $I_{p-1}(\mathbf{r}; \mathbf{r}'; s) = -\delta(\mathbf{r} - \mathbf{r}')$. The PDE satisfied by $\widetilde{G}_\mathrm{H}^+$ is simply

$$\boxed{[P_p(s) + \mathcal{O}_\mathbf{r}]\, \widetilde{G}_\mathrm{H}^+(\mathbf{r}; \mathbf{r}'; s) = \delta(\mathbf{r} - \mathbf{r}').} \tag{3.21}$$

PDE (3.21) shows that the Laplace transform $\widetilde{G}_\mathrm{H}^+(\mathbf{r}; \mathbf{r}'; s)$ is the Green's function $G_\lambda(\mathbf{r}; \mathbf{r}')$ associated with the resolvent $[\lambda + \mathcal{O}_\mathbf{r}]^{-1}$, $\lambda = P_p(s)$ and the same homogeneous boundary conditions on the domain boundary $\partial\mathcal{D}$:

$$\boxed{\widetilde{G}_\mathrm{H}^+(\mathbf{r}; \mathbf{r}'; s) = G_\lambda(\mathbf{r}; \mathbf{r}') \quad \text{with} \quad \lambda = P_p(s).} \tag{3.22}$$

Spectral representation, Fourier and Laplace transforms We first point out the link between the spectral representation (3.15), p. 131, and the Laplace transform. Spectral representation of $G_\lambda(\mathbf{r}, \mathbf{r}')$ (2.77), p. 88 obtained in Chapter 2 leads to

$$\boxed{\widetilde{G}_\mathrm{H}^+(\mathbf{r}; \mathbf{r}'; s) = \sum_n \frac{1}{P_p(s) + \lambda_n} \psi_n(\mathbf{r}) \psi_n^*(\mathbf{r}').} \tag{3.23}$$

One can also obtain this representation by directly taking the Laplace transform of expression (3.15). It then leads to

$$\widetilde{G}_\mathrm{H}^+(\mathbf{r}; \mathbf{r}'; s) = \sum_n \widetilde{Z}_n(s) \psi_n(\mathbf{r}) \psi_n^*(\mathbf{r}'),$$

---

[5]The argument is carried out for a given position $\mathbf{r}'$ of the flash source. Since initial conditions depend on both $\mathbf{r}$ and $\mathbf{r}'$, the polynomial I_{p-1} also depends on these two points.

while combining PDE (3.13) with IC (3.14) satisfied by $Z_n(t - t')$, we can recast $\widetilde{Z}_n(s)$ as

$$\widetilde{Z}_n(s) = \frac{1}{P_p(s) + \lambda_n}.$$

We now turn to the relation between Laplace and Fourier transforms. As causal Green's function $G_{\mathrm{H}}^+(\mathbf{r}; \mathbf{r}'; \tau)$ satisfies $G_{\mathrm{H}}^+(\mathbf{r}; \mathbf{r}'; \tau) = 0$ for $\tau < 0$ we have

$$\int_0^{+\infty} \mathrm{d}\tau\, e^{-s\tau} G_{\mathrm{H}}^+(\mathbf{r}; \mathbf{r}'; \tau) = \int_{-\infty}^{+\infty} \mathrm{d}\tau\, e^{-s\tau} G_{\mathrm{H}}^+(\mathbf{r}; \mathbf{r}'; \tau). \tag{3.24}$$

For $s = -iz$, the previous expression coincides with the definition of the Fourier transform $\widehat{G}_{\mathrm{H}}^+(\mathbf{r}; \mathbf{r}'; z)$,

$$\widetilde{G}_{\mathrm{H}}^+(\mathbf{r}; \mathbf{r}'; -iz) = \widehat{G}_{\mathrm{H}}^+(\mathbf{r}; \mathbf{r}'; z).$$

We then have, using the previous result (3.18),

$$\boxed{\widehat{G}_{\mathrm{H}}^+(\mathbf{r}; \mathbf{r}'; \omega) = \lim_{\epsilon \to 0^+} \widetilde{G}_{\mathrm{H}}^+(\mathbf{r}; \mathbf{r}'; s = -i\omega + \epsilon).} \tag{3.25}$$

3.1.3 *Diffusion equation*

The diffusion equation appears in many areas. It is known as the heat equation in thermodynamics because it governs the evolution of local temperature in materials. It also arises frequently in statistical physics and fluid mechanics in the study of liquid mixtures. It plays an important role in the theory of Brownian motion through the description of the probability of presence induced by a random walk. Its explicit form reads

$$\boxed{\left[\frac{\partial}{\partial t} - D\, \Delta_{\mathbf{r}}\right] \phi(\mathbf{r}, t) = \rho(\mathbf{r}, t).} \tag{3.26}$$

As mentioned above, $\phi(\mathbf{r}, t)$ can represent for instance the temperature or the particles density. In these examples, the source term $\rho(\mathbf{r}, t)$ takes into account a local injection of energy or matter. This term often vanishes and the homogeneous version of PDE (3.26) then describes free diffusion. The underlying operator $\mathcal{O}$ has the structure (3.10), with $\mathcal{O}_t = \partial/\partial t$ and $\mathcal{O}_{\mathbf{r}} = -D\Delta_{\mathbf{r}}$, where the positive constant D is the diffusion coefficient. The methods outlined in the previous section are therefore applicable to the study of PDE (3.26).

◇ Boundary conditions ◇

As $\mathcal{O}_t$ is a first order operator with respect to the time variable, i.e. $p = 1$, the initial conditions $\mathrm{IC}(\phi|t_0)$ reduce to fixing the field value at an initial time t_0 anywhere on the domain,

$$\phi(\mathbf{r}, t_0) = \phi_0(\mathbf{r}) \qquad \text{for all } \mathbf{r} \in \mathcal{D}. \tag{3.27}$$

The static operator $(P_p(s) + \mathcal{O}_{\mathbf{r}})$ reads in Laplace space

$$\boxed{s - D\Delta_{\mathbf{r}}.}$$

It is therefore simply proportional to the Helmholtz operator for all real positive values of s. Then, as shown in Chapter 2, Dirichlet and Neumann boundary conditions are among the various boundary conditions $BC(\phi|\partial\mathcal{D})$ which guarantee uniqueness of the solution of PDE (3.26).

Dirichlet boundary conditions In Laplace space we must set $\widetilde{\phi}(\mathbf{r}, s)$ equal to a given function $\widetilde{D}(\mathbf{r}, s)$ on the boundary $\partial\mathcal{D}$, for all s. By inverse Laplace transform, this amounts to impose

$$\phi(\mathbf{r}, t) = D(\mathbf{r}, t) \qquad \text{for } \mathbf{r} \in \partial\mathcal{D} \text{ and for all } t. \tag{3.28}$$

Note that the boundary function $D(\mathbf{r}, t)$ may vary not only when the position moves along $\partial\mathcal{D}$, but it can also change arbitrarily with time, under the condition that it remains consistent with the IC($\phi|t_0$), namely $D(\mathbf{r}, t_0) = \phi_0(\mathbf{r})$. Boundary conditions defined by the combination of constraints (3.27) and (3.28) uniquely determine the field ϕ.

Neumann boundary conditions By inversion of simple Neumann boundary conditions in the Laplace space, we find the corresponding constraints on $\phi(\mathbf{r}, t)$

$$\mathbf{n} \cdot \boldsymbol{\nabla}\phi(\mathbf{r}, t) = N(\mathbf{r}, t) \qquad \text{for } \mathbf{r} \in \partial\mathcal{D} \text{ and all } t, \tag{3.29}$$

where the given boundary function $N(\mathbf{r}, t)$ varies arbitrarily on $\partial\mathcal{D}$ and in time, while the compatibility condition with IC($\phi|t_0$) becomes here $N(\mathbf{r}, t_0) = \mathbf{n} \cdot \boldsymbol{\nabla}\phi_0(\mathbf{r})$ for $\mathbf{r} \in \partial\mathcal{D}$. Conditions (3.27) and (3.29) insure uniqueness of the field ϕ.

Exercise 3.1 proposes to recover these results about uniqueness of diffusion equation, by manipulations of the first Green's formula, analogous to those used in section 2.1.3. Note also that exercises 3.9 and 3.10 introduce other boundary conditions, called Robin boundary conditions.

◊ Integral equation and Green's functions ◊

The PDE (3.20) on $\widetilde{\phi}(\mathbf{r}, s)$ becomes in Laplace space

$$[s - D\Delta_{\mathbf{r}}] \widetilde{\phi}(\mathbf{r}, s) = \widetilde{\rho}(\mathbf{r}, s) + \phi_0(\mathbf{r}), \tag{3.30}$$

where we used $I_0(\mathbf{r}, s) = -\phi_0(\mathbf{r})$. Equation (3.22) moreover leads to $\widetilde{G}_{\mathrm{H}}^+(\mathbf{r}; \mathbf{r}'; s) = G_s(\mathbf{r}; \mathbf{r}')$, since this Green's function satisfies

$$[s - D\Delta_{\mathbf{r}}]G_s(\mathbf{r}; \mathbf{r}') = \delta(\mathbf{r} - \mathbf{r}').$$

Noting that $[s - D\Delta_{\mathbf{r}}]$ is proportional to Helmholtz operator $[-\Delta_{\mathbf{r}} + s/D]$, we find

$$\widetilde{\phi}(\mathbf{r}, s) = \int_{\mathcal{D}} \mathrm{d}\mathbf{r}' \, \widetilde{\rho}(\mathbf{r}', s)\widetilde{G}_{\mathrm{H}}^+(\mathbf{r}'; \mathbf{r}; s) + \int_{\mathcal{D}} \mathrm{d}\mathbf{r}' \, \phi_0(\mathbf{r}') \, \widetilde{G}_{\mathrm{H}}^+(\mathbf{r}'; \mathbf{r}; s)$$

$$- D\int_{\partial\mathcal{D}} \mathrm{d}\Sigma' \, \widetilde{\phi}(\mathbf{r}', s)\mathbf{n}' \cdot \boldsymbol{\nabla}_{\mathbf{r}'}\widetilde{G}_{\mathrm{H}}^+(\mathbf{r}'; \mathbf{r}; s)$$

$$+ D\int_{\partial\mathcal{D}} \mathrm{d}\Sigma' \, \widetilde{G}_{\mathrm{H}}^+(\mathbf{r}'; \mathbf{r}; s)\mathbf{n}' \cdot \boldsymbol{\nabla}_{\mathbf{r}'}\widetilde{\phi}(\mathbf{r}', s), \tag{3.31}$$

which is analogous to equation (2.27) established on p. 65 for a field solution of Helmholtz PDE (2.42).

Let us take the inverse Laplace transform of integral equation (3.31). This inversion is linear, so that the inverse Laplace transform of a spatial integral is equal to the integral of the inverse transform of the integrand: in other words the inversion operator $\mathcal{L}^{-1}$ can cross the integral sign. In addition, the inverse transform of the product of any two functions $\widetilde{f}_1(s)$ and $\widetilde{f}_2(s)$ is given by the convolution product

$$\mathcal{L}^{-1}[\widetilde{f}_1\widetilde{f}_2](t) = \int_0^{t-t_0} d\tau\, f_1(t_0 + \tau)f_2(t - \tau).$$

Using these calculation rules, we finally obtain

$$
\begin{aligned}
\phi(\mathbf{r}, t) =\ & \int_0^{t-t_0} d\tau \int_{\mathcal{D}} d\mathbf{r}'\, \rho(\mathbf{r}', t - \tau)G_{\mathrm{H}}^+(\mathbf{r}'; \mathbf{r}; \tau) \\
& + \int_{\mathcal{D}} d\mathbf{r}'\, \phi_0(\mathbf{r}')\, G_{\mathrm{H}}^+(\mathbf{r}'; \mathbf{r}; t - t_0) \\
& - D \int_0^{t-t_0} d\tau \int_{\partial\mathcal{D}} d\Sigma'\, \mathbf{n}' \cdot \phi(\mathbf{r}', t - \tau)\boldsymbol{\nabla}_{\mathbf{r}'}G_{\mathrm{H}}^+(\mathbf{r}'; \mathbf{r}; \tau) \\
& + D \int_0^{t-t_0} d\tau \int_{\partial\mathcal{D}} d\Sigma'\, \mathbf{n}' \cdot G_{\mathrm{H}}^+(\mathbf{r}'; \mathbf{r}; \tau)\boldsymbol{\nabla}_{\mathbf{r}'}\phi(\mathbf{r}', t - \tau).
\end{aligned}
\tag{3.32}
$$

The reader is encouraged to find this integral equation without introducing the Laplace space. It requires exploiting the second Green formula (2.20), as in the argument leading to formula (2.27). In addition, it is necessary to carry out a temporal integration of the obtained identities between initial time t_0 and an instant $t_f > t$.

Each term in equation (3.32) has a simple physical interpretation. The first term is the contribution of the superposition of flash sources. In addition to the expected spatial integral over the domain $\mathcal{D}$, it also appears a time integral over the time delay τ of the contribution emitted by each flash source. The density $\rho(\mathbf{r}', t - \tau)$ is to be evaluated at a delayed time, as a consequence of causality. The second term describes the spread of the initial condition ϕ_0. The third and fourth terms are analogous to surface terms already met in formula (2.27). They take into account the contribution of induced surface sources, and their structure includes both diffusion and causality as the first volume term.

Just as formula (2.27) for a static field, expression (3.32) is not completely explicit since the surface term depends on the values of field ϕ and its normal derivative $\mathbf{n} \cdot \boldsymbol{\nabla}\phi$ on the boundary $\partial\mathcal{D}$, which cannot be imposed simultaneously by boundary conditions. However, as in the static case, the freedom in the choice of causal Green's function G^+ makes expression (3.32) very useful as we shall see.

◇ **Causal Green's functions** ◇

Causal Green's functions G^+ are solutions of PDE

$$\left[\frac{\partial}{\partial t} - D\Delta_{\mathbf{r}}\right] G^+(\mathbf{r};\mathbf{r}';t-t') = \delta(\mathbf{r}-\mathbf{r}')\delta(t-t'), \qquad (3.33)$$

and they satisfy the causality condition $G^+(\mathbf{r};\mathbf{r}';t-t') = 0$ for $t < t'$. They differ from each other by the nature of chosen boundary conditions. We will successively consider homogeneous Dirichlet and Neumann boundary conditions on $\partial\mathcal{D}$.

Homogeneous Dirichlet boundary conditions These conditions state that $G_{\mathrm{HD}}^+(\mathbf{r};\mathbf{r}';t-t')$ vanishes identically for $\mathbf{r} \in \partial\mathcal{D}$. As shown in the general section above, $G_{\mathrm{HD}}^+(\mathbf{r};\mathbf{r}';t-t')$ can be decomposed in terms of eigenfunctions ψ_n of operator $-\Delta$ satisfying homogeneous Dirichlet boundary conditions with eigenvalues λ_n. These eigenfunctions are assumed to form an orthonormal basis of the space of functions vanishing on the boundary $\mathbf{r} \in \partial\mathcal{D}$. In the corresponding spectral representation (3.15), the function $Z_n(\tau)$ is the solution of the ordinary differential equation[6]

$$\frac{\mathrm{d}}{\mathrm{d}\tau}Z_n(\tau) + D\,\lambda_n Z_n(\tau) = 0$$

with the initial condition $Z_n(0) = 1$. We then immediately get $Z_n(\tau) = e^{-D\lambda_n\tau}$, so that spectral representation (3.15) reads here

$$\boxed{G_{\mathrm{HD}}^+(\mathbf{r};\mathbf{r}';t-t') = \theta(t-t')\sum_n e^{-D\lambda_n(t-t')}\,\psi_n(\mathbf{r})\psi_n^*(\mathbf{r}').} \qquad (3.34)$$

This formula only becomes fully explicit after determining the Laplace operator spectrum on the finite domain $\mathcal{D}$, which remains a challenging problem as discussed on p. 66. It yet allows to understand some important properties of G_{HD}^+.

At short time, $t \to t'$, the behaviour of G_{HD}^+ is obtained from general formula (3.17) with $p = 1$, i.e.

$$\boxed{\lim_{\tau\to 0^+} G_{\mathrm{HD}}^+(\mathbf{r};\mathbf{r}';\tau) = \delta(\mathbf{r}-\mathbf{r}').} \qquad (3.35)$$

As shown by the spectral representation (3.34), the behaviour of G_{HD}^+ at long times $t \to +\infty$ is controlled by λ_0, the smallest eigenvalue of operator $-\Delta$. The latter being positive definite with homogeneous Dirichlet conditions, λ_0 is strictly positive. Thus $G_{\mathrm{HD}}^+(\mathbf{r};\mathbf{r}';t-t')$ decreases exponentially fast to zero as $t \to +\infty$. Finally, G_{HD}^+ satisfies the reciprocal relation

$$\boxed{G_{\mathrm{HD}}^+(\mathbf{r};\mathbf{r}';t-t') = G_{\mathrm{HD}}^+(\mathbf{r}';\mathbf{r};t-t').} \qquad (3.36)$$

---

[6]It should be noted, in relation to the general discussion, that the eigenvalues of functions $\psi_n(\mathbf{r})$ are defined here as eigenvalues of the operator $\mathcal{O}_{\mathbf{r}}/D$ instead of $\mathcal{O}_{\mathbf{r}}$.

The reader may prove this relation either using the second Green's formula (2.20) or noticing that eigenfunctions of Laplace operator are real.

The above properties can easily be interpreted from a physical point of view. Limit behaviour (3.35) shows that $G_{\mathrm{HD}}^{+}(\mathbf{r};\mathbf{r}';t-t')$ is proportional to a density of diffusing particles, all localised at $\mathbf{r} = \mathbf{r}'$ at initial time $t = t'$. Decay of $G_{\mathrm{HD}}^{+}(\mathbf{r};\mathbf{r}';t-t')$ at long times then leads to a decrease in the number of particles inside the domain $\mathcal{D}$. This is consistent with the interpretation of homogeneous Dirichlet boundary conditions: They describe an absorbing wall since the condition that $G_{\mathrm{HD}}^{+}(\mathbf{r};\mathbf{r}';t-t')$ vanishes on $\partial\mathcal{D}$ means that any particle reaching the border actually disappears! Finally the reciprocal relation expresses the equivalence between diffusions from $\mathbf{r}$ to $\mathbf{r}'$ on one hand, and from $\mathbf{r}'$ to $\mathbf{r}$ on the other hand.

Homogeneous Neumann boundary conditions   These conditions impose that $\mathbf{n}\cdot\boldsymbol{\nabla}_{\mathbf{r}}G_{\mathrm{HN}}^{+}(\mathbf{r};\mathbf{r}';t-t')$ vanishes for all $\mathbf{r}\in\partial\mathcal{D}$. The form of the spectral representation of G_{HN}^{+} is the same as (3.34) for G_{HD}^{+}, now with the eigenfunctions and eigenvalues of operator $-\Delta$ defined by homogeneous Neumann boundary conditions.

Function $G_{\mathrm{HN}}^{+}(\mathbf{r};\mathbf{r}';t-t')$ presents the same asymptotic behaviour (3.35) when $t-t'\to 0^{+}$ as $G_{\mathrm{HD}}^{+}(\mathbf{r};\mathbf{r}';t-t')$. For $t\to+\infty$ however we find

$$\lim_{t\to+\infty}G_{\mathrm{HN}}^{+}(\mathbf{r};\mathbf{r}';t-t') = \frac{1}{V} \tag{3.37}$$

where $V = \int_{\mathcal{D}}\mathrm{d}\mathbf{r}$ is the volume of domain $\mathcal{D}$. This result comes from the existence of the zero eigenvalue $\lambda_0 = 0$ associated with the eigenfunction $\psi_0(\mathbf{r}) = 1/V^{1/2}$ for homogeneous Neumann boundary conditions. All other eigenvalues λ_n are strictly positive, and their contributions in the spectral representation of $G_{\mathrm{HN}}^{+}(\mathbf{r};\mathbf{r}';t-t')$ become exponentially smaller than the constant $1/V$, the contribution from $\lambda_0 = 0$. The spectral representation of G_{HN}^{+} then reads:

$$G_{\mathrm{HN}}^{+}(\mathbf{r};\mathbf{r}';t-t') = \theta(t-t')\left[\frac{1}{V} + \sum_{n\neq 0}e^{-D\lambda_n(t-t')}\,\psi_n(\mathbf{r})\psi_n^{*}(\mathbf{r}')\right]. \tag{3.38}$$

$G_{\mathrm{HN}}^{+}(\mathbf{r};\mathbf{r}';t-t')$ also satisfies the sum rule

$$\int_{\mathcal{D}}\mathrm{d}\mathbf{r}\,G_{\mathrm{HN}}^{+}(\mathbf{r};\mathbf{r}';t-t') = 1 \tag{3.39}$$

for all times $t > t'$. This result can be established by remarking that, for $n\neq 0$,

$$\int_{\mathcal{D}}\mathrm{d}\mathbf{r}\psi_n(\mathbf{r}) = -\frac{1}{\lambda_n}\int_{\mathcal{D}}\mathrm{d}\mathbf{r}\,\Delta\psi_n(\mathbf{r}) = -\frac{1}{\lambda_n}\int_{\partial\mathcal{D}}\mathrm{d}\Sigma\,\mathbf{n}\cdot\boldsymbol{\nabla}\psi_n(\mathbf{r}) = 0,$$

since functions ψ_n satisfy homogeneous Neumann boundary conditions. Finally $G_{HN}^+(\mathbf{r}; \mathbf{r}'; t - t')$ is invariant by exchange of $\mathbf{r}$ and $\mathbf{r}'$, since eigenfunctions of $-\Delta$ are real.

As for Dirichlet case, $G_{HN}^+(\mathbf{r}; \mathbf{r}'; t - t')$ describes the diffusion of particles all initially located at $\mathbf{r} = \mathbf{r}'$. The main difference in the subsequent evolution comes from the completely different nature of boundary conditions. Homogeneous Neumann boundary conditions amount to cancel the component of particle current normal to the wall, because this current is proportional to $\nabla_{\mathbf{r}} G_{HN}^+(\mathbf{r}; \mathbf{r}'; t - t')$ according to Fick's law. They therefore represent a perfect reflecting wall. The diffusion of particles thus remains within domain $\mathcal{D}$. Sum rule (3.39) then expresses the conservation of the total number of particles, while behaviour (3.37) simply reflects equiprobability of presence when $t \to +\infty$.

Note that the behaviours highlighted above are well illustrated by the example of diffusion on a segment, discussed in section 3.2.1, p. 174.

◊ Causal Green's function for an infinite system ◊

Consider the case where the domain $\mathcal{D}$ is now the whole space $\mathbb{R}^d$. Let G_∞^+ be the causal Green's function associated with homogeneous Dirichlet boundary conditions at infinity, i.e. $G_\infty^+(\mathbf{r}; \mathbf{r}'; t - t') \to 0$ when $|\mathbf{r}| \to \infty$. As a result of spatial translation invariance of the diffusion operator, G_∞^+ depends only on the difference $\mathbf{r} - \mathbf{r}'$.

Explicit calculation by Fourier-Laplace transform The Laplace transform of $\widetilde{G}_\infty^+(\mathbf{r} - \mathbf{r}'; s)$ is nothing but the Green's function of the infinite homogeneous system associated with operator $s - D\Delta$. The Fourier transform of the latter is then simply obtained from formula (2.46), p. 74, which leads to

$$\widehat{\widetilde{G}}_\infty^+(\mathbf{k}; s) = \frac{1}{D\mathbf{k}^2 + s}. \tag{3.40}$$

Inverse Laplace transform of expression (3.40) immediately yields

$$\widehat{G}_\infty^+(\mathbf{k}; t - t') = e^{-D\mathbf{k}^2(t-t')}.$$

Inverse Fourier transform of this expression reads

$$G_\infty^+(\mathbf{r} - \mathbf{r}'; t - t') = \frac{1}{(2\pi)^d} \int d\mathbf{k} \, e^{[i\mathbf{k}\cdot(\mathbf{r}-\mathbf{r}')-D\mathbf{k}^2(t-t')]}.$$

The integral over $\mathbf{k}$ is nothing but the Fourier transform of a Gaussian function. As detailed in Appendix E, the result is also a Gaussian function, namely

$$G_\infty^+(\mathbf{r} - \mathbf{r}'; t - t') = \frac{1}{(4\pi D(t - t'))^{d/2}} \exp\left[-\frac{(\mathbf{r} - \mathbf{r}')^2}{4D(t - t')}\right] \tag{3.41}$$

for $t > t'$. Let us note that relation (3.25), p. 136, gives here for the spatio-temporal Fourier transform of G^+_∞,

$$\widehat{G}^+_\infty(\mathbf{k}; \omega) = \frac{1}{D\mathbf{k}^2 - i\omega}.$$

The absence of singularity for ω real comes from the dissipative nature of diffusion, a feature which emerges several times in the following.

Interpretation and sum rules Just as causal Green's functions in a finite domain, G^+_∞ can be seen as proportional to the density of diffusing particles, all initially located at $\mathbf{r} = \mathbf{r}'$. In the same way, G^+_∞ also describes the probability distribution at $\mathbf{r}$ at time t of a Brownian particle initially located in $\mathbf{r}'$ at the initial time t'. It is easy to derive a few simple results from expression (3.41). First we obtain by simple spatial integration of a Gaussian function,

$$\int d\mathbf{r}\, G^+_\infty(\mathbf{r} - \mathbf{r}'; t - t') = 1 \qquad \forall\, t > t', \tag{3.42}$$

which is the exact analogue of sum rule (3.39) for homogeneous Neumann boundary conditions. In fact, for an infinite system, homogeneous Dirichlet and Neumann boundary conditions are equivalent and lead to the same causal Green's function G^+_∞. Formula (3.42) therefore reflects the invariant normalisation of the probability distribution of a Brownian particle.

The second spatial moment of G^+_∞, also called mean square displacement, is simply related to the variance of Gaussian formula (3.41),

$$\langle (\mathbf{r} - \mathbf{r}')^2 \rangle = \int d\mathbf{r}\, (\mathbf{r} - \mathbf{r}')^2\, G^+_\infty(\mathbf{r} - \mathbf{r}'; t - t')$$

and reads

$$\langle (\mathbf{r} - \mathbf{r}')^2 \rangle = 6D(t - t').$$

This major and well-known result means that the average distance of the particle from its starting point only increases as the square root of time (see Figure 3.1).

◇ Solution of diffusion equation ◇

We now come back to the field ϕ induced by an arbitrary source density ρ and satisfying the boundary conditions $\mathrm{BC}(\phi|\partial\mathcal{D})$. We will show that an appropriate choice of causal Green's function in integral equation (3.32) immediately provides an explicit representation of ϕ.

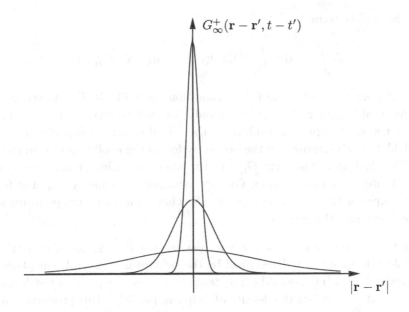

Fig. 3.1   Representation of Green's function (3.41) as a function of $|\mathbf{r}-\mathbf{r}'|$ for various time intervals $t-t'$: the time interval is multiplied by a factor of ten between each curve.

Dirichlet boundary conditions   Assume that $\mathrm{BC}(\phi|\partial\mathcal{D})$ are of Dirichlet type (3.28). Applying formula (3.32) with causal homogeneous Dirichlet Green's function G_{HD}^+ gives[7]

$$
\boxed{
\begin{aligned}
\phi(\mathbf{r},t) &= \int_0^{t-t_0} \mathrm{d}\tau \int_{\mathcal{D}} \mathrm{d}\mathbf{r}'\, \rho(\mathbf{r}',t-\tau)G_{\mathrm{HD}}^+(\mathbf{r};\mathbf{r}';\tau) \\
&\quad + \int_{\mathcal{D}} \mathrm{d}\mathbf{r}'\, \phi_0(\mathbf{r}')\, G_{\mathrm{HD}}^+(\mathbf{r};\mathbf{r}';t-t_0) \\
&\quad -D\int_0^{t-t_0} \mathrm{d}\tau \int_{\partial\mathcal{D}} \mathrm{d}\Sigma'\, D(\mathbf{r}',t-\tau)\mathbf{n}'\cdot\boldsymbol{\nabla}_{\mathbf{r}'}G_{\mathrm{HD}}^+(\mathbf{r}';\mathbf{r};\tau).
\end{aligned}
}
\tag{3.43}
$$

This is of course an explicit expression for the solution of PDE (3.26) up to the knowledge of G_{HD}^+.

Notice that only the diffusion term of the initial condition contributes for $t \to t_0^+$. In addition, relation (3.43) indeed leads to $\phi(\mathbf{r},t) \to \phi_0(\mathbf{r})$ in this limit by virtue of result (3.35), p. 139. This property is the exact analogue of the one discussed in the static case for the Laplacian on p. 70.

In the absence of sources, formula (3.43) with $\rho(\mathbf{r}',t) = 0$ reduces to the diffusion of the initial condition

$$
\int_{\mathcal{D}} \mathrm{d}\mathbf{r}'\, \phi_0(\mathbf{r}')\, G_{\mathrm{HD}}^+(\mathbf{r};\mathbf{r}';t-t_0),
\tag{3.44}
$$

---

[7]We use the symmetry between $\mathbf{r}$ and $\mathbf{r}'$ of Green's function $G_{\mathrm{HD}}^+(\mathbf{r};\mathbf{r}';\tau)$ to write this formula.

plus the surface term

$$-D \int_0^{t-t_0} d\tau \int_{\partial \mathcal{D}} d\Sigma' \, D(\mathbf{r}', t-\tau) \mathbf{n}' \cdot \boldsymbol{\nabla}_{\mathbf{r}'} G_{\text{HD}}^+(\mathbf{r}'; \mathbf{r}; \tau). \tag{3.45}$$

At long time $t \to +\infty$, the initial condition term (3.44) disappears, because of the decay of $G_{\text{HD}}^+(\mathbf{r}; \mathbf{r}'; t-t_0)$. Then only the surface term (3.45) remains. Thus once a transient regime depending on the initial condition has disappeared, only the field $D(\mathbf{r}', t')$ imposed on the boundary forces the evolution of ϕ in the domain $\mathcal{D}$. This forcing diffuses via G_{HD}^+ in the whole domain, in agreement with the general interpretation of causal Green's functions in terms of response functions. An example of this situation is a sample whose boundary temperature is varied through external thermostats.

In the absence of sources, for homogeneous conditions, i.e. with $D(\mathbf{r}', t') = 0$, the field ϕ is completely determined by the term of initial condition (3.44). So for any initial field $\phi_0(\mathbf{r})$, the field $\phi(\mathbf{r}, t)$ tends to zero at any point of the domain when $t \to +\infty$. If ϕ describes the density of diffusing particles, this property reflects the complete escape of all the particles outside the domain, as a result of the absorbing nature of the walls. This is an evidence of the irreversible nature of diffusion.

Neumann boundary conditions   Now imagine that the $\text{BC}(\phi|\partial \mathcal{D})$ are of Neumann type (3.29). Applying formula (3.32), with the causal homogeneous Neumann Green's function of G_{HN}^+, then leads to an explicit expression for ϕ as

$$\phi(\mathbf{r}, t) = \int_0^{t-t_0} d\tau \int_{\mathcal{D}} d\mathbf{r}' \, \rho(\mathbf{r}', t-\tau) G_{\text{HN}}^+(\mathbf{r}; \mathbf{r}'; \tau)$$

$$+ \int_{\mathcal{D}} d\mathbf{r}' \, \phi_0(\mathbf{r}') \, G_{\text{HN}}^+(\mathbf{r}; \mathbf{r}'; t-t_0) \tag{3.46}$$

$$+ D \int_0^{t-t_0} d\tau \int_{\partial \mathcal{D}} d\Sigma' \, G_{\text{HN}}^+(\mathbf{r}'; \mathbf{r}; \tau) N(\mathbf{r}', t-\tau).$$

In the absence of sources, only the initial condition and surface terms contribute to ϕ. Suppose that the Neumann conditions are homogeneous, i.e. $N(\mathbf{r}', t) = 0$, so that only the initial condition term remains. In contrast to the Dirichlet case, this term does not disappear at large times. Indeed, by virtue of limit behaviour (3.37) of $G_{\text{HN}}^+(\mathbf{r}; \mathbf{r}'; t-t_0)$ when $t \to +\infty$, the field $\phi(\mathbf{r}, t)$ tends to the constant

$$\frac{\int_{\mathcal{D}} d\mathbf{r}' \, \phi_0(\mathbf{r}')}{V}. \tag{3.47}$$

When ϕ describes a local density of particles, it renders the homogenisation of density at sufficiently long time, regardless of the form of the initial density. This remarkable behaviour is another manifestation of the irreversibility of diffusion.

Use of causal Green's function for the infinite system Formula (3.32) holds when the causal Green's function of an infinite system G_∞^+ is used. If this formula does not provide an explicit expression for ϕ, the simple Gaussian form (3.41) of $G_\infty^+(\mathbf{r} - \mathbf{r}'; t - t')$ facilitates the analysis of behaviours at large time or far away from the finite domain border. More generally, the use of G_∞^+ helps us to understand the evolution of the different contributions to the field ϕ, provided that the Green's function incorporates the basic mechanisms of diffusion.

To conclude this section on diffusion equation, let us consider again the case without source, i.e. $\rho = 0$, with homogeneous BC($\phi|\partial\mathcal{D}$) of Dirichlet type for example. Since the corresponding PDE is homogeneous, one might wonder why introducing Green's functions in this simple case. As mentioned above, the relation (3.43) however shows that

$$\phi(\mathbf{r}, t) = \int_{\mathcal{D}} d\mathbf{r}' \; \phi_0(\mathbf{r}') \; G_{HD}^+(\mathbf{r}; \mathbf{r}'; t - t_0).$$

So a preliminary calculation of G_{HD}^+ gives access to the solution ϕ. In fact, as discussed on p. 134, the initial condition becomes in the Laplace space a source term associated with the IC In short the knowledge of the causal Green's function is also adapted to solve the homogeneous diffusion equation with IC. This remark will take its full meaning in the next section devoted to Schrödinger equation.

3.1.4 *Schrödinger equation*

Schrödinger equation is the central equation of quantum mechanics. We have already encountered it in Chapter 2, where we have shown in section 2.2.3 how eigenfunctions and eigenvectors of a given Hamiltonian H could be determined from the Green's functions associated with the resolvent $[\lambda + H]^{-1}$. In the following we first consider the non-relativistic Hamiltonian of a particle of mass m evolving in a time-independent potential $V(\mathbf{r})$,

$$H = -\frac{\hbar^2}{2m}\Delta_\mathbf{r} + V(\mathbf{r}). \tag{3.48}$$

The Schrödinger equation

$$i\hbar\frac{\partial\phi}{\partial t} = H\phi \tag{3.49}$$

then generates the time evolution of the particle wave function $\phi(\mathbf{r}, t)$. This equation takes the generic form (3.1) without source, i.e.

$$\left[\frac{\partial}{\partial t} - \frac{H}{i\hbar}\right]\phi(\mathbf{r}, t) = 0. \tag{3.50}$$

The field ϕ thus corresponds to the wave function. As announced at the end of the previous section, we note that this PDE is homogeneous. Moreover, since the potential is time independent, operator $\mathcal{O}$ is indeed of the additive form (3.10) with

$\mathcal{O}_t = (\partial/\partial t)$ and $\mathcal{O}_{\mathbf{r}} = -(H/i\hbar)$. Methods introduced in section 3.1.2 can then be applied: They allow us first to specify the boundary conditions to impose on ϕ to ensure uniqueness of the solution of PDE (3.50). We then study properties of the corresponding causal Green's functions, and consider the case of a free quantum particle. Then comes the standard paragraph where the expression of the solution of the PDE under consideration, here Schrödinger equation, is derived. We finally highlight the interest of Green's functions in the construction of perturbative expansions.

$$\lozenge \textbf{ Boundary conditions } \lozenge$$

Let us therefore consider the Schrödinger equation (3.50) in a finite domain $\mathcal{D}$, and determine the boundary conditions leading to a unique solution in such domain. Since $\mathcal{O}_t$ is a first order operator, the initial conditions $\text{CI}(\phi|t_0)$ consist in setting the value of the wave function at an initial time t_0 at every point of domain, as in the case of the diffusion equation:

$$\phi(\mathbf{r}, t_0) = \phi_0(\mathbf{r}) \qquad \text{for all } \mathbf{r} \in \mathcal{D}. \tag{3.51}$$

It is convenient to go back to Laplace space to identify the boundary conditions $\text{BC}(\phi|\partial\mathcal{D})$ to ensure uniqueness. The static operator $[P_1(s) + \mathcal{O}_{\mathbf{r}}]$, introduced on p. 134, reads here

$$\boxed{[P_1(s) + \mathcal{O}_{\mathbf{r}}] = s - \frac{H}{i\hbar}.}$$

Up to a multiplicative constant, its structure therefore takes the form studied on p. 87, in section 2.1.6 of Chapter 2. In particular, as shown on p. 87 Dirichlet or Neumann boundary conditions are sufficient to ensure uniqueness of ϕ.

Mathematically speaking, the boundary conditions of Dirichlet and Neumann are also acceptable. It turns out that homogeneous Dirichlet conditions are most common in quantum mechanics, since they describe physical situations where the particle remains confined within the domain $\mathcal{D}$ during its evolution. Indeed potential $V(\mathbf{r})$ must enforce a strict confinement within this domain, so that $V(\mathbf{r}) = +\infty$ for $\mathbf{r} \notin \mathcal{D}$. Since the wave function vanishes identically outside $\mathcal{D}$, it is also zero by continuity on the boundary $\partial\mathcal{D}$. Corresponding homogeneous Dirichlet conditions

$$\phi(\mathbf{r}, t) = 0 \qquad \text{for } \mathbf{r} \in \partial\mathcal{D} \text{ and all } t, \tag{3.52}$$

together with initial conditions (3.51) define uniquely the solution of Schrödinger equation in domain $\mathcal{D}$. We will only use these boundary conditions on ϕ in the remaining of this section.

◇ Wave function and Green's functions ◇

Let us start by writing PDE (3.20) for $\widetilde{\phi}(\mathbf{r}, s)$, the Laplace transform of $\phi(\mathbf{r}, t)$. It takes the form

$$\left[s - \frac{H}{i\hbar} \right] \widetilde{\phi}(\mathbf{r}, s) = \phi_0(\mathbf{r}), \tag{3.53}$$

since $\widetilde{\rho} = 0$ and $I_0(\mathbf{r}, s) = -\phi_0(\mathbf{r})$. In addition equation (3.22) leads here to $\widetilde{G}_{\mathrm{H}}^+(\mathbf{r}; \mathbf{r}'; s) = G_s(\mathbf{r}; \mathbf{r}')$. As exposed in section 2.1.6 on p. 86, $\widetilde{\phi}(\mathbf{r}, s)$ then satisfies integral equation

$$\widetilde{\phi}(\mathbf{r}, s) = \int_{\mathcal{D}} \mathrm{d}\mathbf{r}' \, \phi_0(\mathbf{r}') \, \widetilde{G}_{\mathrm{H}}^+(\mathbf{r}'; \mathbf{r}; s) + \int_{\partial \mathcal{D}} \mathrm{d}\Sigma' \, \widetilde{G}_{\mathrm{H}}^+(\mathbf{r}'; \mathbf{r}; s) \mathbf{n}' \cdot \boldsymbol{\nabla}_{\mathbf{r}'} \widetilde{\phi}(\mathbf{r}', s). \tag{3.54}$$

The inverse Laplace transform of the above formula is similar to that carried out on p. 138 in the case of diffusion. We obtain:

$$\phi(\mathbf{r}, t) = \int_{\mathcal{D}} \mathrm{d}\mathbf{r}' \, \phi_0(\mathbf{r}') \, G_{\mathrm{H}}^+(\mathbf{r}'; \mathbf{r}; t - t_0)$$

$$+ \int_0^{t - t_0} \mathrm{d}\tau \int_{\partial \mathcal{D}} \mathrm{d}\Sigma' \, G_{\mathrm{H}}^+(\mathbf{r}'; \mathbf{r}; \tau) \mathbf{n}' \cdot \boldsymbol{\nabla}_{\mathbf{r}'} \phi(\mathbf{r}', t - \tau), \tag{3.55}$$

where $G_{\mathrm{H}}^+(\mathbf{r}; \mathbf{r}', t - t')$ is a homogeneous causal Green's function of operator $[(\partial/\partial t) - (H/i\hbar)]$, solution of the PDE:

$$\left[\frac{\partial}{\partial t} - i \frac{\hbar}{2m} \Delta_{\mathbf{r}} - \frac{V(\mathbf{r})}{i\hbar} \right] G_{\mathrm{H}}^+(\mathbf{r}; \mathbf{r}', t - t') = \delta(\mathbf{r} - \mathbf{r}') \delta(t - t'), \tag{3.56}$$

with the causality condition $G_{\mathrm{H}}^+(\mathbf{r}; \mathbf{r}', t - t') = 0$ for $t < t'$.

The expression (3.55) is similar to that obtained on p. 138 for diffusion. In the case of quantum mechanics, the surface term depending on value of $\phi(\mathbf{r}', t - \tau)$ at the boundary does not appear, in agreement with homogeneous Dirichlet conditions (3.52). On the contrary, the one involving $\boldsymbol{\nabla}_{\mathbf{r}'} \phi(\mathbf{r}', t - \tau)$ is generally present since $G_{\mathrm{H}}^+(\mathbf{r}; \mathbf{r}', \tau)$ may take non-zero values at the domain boundary.

◇ Causal Green's functions ◇

We present here only the causal Green's functions satisfying the same BC as the wave function, i.e. homogeneous Dirichlet boundary conditions.

Case of a finite domain   Green's function $G_{\text{HD}}^+(\mathbf{r};\mathbf{r}';t-t')$ vanishes identically on the domain boundary $\partial \mathcal{D}$. Suppose that the eigenfunctions ψ_n of Hamiltonian H, defined by homogeneous Dirichlet conditions, form an orthonormal basis. In spectral representation (3.15) corresponding to $G_{\text{HD}}^+(\mathbf{r};\mathbf{r}';t-t')$ each function $Z_n(\tau)$ is the solution of ordinary differential equation

$$\frac{\mathrm{d}}{\mathrm{d}\tau}Z_n(\tau) - \frac{E_n}{i\hbar}Z_n(\tau) = 0$$

with initial condition $Z_n(0) = 1$, where E_n is the energy corresponding to the eigenfunction ψ_n. A straightforward integration leads to the pure phase factor $Z_n(\tau) = e^{-iE_n\tau/\hbar}$, so that $G_{\text{HD}}^+(\mathbf{r};\mathbf{r}';t-t')$ reads

$$G_{\text{HD}}^+(\mathbf{r};\mathbf{r}';t-t') = \theta(t-t') \sum_n e^{-iE_n(t-t')/\hbar}\psi_n(\mathbf{r})\psi_n^*(\mathbf{r}'). \qquad (3.57)$$

Note that one can also derive formula (3.57) by taking the inverse Laplace transform of spectral representation

$$\widetilde{G}_{\text{HD}}^+(\mathbf{r};\mathbf{r}';s) = G_s(\mathbf{r};\mathbf{r}') = \sum_n \frac{\psi_n(\mathbf{r})\psi_n^*(\mathbf{r}')}{s - \frac{E_n}{i\hbar}}. \qquad (3.58)$$

As shown in section 2.1.6, p. 87, from an operator point of view, $G_s(\mathbf{r};\mathbf{r}')$ is a matrix element of resolvent $[s - (H/i\hbar)]^{-1}$,

$$G_s(\mathbf{r};\mathbf{r}') = \langle \mathbf{r}|\frac{1}{s - (H/i\hbar)}|\mathbf{r}'\rangle.$$

Representation (3.57) can then be written as

$$G_{\text{HD}}^+(\mathbf{r};\mathbf{r}';t-t') = \theta(t-t')\,\langle\mathbf{r}|e^{-iH(t-t')/\hbar}|\mathbf{r}'\rangle. \qquad (3.59)$$

So causal Green's function $G_{\text{HD}}^+(\mathbf{r};\mathbf{r}';t-t')$, for $t > t'$, is nothing but the matrix element of the evolution operator $e^{-iH(t-t')/\hbar}$ between bra $\langle\mathbf{r}|$ and ket $|\mathbf{r}'\rangle$. In other words $G_{\text{HD}}^+(\mathbf{r};\mathbf{r}';t-t')$ is the probability amplitude for the particle to go from position $\mathbf{r}'$ at t' to position $\mathbf{r}$ at $t > t'$. This interpretation is consistent with the short time behaviour

$$\lim_{\tau\to 0^+} G_{\text{HD}}^+(\mathbf{r};\mathbf{r}';\tau) = \delta(\mathbf{r} - \mathbf{r}'). \qquad (3.60)$$

Case of the entire space   For homogeneous causal Dirichlet Green's function $G_\infty^+(\mathbf{r};\mathbf{r}';t-t')$ in $\mathbb{R}^d$, we use a limit process where a finite domain is introduced and its size is then sent to infinity. Once this limit has been taken, the Hamiltonian spectrum is generally composed of a discrete part and of a continuum part. Eigenfunctions in the discrete part of the spectrum are square summable, i.e. $\int_{\mathcal{D}} \mathrm{d}\mathbf{r}\,|\psi_n(\mathbf{r})|^2 < \infty$, and describe localised states such that $\psi_n(\mathbf{r}) \to 0$ when $|\mathbf{r}| \to \infty$. The spectral representation of $G_\infty^+(\mathbf{r};\mathbf{r}';t-t')$ takes the form (3.57), where the summation over continuum states reduces to an integral over the continuous variable E_n involving the density of states. An explicit calculation of $G_\infty^+(\mathbf{r};\mathbf{r}';t-t')$ is not possible for a general Hamiltonian. We return to this issue later in the paragraph on perturbative expansions. First we present below the case of a free particle.

◇ Quantum free particle ◇

Case of the whole $\mathbb{R}^3$ For a free particle of Hamiltonian $H = -\frac{\hbar^2}{2m}\Delta$ the calculation of G_∞^+, solution of equation (3.56) is elementary. There is no localised state since potential $V(\mathbf{r})$ is identically zero. The spectrum consists only of a continuous part, where n can be replaced by the wave number $\mathbf{k}$, and the corresponding energy E_n becomes $E(\mathbf{k}) = \hbar^2 \mathbf{k}^2/(2m)$. Furthermore, the wave functions ψ_n are linear combinations of plane waves $e^{\pm i\mathbf{k}\cdot\mathbf{r}}$, and the sum over n becomes an integral over $\mathbf{k}$ with a density of states independent of $\mathbf{k}$. We thus obtain from equation (3.57) in three dimensions

$$G_\infty^+(\mathbf{r} - \mathbf{r}'; t - t') = \frac{1}{(2\pi)^3} \int d\mathbf{k} \, \exp\left[i\mathbf{k}\cdot(\mathbf{r}-\mathbf{r}') - i\frac{\hbar\mathbf{k}^2}{2m}(t-t')\right] \qquad (3.61)$$

for $t > t'$. Just as in the case of Green's function for free diffusion, we are left with decoupled Gaussian integrals but here with purely imaginary arguments. We then obtain, by using the method described in Appendix E and a simple change of variables:

$$\boxed{\textbf{Free particle:} \qquad G_\infty^+(\mathbf{r} - \mathbf{r}'; t - t') = \left(\frac{m}{2\pi i\hbar(t-t')}\right)^{3/2} e^{\frac{im(\mathbf{r}-\mathbf{r}')^2}{2\hbar(t-t')}}.} \qquad (3.62)$$

The analytic function $Z^{3/2}$ appearing in formula (3.62) is defined with the choice of determination such that

$$Z^{3/2} = |Z|^{3/2} e^{i\frac{3}{2}\arg Z} \qquad \text{with } \arg Z \in \,]-\pi, \pi[,$$

where the negative real axis is a cut starting from the singular point $Z = 0$.

Expression (3.62) is similar to equation (3.41) obtained on p. 141 for diffusion. In fact, to go from one to the other, it is sufficient to make the substitution $D \to i(\hbar/2m)$, as suggested by examination of operators defining the respective Green's functions. In other words, the free particle problem in quantum mechanics is equivalent to free diffusion, with a purely imaginary diffusion constant. The analogue of relation (3.40) is then

$$\boxed{\widehat{\widetilde{G}}_\infty^+(\mathbf{k}; s) = \frac{1}{\frac{i\hbar}{2m}\mathbf{k}^2 + s}.} \qquad (3.63)$$

We shall return in detail to this analogy later on. Note that expression (3.62) is an essential ingredient for the path integral represention of $G_\infty^+(\mathbf{r}; \mathbf{r}'; t - t')$ in the presence of a non-vanishing potential $V(\mathbf{r})$, as discussed in Chapter 4. In the case of free quantum particle, rule (3.25), p. 136, applied to result (3.63) then leads to

$$\boxed{\widehat{G}_\infty^+(\mathbf{k}; \omega) = \lim_{\epsilon \to 0^+} \frac{1}{\frac{i\hbar}{2m}\mathbf{k}^2 - i\omega + \epsilon}.}$$

So here, unlike in the case of diffusion, the spatio-temporal Fourier transform of G_∞^+ is singular at $\omega = \hbar \mathbf{k}^2/(2m)$ which is a simple pole[✠]. This singularity comes from the oscillating nature without decay of $\widehat{G}_\infty^+(\mathbf{k}; \tau)$ when $\tau \to \infty$. This is not surprising since Schrödinger equation leads to a conservative evolution without dissipation.

✠ Comment: For a system of N interacting particles described by a time-independent Hamiltonian H_N, the N-body Green's function can be defined as the matrix element of evolution operator $e^{-iH_N(t-t')/\hbar}$ in the configuration space of positions. We then introduce the one-particle Green's function by integrating over all particle positions but one. This quantity describes the propagation of a particle interacting with all the others. Analytic properties of its double Fourier-Laplace transform are similar to those of $\widetilde{\widehat{G}}_\infty(\mathbf{k}; -iz)$. In general, interactions blur the singularity of the free case on the real axis at $\omega = \hbar \mathbf{k}^2/(2m)$. If interactions are weak, it appears instead a very sharp peak of finite width. This allows to define the notion of quasi-particle, identifying its effective mass from the peak position and its lifetime from its width. The concept of one-particle Green's function plays a fundamental role in the study of quantum many-body problem.

Comparison with diffusion   Let us continue with the comparison between diffusion and a free quantum particle. While the corresponding Schrödinger equation

$$\frac{\partial}{\partial t}\phi(\mathbf{r}, t) = i\frac{\hbar}{2m}\Delta_{\mathbf{r}}\phi(\mathbf{r}, t),$$

takes the same form as the diffusion equation for a density $n(\mathbf{r}, t)$ of classical particles

$$\frac{\partial}{\partial t}n(\mathbf{r}, t) = D\Delta_{\mathbf{r}}n(\mathbf{r}, t),$$

the respective evolutions of $\phi(\mathbf{r}, t)$ and $n(\mathbf{r}, t)$ are very different! First of all, note that the boundary conditions have opposed meanings. Indeed homogeneous Dirichlet conditions reflect the confinement of the quantum particle in the box, while they cause the escape of diffusing particles! In order to confine such diffusing particles, one must actually impose homogeneous Neumann boundary conditions.

As for the evolution itself, if the initial wave function $\phi(\mathbf{r}, t_0) = \phi_0(\mathbf{r})$ is an eigenfunction $\psi_n(\mathbf{r})$ of Hamiltonian with eigenenergy E_n, then

$$\phi(\mathbf{r}, t) = e^{-iE_n(t-t_0)/\hbar}\psi_n(\mathbf{r}),$$

and $|\phi(\mathbf{r}, t)|^2$ remains constant over time at any point in the field, and equal to $|\psi_n(\mathbf{r})|^2$. There is therefore an infinite number of initial configurations of the quantum probability density $|\phi(\mathbf{r}, t)|^2$ remaining invariant during the evolution! This is not the case with diffusion as may already be seen on the limit behaviour

$$\lim_{t \to +\infty} \phi(\mathbf{r}, t) = \frac{\int \mathrm{d}\mathbf{r}\, \phi_0(\mathbf{r})}{V},$$

obtained on p. 144 for homogeneous Neumann boundary conditions.

In fact, these very different evolutions are related to invariance properties of the corresponding equations with respect to time reversal, i.e. the transformation $t \to -t$. For Schrödinger equation, this transformation is equivalent to consider the evolution of ϕ^*, which is of course identical to that of ϕ: this reflects the reversibility of quantum evolution. On the contrary diffusion equation is not invariant under the transformation $t \to -t$, hence it is irreversible!

These very different behaviours actually come from the complex number i. In conclusion, if Schrödinger equation for a free particle can be formally derived from diffusion equation through the simple substitution $D \to i\hbar/(2m)$, the corresponding analytic continuation is not trivial and involves many surprises.

◇ Solution of Schrödinger equation ◇

Let us return to the case of an arbitrary potential $V(\mathbf{r})$ and to expression (3.55). Using $G_{\mathrm{HD}}^+(\mathbf{r}; \mathbf{r}'; t - t')$ in this equation leads to the sought wave function, while the introduction of $G_\infty^+(\mathbf{r}; \mathbf{r}'; t - t')$ provides a better understanding of boundary effects, as we shall now see.

Explicit form of the wave function As $G_{\mathrm{HD}}^+(\mathbf{r}; \mathbf{r}'; t-t')$ vanishes on the domain boundary $\partial \mathcal{D}$, the surface term in the integral equation (3.55) vanishes, leading to the explicit form

$$\boxed{\phi(\mathbf{r}, t) = \int_{\mathcal{D}} d\mathbf{r}' \; G_{\mathrm{HD}}^+(\mathbf{r}; \mathbf{r}'; t - t_0) \; \phi_0(\mathbf{r}').} \tag{3.64}$$

Note that by using expression (3.59) of $G_{\mathrm{HD}}^+(\mathbf{r}; \mathbf{r}'; t - t')$ in terms of matrix element of evolution operator given on p. 148, we find the well-known formula,

$$\phi(\mathbf{r}, t) = \langle \mathbf{r} | e^{-iH(t-t_0)/\hbar} | \phi_0 \rangle,$$

which can be obtained by formal integration of Schrödinger equation rewritten as

$$\frac{\partial}{\partial t} |\phi\rangle = -\frac{iH}{\hbar} |\phi\rangle.$$

Formula (3.64) allows us to complete the interpretation of Green's function $G_{\mathrm{HD}}^+(\mathbf{r}; \mathbf{r}'; t - t')$. First of all, according to the analysis of the previous paragraph, we check that if the particle is strictly localised at $\mathbf{r}_0$ at time t_0, that is to say if $\phi_0(\mathbf{r}') = \delta(\mathbf{r}' - \mathbf{r}_0)$, then $\phi(\mathbf{r}, t) = G_{\mathrm{HD}}^+(\mathbf{r}; \mathbf{r}_0; t - t_0)$. For an arbitrary initial wave function, it is necessary to linearly combine probability amplitudes to go from $\mathbf{r}'$ to $\mathbf{r}$, $G_{\mathrm{HD}}^+(\mathbf{r}; \mathbf{r}'; t - t_0)$, weighted by the initial distribution $\phi_0(\mathbf{r}')$. If short time behaviour (3.60) guarantees that ϕ reduces to the initial condition ϕ_0, the analysis of long time behaviour is much more difficult. This difficulty arises from the summation of factors with oscillating phase contained in $G_{\mathrm{HD}}^+(\mathbf{r}; \mathbf{r}'; t - t_0)$, which interfere in a highly complex way even in the simplest case of a free particle.

Boundary effects   Integral equation (3.55) becomes, using $G_\infty^+(\mathbf{r};\mathbf{r}';t-t')$,

$$
\begin{aligned}
\phi(\mathbf{r},t) = &\int_{\mathcal{D}} d\mathbf{r}'\, \phi_0(\mathbf{r}')\, G_\infty^+(\mathbf{r};\mathbf{r}';t-t_0) \\
&+i\frac{\hbar}{2m}\int_0^{t-t_0} d\tau\, \int_{\partial\mathcal{D}} d\Sigma'\, G_\infty^+(\mathbf{r};\mathbf{r}';\tau)\mathbf{n}'\cdot\boldsymbol{\nabla}_{\mathbf{r}'}\phi(\mathbf{r}',t-\tau).
\end{aligned}
\tag{3.65}
$$

The volume integral describes the intrinsic evolution without boundary. The surface integral takes into account all reflections on the boundary between initial time t_0 and time t, while these reflections propagate inside the volume via the intrinsic evolution governed by $G_\infty^+(\mathbf{r};\mathbf{r}';\tau)$. Again, the interferences of these reflected waves are rather difficult to handle.

Iteration of integral equation (3.65) provides a perturbative expansion of $\phi(\mathbf{r},t)$ whose p-th term may be understood as describing interferences induced by p reflections on the domain boundary $\partial\mathcal{D}$. An estimation of each of these terms requires knowledge of $G_\infty^+(\mathbf{r};\mathbf{r}';t-t')$, whose explicit calculation is already a problem in itself! In fact, even in the case of the free particle where simple expression (3.62) is available, it still remains to determine difficult surface integrals due to the presence of oscillating terms.

◇ **Perturbative expansions** ◇

As mentioned earlier, it is unrealistic to think it is possible to accurately calculate Green's functions for any arbitrary potential $V(\mathbf{r})$. Under certain circumstances the potential V can be considered as close[8] to a reference potential $V^{(0)}$, for which it is possible to get relatively simple analytic representations of the corresponding Green's functions. It is then sensible to make a perturbative expansion of the difference $W = V - V^{(0)}$. In a first step we establish the perturbative expansion of the wave function $\phi(\mathbf{r},t)$, obtained by inserting the expansion of the causal Green's function in equation (3.64). We then show that the structure of this expansion remains unchanged if the perturbation W depends explicitly on time.

We shall build perturbative expansions in the case where the domain is the whole space. Extension of the latter to a finite domain is immediate. We omit the index ∞ in the notation for causal Green's function of the infinite system, in order to avoid heavy notations.

Use of the resolvent   First we derive the perturbative expansion of $G^+(\mathbf{r};\mathbf{r}';t-t')$ in powers of W. It can be achieved by starting from expansion (2.81), p. 90 for

---

[8]This hypothesis must be checked a *posteriori* by comparing the corrections generated by the perturbation $V - V^{(0)}$ to reference quantities.

matrix elements of resolvent $[\lambda + H^{(0)} + W]^{-1}$,

$$G_\lambda(\mathbf{r};\mathbf{r}') = G_\lambda^{(0)}(\mathbf{r};\mathbf{r}') - \int d\mathbf{r}_1\, G_\lambda^{(0)}(\mathbf{r};\mathbf{r}_1)\, W(\mathbf{r}_1)\, G_\lambda^{(0)}(\mathbf{r}_1;\mathbf{r}')$$

$$+ \int d\mathbf{r}_1 \int d\mathbf{r}_2\, G_\lambda^{(0)}(\mathbf{r};\mathbf{r}_1)\, W(\mathbf{r}_1)\, G_\lambda^{(0)}(\mathbf{r}_1;\mathbf{r}_2)\, W(\mathbf{r}_2)\, G_\lambda^{(0)}(\mathbf{r}_2;\mathbf{r}')$$

$$+ \cdots .$$

The perturbative expansion of G^+ is easily obtained using identity $G^+(\mathbf{r};\mathbf{r}';t-t') = -i\hbar\mathcal{L}^{-1}[G_\lambda(\mathbf{r};\mathbf{r}')]$ with $\lambda = -i\hbar s$. We then find[a], using the convolution formula for inverse Laplace transform of a product,

$$G^+(\mathbf{r};\mathbf{r}';\tau) = G^{+(0)}(\mathbf{r};\mathbf{r}';\tau)$$

$$- \frac{i}{\hbar} \int d\mathbf{r}_1 \int_0^\tau d\tau_1\, G^{+(0)}(\mathbf{r};\mathbf{r}_1;\tau - \tau_1)\, W(\mathbf{r}_1)\, G^{+(0)}(\mathbf{r}_1;\mathbf{r}';\tau_1)$$

$$- \frac{1}{\hbar^2} \int d\mathbf{r}_1 d\mathbf{r}_2 \int_0^\tau d\tau_1 \int_0^{\tau_1} d\tau_2\, G^{+(0)}(\mathbf{r};\mathbf{r}_1;\tau - \tau_1)\, W(\mathbf{r}_1)$$

$$\times G^{+(0)}(\mathbf{r}_1;\mathbf{r}_2;\tau_1 - \tau_2)\, W(\mathbf{r}_2)\, G^{+(0)}(\mathbf{r}_2;\mathbf{r}';\tau_2) + \cdots . \qquad (3.66)$$

This perturbative series can also be inferred from identity (3.59), where one can apply Dyson's formula for the expansion of evolution operator $e^{-i\tau(H_0+W)/\hbar}$ in powers of W. The expansion of $\phi(\mathbf{r},t)$ is immediately obtained by inserting the perturbative series (3.66) in formula (3.64) leading to

$$\phi(\mathbf{r},t) = \int d\mathbf{r}_1\, G^{+(0)}(\mathbf{r};\mathbf{r}_1;t - t_0)\, \phi_0(\mathbf{r}_1)$$

$$- \frac{i}{\hbar} \int d\mathbf{r}_1 d\mathbf{r}_2 \int_0^{t-t_0} d\tau_1\, G^{+(0)}(\mathbf{r};\mathbf{r}_1;\tau_1) W(\mathbf{r}_1)$$

$$\times G^{+(0)}(\mathbf{r}_1;\mathbf{r}_2;t - t_0 - \tau_1)\phi_0(\mathbf{r}_2) + \cdots . \qquad (3.67)$$

The first term, of order zero in W, describes the evolution under the action of reference Hamiltonian $H^{(0)}$. The second term, linear in W, represents the first correction to this reference evolution. The sole consideration of the latter is often sufficient to capture interesting effects, as is shown in section 2.2.4 dedicated to diffusion by a repulsive potential.

Time-dependent perturbation Consider now the case of an explicitly time-dependent perturbation $W(\mathbf{r},t)$. Hamiltonian H can then be written as $H = H^{(0)} + W$, where $H^{(0)}$ is a reference Hamiltonian of the form (3.48) with a static potential $V(\mathbf{r})$. Note that operator

$$\mathcal{O} = \frac{\partial}{\partial t} + \frac{i}{\hbar} H$$

associated with Schrödinger equation

$$i\hbar \frac{\partial}{\partial t}\phi = (H^{(0)} + W)\phi, \qquad (3.68)$$

no longer takes the translation invariant additive form (3.10)! Above methods using the resolvent are therefore not applicable here. We draw inspiration from what has been done on p. 88 to overcome this difficulty, and rewrite Schrödinger equation (3.68) as

$$\left[\frac{\partial}{\partial t} - i\frac{\hbar}{2m}\Delta_{\mathbf{r}} + \frac{i}{\hbar}V(\mathbf{r})\right]\phi(\mathbf{r}, t) = -\frac{i}{\hbar}W(\mathbf{r}, t)\phi(\mathbf{r}, t). \qquad (3.69)$$

The introduction of

$$\mathcal{O}^{(0)} = \frac{\partial}{\partial t} + \frac{i}{\hbar}H^{(0)},$$

the operator associated with Hamiltonian $H^{(0)}$, brings us back to the study of PDE

$$\mathcal{O}^{(0)}\phi(\mathbf{r}, t) = \rho(\mathbf{r}, t), \qquad (3.70)$$

with a source term $\rho(\mathbf{r}, t) = -(i/\hbar)W(\mathbf{r}, t)\phi(\mathbf{r}, t)$.

To solve PDE (3.70) with the initial condition $\phi(\mathbf{r}, t_0) = \phi_0(\mathbf{r})$ and boundary conditions $\phi(\mathbf{r}, t) \to 0$ when $|\mathbf{r}| \to \infty$, one just applies the methods introduced for diffusion equation on the one hand, and Schrödinger equation with $H^{(0)}$ on the other hand. This provides

$$\phi(\mathbf{r}, t) = \int d\mathbf{r}_1\, G^{+(0)}(\mathbf{r}; \mathbf{r}_1; t - t_0)\, \phi_0(\mathbf{r}_1)$$

$$+ \int d\mathbf{r}_1 \int_0^{t-t_0} d\tau_1\, G^{+(0)}(\mathbf{r}; \mathbf{r}_1; \tau_1)\, \rho(\mathbf{r}_1, t - \tau_1), \quad (3.71)$$

where $G^{+(0)}$ is once again the causal Green's function associated with $H^{(0)}$. Replacing the source density $\rho(\mathbf{r}, \tau_1)$ by $-(i/\hbar)W(\mathbf{r}, \tau_1)\phi(\mathbf{r}, \tau_1)$, expression (3.71) becomes an integral equation for the wave function,

$$\phi(\mathbf{r}, t) = \int d\mathbf{r}_1\, G^{+(0)}(\mathbf{r}; \mathbf{r}_1; t - t_0)\, \phi_0(\mathbf{r}_1)$$

$$- \frac{i}{\hbar}\int d\mathbf{r}_1 \int_0^{t-t_0} d\tau_1\, G^{+(0)}(\mathbf{r}; \mathbf{r}_1; \tau_1)\, W(\mathbf{r}_1, t - \tau_1)\phi(\mathbf{r}_1, t - \tau_1). \quad (3.72)$$

Of course the integral equation (3.72) is not easier to solve than the original Schrödinger equation (3.68). Its iteration easily provides however a perturbative expansion of the wave function in powers of W. More precisely, replacing $\phi(\mathbf{r}_1, t - \tau_1)$ by its full expression in the above equation, we still get an integral equation but where linear and higher order terms in W are now clearly separated. We then repeat this iteration $n + 1$ times and keep all terms except the last one to get an expansion at n-th order in W. We then find, at first order in W,

$$\phi(\mathbf{r}, t) = \int d\mathbf{r}_1\, G^{+(0)}(\mathbf{r}; \mathbf{r}_1; t - t_0)\, \phi_0(\mathbf{r}_1)$$

$$- \frac{i}{\hbar}\int d\mathbf{r}_1 d\mathbf{r}_2 \int_0^{t-t_0} d\tau_1\, G^{+(0)}(\mathbf{r}; \mathbf{r}_1; \tau_1)W(\mathbf{r}_1, t - \tau_1)$$

$$\times G^{+(0)}(\mathbf{r}_1; \mathbf{r}_2; t - \tau_1 - t_0)\phi_0(\mathbf{r}_2) + \cdots. \quad (3.73)$$

Note that the expansion (3.73) for a time-dependent perturbation has the same structure as expansion (3.67) for a purely static perturbation! It is sufficient to replace $W(\mathbf{r}_i)$ by $W(\mathbf{r}_i, t - \tau_i)$ in each term of this latter series. The use of causal Green's function $G^{+(0)}$ associated with reference Hamiltonian $H^{(0)}$ thus provides a unified and elegant framework for the general perturbation theory.

Expansion (3.73) is particularly useful and can be applied with good accuracy in many situations. We used it for instance in section 1.2.5 to establish Kubo formula✠. It is also applied at section 3.2.6 to a Hydrogen atom submitted to an oscillating electric field. Perturbative series are obviously not always sufficient to capture all the physics in the system under consideration[9].

✠ **Comment:** It is useful for this example to rewrite this expansion in terms of kets. We indicate here for convenience the analogue of equation (3.72),

$$|\phi(t)\rangle = e^{-i\frac{H^{(0)}(t-t_0)}{\hbar}}|\phi_0\rangle$$
$$- \frac{i}{\hbar}\int_0^{t-t_0}\mathrm{d}\tau_1 e^{-i\frac{H^{(0)}\tau_1}{\hbar}}W(t-\tau_1)|\phi(t-\tau_1)\rangle, \quad (3.74)$$

and equation (3.73)

$$|\phi(t)\rangle \simeq e^{-i\frac{H^{(0)}(t-t_0)}{\hbar}}|\phi_0\rangle$$
$$- \frac{i}{\hbar}\int_0^{t-t_0}\mathrm{d}\tau_1 e^{-i\frac{H^{(0)}\tau_1}{\hbar}}W(t-\tau_1)e^{-i\frac{H^{(0)}(t-\tau_1-t_0)}{\hbar}}|\phi_0\rangle, \quad (3.75)$$

where we used relation (3.59), p. 148. Note that these relations remain valid in the more general case where operator W is not diagonal in the space of positions. More generally, there are similar relations for operators that are not separable but take the form $\mathcal{O} = \mathcal{O}^{(0)} + \mathcal{P}$, where $\mathcal{P}$ is a small perturbation.

3.1.5 *Bloch equation*

In this section we return to the case of a quantum particle with a time-independent Hamiltonian H, confined inside a domain $\mathcal{D}$. Gibbs operator $e^{-H/(k_\mathrm{B}T)}$ is the fundamental object in statistical physics as it determines thermodynamic quantities at equilibrium. It turns out that this operator formally follows from evolution operator $e^{-itH/\hbar}$ through the substitution $t \to -i\beta\hbar$ with inverse temperature $\beta = 1/(k_\mathrm{B}T)$. Operator $e^{-\beta H}$ is therefore said to describe the evolution in imaginary time. Its matrix elements define the thermal propagator

$$\boxed{\rho(\mathbf{r}; \mathbf{r}'; \beta) = \langle \mathbf{r}|e^{-\beta H}|\mathbf{r}'\rangle.} \quad (3.76)$$

Here we establish the evolution equation of this thermal propagator with respect to variable β: it is similar to a diffusion equation, as one might expect following

---

[9]See for instance the case of perturbation series for the Green's function associated with the resolvent, illustrated by the example discussed in section 2.2.3.

considerations of the previous paragraph. We then show that $\rho(\mathbf{r};\mathbf{r}';\beta)$ identifies with a Green's function, and we list some essential properties.

◊ Thermal propagator and Green's functions ◊

Here we take β as variable and focus on the corresponding evolution of $\rho(\mathbf{r};\mathbf{r}';\beta)$. We use calculation rules from operatorial formalism summarised in Appendix D to get

$$\boxed{\left[\frac{\partial}{\partial\beta} - \frac{\hbar^2}{2m}\Delta_{\mathbf{r}} + V(\mathbf{r})\right]\rho(\mathbf{r};\mathbf{r}';\beta) = 0.} \tag{3.77}$$

This PDE is called Bloch equation and can be seen as an evolution equation for the thermal propagator with respect to pseudo-time $\beta > 0$. By construction $\rho(\mathbf{r};\mathbf{r}';\beta)$ satisfies the initial condition

$$\lim_{\beta\to 0^+}\rho(\mathbf{r};\mathbf{r}';\beta) = \delta(\mathbf{r}-\mathbf{r}'), \tag{3.78}$$

and homogeneous Dirichlet boundary conditions

$$\rho(\mathbf{r};\mathbf{r}';\beta) = 0 \; \forall \mathbf{r}\in\partial\mathcal{D}. \tag{3.79}$$

PDE (3.77) is of the same type as diffusion equation (3.26). For a free particle, we have the correspondence $\beta\leftrightarrow t$ and $\hbar^2/(2m)\leftrightarrow D$ between real quantities without further complications arising from the imaginary number i. Evolution of the free thermal propagator $\rho^{(0)}(\mathbf{r};\mathbf{r}';\beta)$ with β is thus identical to that of $G_{\mathrm{HD}}^+(\mathbf{r};\mathbf{r}';t)$ given by equation (3.34) with respect to t with homogeneous Dirichlet boundary conditions. In the presence of a non-zero potential $V(\mathbf{r})$, everything happens as if free diffusion was altered by local injection ($V(\mathbf{r}) < 0$) or absorption ($V(\mathbf{r}) > 0$) of particles. The consequences of this additional process on further evolution are discussed in the following paragraph.

A simple adaptation of the argument proves uniqueness of the solution of diffusion equation. First, one shows that PDE (3.77) with boundary conditions (3.78) and (3.79) uniquely defines the thermal propagator for any $\beta > 0$. Since initial condition (3.78) is the same as condition (3.17) on causal homogeneous Dirichlet Green's function of operator

$$\mathcal{O} = \frac{\partial}{\partial\beta} + H,$$

$\rho(\mathbf{r};\mathbf{r}';\beta)$ reduces to the Green's function.

Link with the resolvent associated with the Hamiltonian   Starting from the previous identification of $\rho(\mathbf{r};\mathbf{r}';\beta)$ and applying general results of section 3.1.2, we find that the Laplace transform with respect to β is equal to homogeneous Dirichlet Green's function associated with resolvent $[s + H]^{-1}$,

$$\widetilde{\rho}(\mathbf{r};\mathbf{r}';s) = \langle\mathbf{r}|\frac{1}{s+H}|\mathbf{r}'\rangle. \tag{3.80}$$

This expression can also be obtained by taking the matrix elements of operatorial identity

$$\int_0^\infty ds\, e^{-\beta s}\, e^{-\beta H} = \frac{1}{s+H}.$$

Note that the integral over β defining the Laplace transform is convergent only if $\mathrm{Re}\, s > -E_0$ where E_0 is the ground state energy of the Hamiltonian. It is assumed here that H is bounded from below, i.e. that $E_0 > -\infty$, which is often the case in practice. Expression (3.80) is valid by analytic continuation for any value of s, except at $s = -E_n$ where $\widetilde{\rho}(\mathbf{r}; \mathbf{r}'; s)$ has a singularity. This singularity of the resolvent matrix elements has already been mentioned in Chapter 2 through the general discussion in section 2.1.6 p. 88, and the example in section 2.2.3.

◇ Fundamental properties of the thermal propagator ◇

Let us apply formula (3.15) specific to the causal Green's function and evaluated here at β instead of $t - t'$. An elementary calculation gives $Z_n(\beta) = e^{-\beta E_n}$, or

$$\boxed{\rho(\mathbf{r}; \mathbf{r}'; \beta) = \sum_n e^{-\beta E_n} \psi_n(\mathbf{r}) \psi_n^*(\mathbf{r}').} \qquad (3.81)$$

This spectral representation is easily found by inserting the completeness relation on $\{\psi_n\}$ in definition (3.76) for the thermal propagator.

Since the functions ψ_n are real, $\rho(\mathbf{r}; \mathbf{r}'; \beta)$ is real and symmetric in the exchange of $\mathbf{r}$ and $\mathbf{r}'$. For small β, i.e. at high temperature, the behaviour of $\rho(\mathbf{r}; \mathbf{r}'; \beta)$ is given by the limit (3.78). At large β, i.e. at low temperature, the dominant contribution to spectral representation (3.81) comes from the ground state. Assuming that it is non-degenerate and that there is a finite gap $E_1 - E_0 > 0$ with the first excited state, we find

$$\rho(\mathbf{r}; \mathbf{r}'; \beta) \simeq e^{-\beta E_0} \psi_0(\mathbf{r}) \psi_0^*(\mathbf{r}') \qquad \text{when } \beta \to \infty. \qquad (3.82)$$

If $E_0 > 0$, $\rho(\mathbf{r}; \mathbf{r}'; \beta)$ vanishes exponentially fast. In terms of diffusion of classical particles, it means that escape through the boundary $\partial\mathcal{D}$ outweighs any other process. In particular, as $E_0 > 0$ in the free case where $V(\mathbf{r}) = 0$, we indeed recover the results of section 3.1.3. On the contrary, if the potential $V(\mathbf{r})$ is negative in part of domain $\mathcal{D}$, E_0 may be strictly negative and then $\rho(\mathbf{r}; \mathbf{r}'; \beta)$ increases exponentially fast at large β. So in terms of diffusion, particle injection in regions where $V(\mathbf{r}) < 0$ outweighs the escape at the boundaries!

It is possible to construct perturbative expansions for the density matrix, for $H = H^{(0)} + W$ where $W(\mathbf{r})$ is supposed to be a weak potential, in analogy with the case of the propagator in real time. We leave the reader to derive the expansion

$$\rho(\mathbf{r}; \mathbf{r}'; \beta) = \rho^{(0)}(\mathbf{r}; \mathbf{r}'; \beta)$$

$$- \int d\mathbf{r}_1 \int_0^\beta d\beta_1\, \rho^{(0)}(\mathbf{r}; \mathbf{r}_1; \beta - \beta_1)\, W(\mathbf{r}_1)\, \rho^{(0)}(\mathbf{r}_1; \mathbf{r}'; \beta_1) + \cdots \qquad (3.83)$$

which is the strict analog of series (3.66). Frequently the free Hamiltonian is chosen as a reference, i.e. $H^{(0)} = -\hbar^2/(2m)\Delta$, and the domain under consideration is the whole space. The essential ingredient is then the free thermal propagator of infinite system[10]

$$\rho^{(0)}(\mathbf{r}; \mathbf{r}'; \beta) = \left(\frac{m}{2\pi\beta\hbar^2}\right)^{d/2} \exp\left[-\frac{m(\mathbf{r} - \mathbf{r}')^2}{2\beta\hbar^2}\right], \tag{3.84}$$

obtained by simple substitutions in formula (3.41) of diffusion constant D by $\hbar^2/(2m)$ on the one hand, and $t - t'$ by β on the other hand. Perturbative expansions of the thermal propagator are of great interest, especially for the N-body problem✠.

✠ **Comment:**   Consider again a system of N interacting particles described by a time-independent Hamiltonian H_N. The N-body thermal propagator is the matrix element of evolution operator $e^{-\beta H_N}$. One then introduces the density matrix reduced to p particles by integrating on positions of $N - p$ particles. This leads to thermodynamic functions as well as correlation functions. They can be represented by perturbative expansions similar to expansion (3.83). Their construction is simplified by introducing the second quantisation formalism, which automatically takes into account the bosonic or fermionic statistics of the particles [21, 61]. Basic ingredients are fermionic/bosonic propagators, similar to the free thermal propagator (3.84) for a single particle.

3.1.6   *Wave equation*

The wave equation is the simplest equation describing propagation without attenuation in a homogeneous medium. It is ubiquitous in various fields ranging from electrodynamics to elasticity theory and fluid mechanics, including wave optics and acoustics for example. It also appears in the context of relativistic quantum theories, under the name of Klein-Gordon equation, in the description of spinless and massless particles. Its general form reads

$$\left[\frac{1}{c^2}\frac{\partial^2}{\partial t^2} - \Delta\right]\phi(\mathbf{r}, t) = \rho(\mathbf{r}, t), \tag{3.85}$$

where c is a given velocity, depending on the environment and on the nature of field ϕ under consideration. As we shall see later, c is the propagation speed of ϕ. For example, the propagation speed of the electromagnetic field in the vacuum is nothing but the famous speed of light.

---

[10]The free expression (3.84) plays a crucial role in the construction of a representation of $\rho(\mathbf{r}; \mathbf{r}'; \beta)$ in terms of path integrals, as discussed in Chapter 4. It also helps to determine so-called Wigner-Kirkwood expansions in powers of $\hbar^2$ of the diagonal part $\rho(\mathbf{r}; \mathbf{r}; \beta)$ in the vicinity of classical Boltzmann factor $e^{-\beta V(\mathbf{r})}$ [52].

Operator $\mathcal{O}$ is here

$$\mathcal{O} = \left[\frac{1}{c^2}\frac{\partial^2}{\partial t^2} - \Delta\right] = -\Box \tag{3.86}$$

where we used the notation

$$\Box = \Delta - \frac{1}{c^2}\frac{\partial^2}{\partial t^2}.$$

D'Alembert operator $\Box$ is also called d'Alembertian. The operator $\mathcal{O}$ again takes the additive structure (3.10) with

$$\mathcal{O}_t = \frac{1}{c^2}\frac{\partial^2}{\partial t^2} \tag{3.87}$$

and

$$\mathcal{O}_\mathbf{r} = -\Delta_\mathbf{r}. \tag{3.88}$$

Unlike operators involved in diffusion or Schrödinger equation, $\mathcal{O}_t$ is here of second order with respect to time. We shall later see that this difference has major consequences on the behaviours induced by the evolution.

As in previous sections, we begin with the study of boundary conditions and the transformation of PDE (3.85) into an integral equation involving causal Green's functions. For homogeneous boundary conditions, we determine spectral representations of causal Green's functions, using the general method described in section 3.1.2. Their structure is analysed and compared with that of the corresponding functions obtained for diffusion or Schrödinger equation. We then give the solution of the wave equation in terms of Green's functions and discuss consequences of this expression. We then compute the Green's function of the d'Alembertian for an infinite system. It leads in particular to the concept of travelling wave moving at finite speed c.

$$\Diamond \text{ **Boundary conditions** } \Diamond$$

Let us start with boundary conditions. Here operator $\mathcal{O}_t$ is of second order with respect to the time variable, i.e. $p = 2$ using the notation of p. 130. At the initial time t_0 we must therefore specify both $\phi(\mathbf{r}, t_0)$ and its time derivative $(\partial/\partial t)\phi(\mathbf{r}, t_0)$ at each point of the domain, namely

$$\phi(\mathbf{r}, t_0) = \phi_0(\mathbf{r}) \text{ and } \frac{\partial}{\partial t}\phi(\mathbf{r}, t_0) = \pi_0(\mathbf{r}) \ \forall \ \mathbf{r} \in \mathcal{D}. \tag{3.89}$$

We must add boundary conditions to these initial conditions. They can be readily determined in Laplace space as argued in section 3.1.2. The corresponding static operator $[P_2(s) + \mathcal{O}_\mathbf{r}]$ is

$$\frac{s^2}{c^2} - \Delta_\mathbf{r}, \tag{3.90}$$

and thus reduces simply to the Helmholtz operator with $m = s/c$. Then, according to the results established in Chapter 2, Dirichlet or Neumann boundary conditions are sufficient to ensure uniqueness of the solution of Helmholtz equation or in other words here, uniqueness of the field $\widetilde{\phi}(\mathbf{r}, s)$. The relevant conditions for the field $\phi(\mathbf{r}, t)$ are immediately obtained by inverse Laplace transform, leading to

$$\phi(\mathbf{r}, t) = D(\mathbf{r}, t) \; \forall \mathbf{r} \in \partial \mathcal{D} \text{ and } \forall t,$$

for Dirichlet boundary conditions, or to

$$\mathbf{n} \cdot \boldsymbol{\nabla} \phi(\mathbf{r}, t) = N(\mathbf{r}, t) \; \forall \mathbf{r} \in \partial \mathcal{D} \text{ and } \forall t,$$

for Neumann boundary conditions. Each of these boundary conditions $\mathrm{BC}(\phi | \partial \mathcal{D})$, combined with the initial conditions (3.89) determine uniquely the field $\phi(\mathbf{r}, t)$.

Note that above boundary conditions are identical to conditions (3.28) and (3.29) introduced for diffusion. Naturally, the boundary functions $D(\mathbf{r}, t)$ and $N(\mathbf{r}, t)$ must be compatible with initial conditions (3.89): There is no solution otherwise! Finally, as proposed in the exercise 3.1, p. 206, one can check that previous boundary conditions indeed guarantee uniqueness, without going in the Laplace space and drawing from the demonstration established for the Laplacian operator in section 2.1.3.

◊ Causal Green's functions and integral equation ◊

As for diffusion or Schrödinger equation, we will transform the original PDE satisfied by $\phi(\mathbf{r}, t)$ into an integral equation involving a causal Green's function G_{H}^{+}, also called delayed Green's function, solution of the PDE

$$\left[\frac{1}{c^2} \frac{\partial^2}{\partial t^2} - \Delta_{\mathbf{r}} \right] G_{\mathrm{H}}^{+}(\mathbf{r}; \mathbf{r}'; t, t') = \delta(\mathbf{r} - \mathbf{r}')\delta(t - t'), \tag{3.91}$$

with causality condition $G_{\mathrm{H}}^{+}(\mathbf{r}; \mathbf{r}'; t, t') = 0$ for $t < t'$. It would be possible *a priori* to obtain an integral equation for any causal Green's function, regardless of the defining boundary conditions. Here we restrict ourselves to homogeneous causal Green functions, which are easier and more transparent to use. Let us recall especially their time-invariance property established in section 3.1.1, i.e. $G_{\mathrm{H}}^{+}(\mathbf{r}; \mathbf{r}'; t, t') = G_{\mathrm{H}}^{+}(\mathbf{r}; \mathbf{r}'; t - t')$. In addition, their Laplace transform $\widetilde{G}_{\mathrm{H}}^{+}(\mathbf{r}; \mathbf{r}'; s)$ is a Green's function of the static Helmholtz operator (3.90) with the same homogeneous boundary conditions.

As for diffusion or Schrödinger equation, the sought equation is easily obtained by working in Laplace space. The starting point is still integral equation (2.27), p. 65, established in Chapter 2 for a field solution of Helmholtz PDE (2.42). Here we can apply this equation to $\widetilde{\phi}(\mathbf{r}, s)$. The only difference with diffusion now lies in the structure of the effective source $I_1(\mathbf{r}, s)$ taking into account the initial conditions,

$$I_1(\mathbf{r}, s) = -\frac{1}{c^2} \pi_0(\mathbf{r}) - \frac{s}{c^2} \phi_0(\mathbf{r}).$$

To go back to field $\phi(\mathbf{r}, t)$, inverse Laplace transformations of the various terms are identical to those performed in section 3.1.3. The inverse transform of initial condition term shows in particular

$$\mathcal{L}^{-1}[s\widetilde{G}_H^+(\mathbf{r}'; \mathbf{r}; s)] = \frac{\partial}{\partial\tau}G_H^+(\mathbf{r}'; \mathbf{r}; \tau),$$

where we used initial condition $G_H^+(\mathbf{r}; \mathbf{r}'; 0^+) = 0$, which is a consequence of the causality condition as shown in section 3.1.2 on p. 132. The final result reads

$$
\begin{aligned}
\phi(\mathbf{r}, t) = {} & \int_0^{t-t_0} d\tau \int_D d\mathbf{r}' \, \rho(\mathbf{r}', t-\tau) G_H^+(\mathbf{r}'; \mathbf{r}; \tau) \\[4pt]
& + \frac{1}{c^2} \int_D d\mathbf{r}' \, \pi_0(\mathbf{r}') \, G_H^+(\mathbf{r}'; \mathbf{r}; t-t_0) \\[4pt]
& + \frac{1}{c^2} \int_D d\mathbf{r}' \, \phi_0(\mathbf{r}') \, \frac{\partial}{\partial t} G_H^+(\mathbf{r}'; \mathbf{r}; t-t_0) \\[4pt]
& - \int_0^{t-t_0} d\tau \int_{\partial D} d\Sigma' \, \phi(\mathbf{r}', t-\tau) \, \mathbf{n}' \cdot \boldsymbol{\nabla}_{\mathbf{r}'} G_H^+(\mathbf{r}'; \mathbf{r}; \tau) \\[4pt]
& + \int_0^{t-t_0} d\tau \int_{\partial D} d\Sigma' \, G_H^+(\mathbf{r}'; \mathbf{r}; \tau) \, \mathbf{n}' \cdot \boldsymbol{\nabla}_{\mathbf{r}'} \phi(\mathbf{r}', t-\tau).
\end{aligned}
\tag{3.92}
$$

Source and surface terms in integral equation (3.92) have exactly the same structure as in equation (3.32) for diffusion. Therefore they can be interpreted along similar lines. The only difference between these two equations comes from the form of the propagation of the initial condition. Here the initial field ϕ_0 propagates via the partial time derivative $(\partial/\partial t)G_H^+$, while the initial partial derivative π_0 propagates via G_H^+ itself. Note that one can also prove equation (3.92) by exploiting the second Green's formula.

◊ Homogeneous causal Green's functions on a finite domain ◊

This paragraph is devoted to the study of causal Green's functions, defined by homogeneous Dirichlet boundary conditions

$$G_{HD}^+(\mathbf{r}; \mathbf{r}'; \tau) = 0 \text{ for } \mathbf{r} \in \partial D \text{ and all } \tau,$$

or homogeneous Neumann boundary conditions

$$\mathbf{n} \cdot \boldsymbol{\nabla}_{\mathbf{r}} G_{HN}^+(\mathbf{r}; \mathbf{r}'; \tau) = 0 \text{ for } \mathbf{r} \in \partial D \text{ and all } \tau.$$

We first determine their spectral representations, and we briefly review their fundamental properties.

Spectral representation Let us indicate here how to obtain the spectral representation of each homogeneous causal Green's function, following the general analysis presented in section 3.1.2. To this end we choose an orthonormal basis of eigenfunctions ψ_n of operator $\mathcal{O}_{\mathbf{r}} = -\Delta_{\mathbf{r}}$, and satisfying the same boundary conditions as the homogeneous Green's function under study. Hermiticity of Laplace operator in the space of functions satisfying homogeneous Dirichlet or Neumann boundary conditions guarantees that eigenvalues λ_n are real. As already noted for diffusion, these eigenvalues are all positive or zero, allowing to define $\sqrt{\lambda_n}$.

Each G_H^+ can be decomposed according to the generic spectral representation (3.15), where $Z_n(\tau)$ are here solutions of the ordinary differential equation

$$\frac{1}{c^2}\frac{d^2}{d\tau^2}Z_n(\tau) + \lambda_n Z_n(\tau) = 0$$

with the initial conditions $Z_n(0) = 0$ and $(dZ_n/d\tau)(0) = c^2$. An elementary calculation gives

$$Z_n(\tau) = c\frac{\sin\left(c\sqrt{\lambda_n}\tau\right)}{\sqrt{\lambda_n}}.$$

The application of spectral formula (3.15) yields

$$G_H^+(\mathbf{r};\mathbf{r}';t-t') = \theta(t-t')\sum_n c\frac{\sin\left(c\sqrt{\lambda_n}(t-t')\right)}{\sqrt{\lambda_n}}\,\psi_n(\mathbf{r})\psi_n^*(\mathbf{r}'). \qquad (3.93)$$

The determination of the spectrum of Laplace operator in an arbitrary domain remains a difficult problem, as already stressed repeatedly. Nevertheless, representation (3.93) is useful since it even allows to explicitly compute functions G_H^+ in some simple geometries. For example, spectral formula (3.93) suggests that it is possible to compute G_H^+ for a semi-infinite domain bounded by a plane wall by the method of images, as presented for example in section 3.2.2 on the Fraunhofer diffraction.

General properties   Causal homogeneous Green's functions G_H^+ satisfy the reciprocity relation $G_H^+(\mathbf{r};\mathbf{r}';t-t') = G_H^+(\mathbf{r}';\mathbf{r};t-t')$. This relation is an immediate consequence of formula (3.93), since eigenfunctions ψ_n of the Laplace operator are real-valued functions. It can also be established using the second Green's formula, as proposed in exercise 3.2. Note also that G_H^+ takes real values.

Functions G_{HD}^+ and G_{HN}^+ have the same short time behaviour determined by the initial conditions set out in section 3.1.2, p. 132, namely

$$G_H^+(\mathbf{r};\mathbf{r}';0^+) = 0 \quad\text{and}\quad \frac{\partial G_H^+}{\partial\tau}(\mathbf{r};\mathbf{r}';0^+) = c^2\delta(\mathbf{r}-\mathbf{r}'). \qquad (3.94)$$

The condition on the time derivative is obtained from formula (3.17) with $p = 2$ and $a_2 = 1/c^2$. Unlike in the case of diffusion, it is not $G_H^+(\mathbf{r};\mathbf{r}';\tau)$ itself but its time derivative which reduces to a Dirac distribution when $\tau \to 0^+$, up to a factor c^2.

The situation at large time is similar to the case of Schrödinger equation. The evolution of each function G_H^+ is complex because it involves sums of oscillating factors. As we shall later see, this follows from multiple reflections on the boundaries, leading to interferences very difficult to analyse in the domain. Note at this point that a significant difference between G_{HD}^+ and G_{HN}^+ appears. If eigenvalues λ_n are all strictly positive with homogeneous Dirichlet boundary conditions, the smallest eigenfunction λ_0 vanishes with homogeneous Neumann boundary conditions. We have seen in the case of diffusion that this peculiarity induces very different long time behaviours depending on the nature of boundary conditions. Herein the term $n = 0$ in the spectral representation of $G_{HN}^+(\mathbf{r}; \mathbf{r}'; \tau)$ reduces to $\frac{c^2\tau}{V}$. Unlike other terms of formula (3.93), the ground state contribution does not oscillate but diverges when $\tau \to +\infty$! We must therefore expect a singular behaviour of G_{HN}^+ at long times. This is related to the non-existence of a static homogeneous Neumann Green's function for Laplace operator as argued in the following paragraph.

$$\Diamond \ \textbf{Solution of the wave equation} \ \Diamond$$

In most physical situations, one has to consider Dirichlet or Neumann boundary conditions on $\phi(\mathbf{r}, t)$. In each case, the solution of the wave equation is explicitly written in terms of the adequate homogeneous causal Green's function. For Dirichlet boundary conditions, we study some particular forms of the resulting field, corresponding to standing waves on the one hand, and to adiabatic connection of a static source on the other hand.

Dirichlet boundary conditions Let us fix Dirichlet boundary conditions (3.28) on $\phi(\mathbf{r}, t)$. It is then suitable to insert the causal homogeneous Green's function G_{HD}^+ in equation (3.92). We then obtain, using again the symmetry property of $G_{HD}^+(\mathbf{r}; \mathbf{r}'; \tau)$ in $\mathbf{r}$ and $\mathbf{r}'$,

$$
\begin{aligned}
\phi(\mathbf{r}, t) = & \int_0^{t-t_0} d\tau \ \int_{\mathcal{D}} d\mathbf{r}' \ \rho(\mathbf{r}', t - \tau) G_{HD}^+(\mathbf{r}; \mathbf{r}'; \tau) \\[1mm]
& + \frac{1}{c^2} \int_{\mathcal{D}} d\mathbf{r}' \ \pi_0(\mathbf{r}') \ G_{HD}^+(\mathbf{r}; \mathbf{r}'; t - t_0) \\[1mm]
& + \frac{1}{c^2} \int_{\mathcal{D}} d\mathbf{r}' \ \phi_0(\mathbf{r}') \ \frac{\partial}{\partial t} G_{HD}^+(\mathbf{r}; \mathbf{r}'; t - t_0) \\[1mm]
& - \int_0^{t-t_0} d\tau \ \int_{\partial\mathcal{D}} d\Sigma' \ D(\mathbf{r}', t - \tau) \ \mathbf{n}' \cdot \boldsymbol{\nabla}_{\mathbf{r}'} G_{HD}^+(\mathbf{r}; \mathbf{r}'; \tau).
\end{aligned}
\tag{3.95}
$$

This representation of $\phi(\mathbf{r}, t)$ is explicit, up to the knowledge of G_{HD}^+, as it only involves given initial conditions ϕ_0 and π_0, boundary function $D(\mathbf{r}, t)$ and source $\rho(\mathbf{r}, t)$, which are given parameters for the problem under consideration.

As in the static case of the Laplace operator, boundary effects implicitly contribute to each term of expression (3.95) via the function G_{HD}^+, itself already incorporating surface contributions. We will later analyse these effects, including in particular reflections on the boundary $\partial\mathcal{D}$.

Eigenmodes and standing waves   Consider a situation without any source with an identically zero boundary function, i.e. $\rho(\mathbf{r}, t) = 0$ and $D(\mathbf{r}, t) = 0$ for $t \geq t_0$. Let us further assume that at the initial time t_0, the field and its time derivative are proportional to an eigenfunction ψ_l of the Laplace operator. In other words decompositions of functions ϕ_0 and π_0 on the basis of functions ψ_n have only a single component along ψ_l,

$$\phi_0(\mathbf{r}) = a\psi_l(\mathbf{r}) \qquad \text{and} \qquad \pi_0(\mathbf{r}) = b\psi_l(\mathbf{r}).$$

The source and surface terms disappear in expression (3.95) and only the terms of initial conditions evolution remain. In these terms, let us replace G_{HD}^+ by its spectral representation (3.93) and permute the integral over $\mathbf{r}$ and the sum over n. Using the orthonormality condition of eigenfunctions,

$$\int_{\mathcal{D}} d\mathbf{r}\ \psi_l(\mathbf{r})\psi_n^*(\mathbf{r}) = \delta_{ln},$$

we find

$$\phi(\mathbf{r}, t) = \left[c^2 a \cos\left(c\sqrt{\lambda_l}(t - t_0) \right) + \frac{cb}{\sqrt{\lambda_l}} \sin\left(c\sqrt{\lambda_l}(t - t_0) \right) \right] \psi_l(\mathbf{r}). \qquad (3.96)$$

Expression (3.96) describes a standing wave, with the usual factorisation of the temporal and spatial parts. There is an infinite number of standing waves, associated with the various eigenmodes l. The situation is then quite similar to the case of Schrödinger equation. Eigenmodes of frequencies $\omega_l = c\sqrt{\lambda_l}$ can only take discrete values just as in quantum mechanics.

Adiabatic connection of a static source   Now imagine that a static source $\rho_S(\mathbf{r})$ was connected adiabatically at $t_0 = -\infty$. In other words, we set $\rho(\mathbf{r}, t) = e^{\epsilon t}\rho_S(\mathbf{r})$ with $\epsilon > 0$ fixed. Assume that the boundary function is always identically zero, $D(\mathbf{r}, t) = 0$, and that initially there is no field present, i.e.

$$\phi_0(\mathbf{r}) = 0 \qquad \text{and} \qquad \pi_0(\mathbf{r}) = 0.$$

Only the volume term contributes in expression (3.95), leading to

$$\phi(\mathbf{r}, t) = e^{\epsilon t} \int_{\mathcal{D}} d\mathbf{r}'\ \rho_S(\mathbf{r}') \int_0^\infty d\tau\ e^{-\epsilon\tau}\ G_{\mathrm{HD}}^+(\mathbf{r}; \mathbf{r}'; \tau). \qquad (3.97)$$

Structure of formula (3.97) is in agreement with the general interpretation of causal homogeneous Green's functions as response functions. Here one finds the static susceptibility at zero frequency by taking the limit of the adiabatic connection $\epsilon \to 0^+$,

$$\lim_{\epsilon \to 0^+} \int_0^\infty d\tau\ e^{-\epsilon\tau}\ G_{\mathrm{HD}}^+(\mathbf{r}; \mathbf{r}'; \tau). \qquad (3.98)$$

The spectral representation of this susceptibility is obtained using formula (3.93) for G_{HD}^+. Integration over time is immediate for each mode after swapping the integral over τ and the sum over n. We then recognise the spectral representation established in Chapter 2 for the homogeneous Dirichlet Green's function for the Laplacian $G_{\mathrm{HD}}(\mathbf{r};\mathbf{r}')$. One thus gets the sum rule

$$\lim_{\epsilon\to 0^+}\int_0^\infty \mathrm{d}\tau\, e^{-\epsilon\tau}\, G_{\mathrm{HD}}^+(\mathbf{r};\mathbf{r}';\tau) = G_{\mathrm{HD}}(\mathbf{r};\mathbf{r}'), \tag{3.99}$$

and expression (3.97) gives back the static solution for Poisson equation with homogeneous Dirichlet conditions, in the limit $\epsilon\to 0^+$,

$$\phi_S(\mathbf{r}) = \int_{\mathcal{D}} \mathrm{d}\mathbf{r}'\, \rho_S(\mathbf{r}')\, G_{\mathrm{HD}}(\mathbf{r};\mathbf{r}').$$

Neumann boundary conditions   Let us now impose Neumann boundary conditions (3.29) on $\phi(\mathbf{r},t)$. Using homogeneous causal Green's function G_{HN}^+ in equation (3.92) and its symmetry property, we find

$$\begin{aligned}
\phi(\mathbf{r},t) = {} & \int_0^{t-t_0}\mathrm{d}\tau\,\int_{\mathcal{D}}\mathrm{d}\mathbf{r}'\,\rho(\mathbf{r}',t-\tau)G_{\mathrm{HN}}^+(\mathbf{r};\mathbf{r}';\tau) \\[4pt]
& +\tfrac{1}{c^2}\int_{\mathcal{D}}\mathrm{d}\mathbf{r}'\,\pi_0(\mathbf{r}')\,G_{\mathrm{HN}}^+(\mathbf{r};\mathbf{r}';t-t_0) \\[4pt]
& +\tfrac{1}{c^2}\int_{\mathcal{D}}\mathrm{d}\mathbf{r}'\,\phi_0(\mathbf{r}')\,\tfrac{\partial}{\partial t}G_{\mathrm{HN}}^+(\mathbf{r};\mathbf{r}';t-t_0) \\[4pt]
& +\int_0^{t-t_0}\mathrm{d}\tau\,\int_{\partial\mathcal{D}}\mathrm{d}\Sigma'\,G_{\mathrm{HN}}^+(\mathbf{r};\mathbf{r}';\tau)\,N(\mathbf{r}',t-\tau).
\end{aligned} \tag{3.100}$$

This representation explicitly provides the solution of the wave equation, as in the Dirichlet case, provided that G_{HN}^+ is known.

In the absence of sources, and under homogeneous boundary conditions, standing waves may exist inside the domain. They are again associated with eigenmodes, i.e. their spatial part is proportional to $\psi_l(\mathbf{r})$ and their temporal part oscillates at frequency $\omega_l = c\sqrt{\lambda_l}$. Note however that the set of possible discrete frequencies is different from the one associated with homogeneous Dirichlet boundary conditions. In addition there is a zero mode frequency associated with the null eigenvalue of the Laplacian, $\lambda_0 = 0$. In fact the field is not stationary, as its temporal part can grow linearly with time!

The study of the possible convergence of $\phi(\mathbf{r},t)$ towards a static solution of Poisson equation highlights specific features of homogeneous Neumann boundary conditions. For example, the equivalent of static susceptibility (3.98), introduced in the case of Dirichlet boundary conditions, diverges here because of the contribution of the zero-frequency fundamental mode! The appearance of this singularity is a manifestation of the non-existence of homogeneous Neumann Green's function for the Laplacian, as discussed in Chapter 2, p. 67.

◇ Causal Green's functions for an infinite system ◇

We continue and conclude this general study of Green's functions for d'Alembert operator by the case where the domain $\mathcal{D}$ is the whole space $\mathbb{R}^d$. We first compute the causal Green function G_∞^+ with homogeneous Dirichlet boundary conditions at infinity,

$$G_\infty^+(\mathbf{r};\mathbf{r}';t-t') \to 0 \quad \text{when } |\mathbf{r}| \to \infty.$$

Since both these boundary conditions and d'Alembert operator are invariant under any spatial translation, G_∞^+ depends spatially only on the difference $\mathbf{r} - \mathbf{r}'$. We restrict ourselves to the three-dimensional case, the calculations in dimension $d = 1$ and $d = 2$ being respectively considered in exercises 3.6 and 3.5.

Calculation using Laplace transform　In Laplace space, the Laplace transform $\widetilde{G}_\infty^+(\mathbf{r} - \mathbf{r}'; s)$ is the homogeneous Green's function of Helmholtz operator $\frac{s^2}{c^2} - \Delta_\mathbf{r}$, in space $\mathbb{R}^d$. This function has been determined in Chapter 2. In three dimensions, application of formula (2.48), p. 74, with $m = s/c$ for $s > 0$, leads to

$$\widetilde{G}_\infty^+(\mathbf{r} - \mathbf{r}'; s) = \frac{e^{-s|\mathbf{r}-\mathbf{r}'|/c}}{4\pi|\mathbf{r} - \mathbf{r}'|}.$$

Inverse Laplace transform of this expression is elementary, thanks to identity

$$\mathcal{L}^{-1}[e^{-as}](\tau) = \delta(\tau - a),$$

where a is a strictly positive constant. We then find

$$\boxed{G_\infty^+(\mathbf{r} - \mathbf{r}'; t - t') = \frac{c}{4\pi|\mathbf{r} - \mathbf{r}'|}\, \delta(c(t - t') - |\mathbf{r} - \mathbf{r}'|).} \tag{3.101}$$

Calculation by Fourier transform　Expression (3.101) can be found by another method, based on the introduction of the Fourier transform in the space-time[✠] of $3 + 1$ dimensions,

$$\widehat{G}_\infty^+(\mathbf{k}, z) = \int d\mathbf{r} \int_{-\infty}^{\infty} dt\; e^{-i\mathbf{k}\cdot(\mathbf{r}-\mathbf{r}')+iz(t-t')}\; G_\infty^+(\mathbf{r} - \mathbf{r}'; t - t').$$

As discussed on p. 136, it is crucial to introduce an imaginary part for the complex frequency z, i.e. $z = \omega + i\epsilon$ with $\epsilon > 0$, in order to ensure convergence of the integral over t when $t \to +\infty$. Divergence of the integrand is avoided by the causality condition, ensuring that $G_\infty^+(\mathbf{r}-\mathbf{r}'; t-t')$ vanishes for time $t < t'$.

> [✠] **Comment:**　This transformation is natural in electromagnetism because of the space-time structure defined by the theory of relativity. Note in this regard that d'Alembert operator is the most simple scalar operator remaining invariant under Lorentz transformations associated with changes of inertial reference frame. D'Alembert operator is then the analogue for Minkowsky space-times of Laplace operator for an Euclidean space.

Let us then compute the Fourier transform of each side of PDE

$$-\Box_{\mathbf{r},t} G_\infty^+(\mathbf{r} - \mathbf{r}'; t - t') = \delta(\mathbf{r} - \mathbf{r}')\delta(t - t').$$

We then get:

$$\left(-\frac{z^2}{c^2} + \mathbf{k}^2\right)\widehat{G}_\infty^+(\mathbf{k}, z) = 1.$$

Singularities at $\omega = \pm c|\mathbf{k}| - i\epsilon$ arise from the eigenmodes and the lack of dissipation. We also check that these singularities are in the lower half complex plane, in agreement with the results of Chapter 1. We then have at this point

$$\boxed{\widehat{G}_\infty^+(\mathbf{k}, \omega) = \lim_{\epsilon \to 0^+} \frac{1}{\mathbf{k}^2 - \frac{(\omega + i\epsilon)^2}{c^2}}.}$$

We should then take the inverse Fourier transform and compute

$$G_\infty^+(\mathbf{r} - \mathbf{r}', t - t') = \lim_{\epsilon \to 0^+} \frac{1}{(2\pi)^4} \int d\mathbf{k} \int_{-\infty}^{\infty} d\omega \frac{e^{i\mathbf{k}\cdot(\mathbf{r} - \mathbf{r}') - i\omega(t - t')}}{(\mathbf{k}^2 - (\omega + i\epsilon)^2/c^2)}. \tag{3.102}$$

The integral over ω is performed combining Jordan's lemma with the residue theorem. Figure 3.2 represents the contour used for $t - t' < 0$. In agreement with

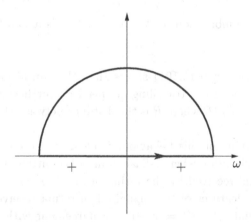

Fig. 3.2   Contour used to perform the integral over ω in the case where $t - t' < 0$. Poles of integrand $\omega = \pm|\mathbf{k}|c - i\epsilon$, represented by crosses, are outside the contour.

the causality condition, integral (3.102) vanishes since the singularities are in the lower complex half-plane. The contour used for $t - t' > 0$ is presented in Figure 3.3. Calculation of the residues at $\omega = \pm|\mathbf{k}|c - i\epsilon$ then gives, in the limit $\epsilon \to 0^+$,

$$G_\infty^+(\mathbf{r} - \mathbf{r}'; t - t') = \frac{c}{(2\pi)^3} \int d\mathbf{k} \frac{\sin(kc(t - t'))}{k} e^{i\mathbf{k}\cdot(\mathbf{r} - \mathbf{r}')}.$$

The integral over $\mathbf{k}$ is then easy to compute. We leave it to the reader to check that we then retrieve formula (3.101).

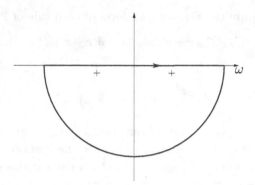

Fig. 3.3 Contour used to perform the integral over ω for $t - t' > 0$, with the same symbols as in Figure 3.2.

The anti-causal homogeneous Green's function (also called advanced Green's function) $G_\infty^-(\mathbf{r} - \mathbf{r}'; t - t')$ is defined by anti-causal condition $G_\infty^-(\mathbf{r} - \mathbf{r}'; t - t') = 0$ for $t > t'$, with the same homogeneous Dirichlet boundary conditions as G_∞^+. One can then use the Fourier transform method in space-time with $\epsilon < 0$ to get

$$G_\infty^-(\mathbf{r} - \mathbf{r}'; t - t') = \frac{c}{4\pi|\mathbf{r} - \mathbf{r}'|}\,\delta(c(t - t') + |\mathbf{r} - \mathbf{r}'|).$$

Note that any linear combination $\alpha G_\infty^+ + (1 - \alpha)G_\infty^-$ is also a Green's function of operator $-\Box$.

Interpretation Expression (3.101) shows that the causal Green's function reduces to a spherical wavefront, travelling at speed c. Furthermore, the magnitude of this front decreases as $1/R$ where R is the distance between the observation point and the source.

The physical meaning of this behaviour is clear in the context of electromagnetism. Imagine the situation where a flash of light is emitted at a time t' at a point $\mathbf{r}'$. What is the signal received by the retina of an observer at a point $\mathbf{r}$? Result (3.101) gives the two features of this signal. The retina receives no signal except at the time t such that $c(t - t') = |\mathbf{r} - \mathbf{r}'|$, in agreement with the propagation of light at speed c. Then the more distant from the emission point the observer is, the weaker the intensity of the received signal is, as everyone can experience every day... We might tend to give a wrong interpretation of the first characteristic by thinking that the signal emitted by a point source in space-time is itself necessarily located on a spherical surface moving at speed c. Such an interpretation is belied by an explicit calculation of G_∞^+ in lower dimension, like in $2 + 1$ and $1 + 1$ dimensions, as proposed to the reader's sagacity in exercises 3.5 and 3.6. The only correct statement concerns the forward wavefront that spreads at the velocity c, so that the retina receives no signal before time t such that $c(t - t') = |\mathbf{r} - \mathbf{r}'|$. But in general the wave described by G_∞^+ may spread backwards with respect to the direction of propagation and presents an extended tail.

The above considerations are of course valid in any context involving the d'Alembert operator. Let us mention the emission of sound waves in hydrodynamics, which is also an example discussed in the second part of this chapter, p. 183. Structural properties of G_∞^+ however take a particular importance in electromagnetism, in relation to the theory of relativity ✠.

✠ **Comment:**   In the space-time of dimension $3+1$, the surface where G_∞^+ is located is nothing but the light cone of equation $c(t-t') = |\mathbf{r}-\mathbf{r}'|$. It is quite natural to see an elementary electromagnetic entity that moves at the speed c underlying the behaviour of G_∞^+. This interpretation takes all its strength with the reformulation of Maxwell's equations in terms of a field theory, namely with classical electrodynamics [49]. Electromagnetic field is then seen as an entity by itself, albeit coupled to the charges present in matter but which can "live its life in all independence". In this construction, homogeneous causal Green's function G_∞^+ characterises the propagation of the electromagnetic entity from one point to another in a given time. This point of view paves the way to quantum electrodynamics: this basic electromagnetic entity then becomes a photon! In particular, eigenmodes highlighted on p. 167 give the photon dispersion relation $\omega^2 = c^2\mathbf{k}^2$. We can then introduce the Green's function describing the propagation of a photon in vacuum as, roughly speaking, the probability of the photon to go from a point $\mathbf{r}'$ at time t' to another point $\mathbf{r}$ at time t. Such Green's function is similar to the one introduced in section 3.1.4 to describe the propagation of a quantum particle according to Schrödinger equation. It involves by construction G_∞^+ [30, 51], which is therefore called upon to play a fundamental role in quantum electrodynamics too!

◊ Remarkable properties of propagation ◊

Consider a domain $\mathcal{D}$ in a three dimensional space. Inserting the causal Green's function G_∞^+ in equation (3.92) yields

$$\phi(\mathbf{r}, t) = \int_{\mathcal{D}} d\mathbf{r}' \; \frac{\rho(\mathbf{r}', t-|\mathbf{r}-\mathbf{r}'|/c)}{4\pi|\mathbf{r}-\mathbf{r}'|} \; \theta(t - t_0 - |\mathbf{r}-\mathbf{r}'|/c)$$

$$+ \frac{1}{c^2} \int_{\mathcal{D}} d\mathbf{r}' \; \pi_0(\mathbf{r}') \; \frac{\delta(t-t_0-|\mathbf{r}-\mathbf{r}'|/c)}{4\pi|\mathbf{r}-\mathbf{r}'|}$$

$$+ \frac{1}{c^2} \int_{\mathcal{D}} d\mathbf{r}' \; \phi_0(\mathbf{r}') \; \frac{\delta'(t-t_0-|\mathbf{r}-\mathbf{r}'|/c)}{4\pi|\mathbf{r}-\mathbf{r}'|} \tag{3.103}$$

$$- \int_0^{t-t_0} d\tau \int_{\partial\mathcal{D}} d\Sigma' \; \phi(\mathbf{r}', t-\tau) \; \mathbf{n}' \cdot \mathbf{\nabla}_{\mathbf{r}'} \frac{\delta(\tau-|\mathbf{r}-\mathbf{r}'|/c)}{4\pi|\mathbf{r}-\mathbf{r}'|}$$

$$+ \int_0^{t-t_0} d\tau \int_{\partial\mathcal{D}} d\Sigma' \; \frac{\delta(\tau-|\mathbf{r}-\mathbf{r}'|/c)}{4\pi|\mathbf{r}-\mathbf{r}'|} \; \mathbf{n}' \cdot \mathbf{\nabla}_{\mathbf{r}'} \phi(\mathbf{r}', t-\tau),$$

after replacing[11] $G_\infty^+(\mathbf{r}-\mathbf{r}'; \tau)$ by its expression (3.101) using the notation $\delta'(\xi) = (d/d\xi)\delta(\xi)$. This formula is particularly interesting because it leads to simple inter-

---

[11]Recall that $\delta(ct) = (1/c)\delta(t)$. Note also that in surface terms, integration over τ cannot be performed immediately, due to the action of the operator $\mathbf{n}' \cdot \mathbf{\nabla}_{\mathbf{r}'}$.

pretations while highlighting some remarkable properties of propagation governed by d'Alembert operator. It is extremely useful in practice too, as shown in the second part of this chapter.

Propagation of initial conditions The contribution

$$\frac{1}{c^2} \int_{\mathcal{D}} d\mathbf{r}' \, \pi_0(\mathbf{r}') \, \frac{\delta(t - t_0 - |\mathbf{r} - \mathbf{r}'|/c)}{4\pi|\mathbf{r} - \mathbf{r}'|}$$

$$+ \frac{1}{c^2} \int_{\mathcal{D}} d\mathbf{r}' \, \phi_0(\mathbf{r}') \, \frac{\delta'(t - t_0 - |\mathbf{r} - \mathbf{r}'|/c)}{4\pi|\mathbf{r} - \mathbf{r}'|} \quad (3.104)$$

describes the propagation of initial field ϕ_0 and its time derivative π_0, carried out by G_∞^+ and its time derivative $(\partial/\partial t)G_\infty^+$ which also has a localised front-like structure moving forward at velocity c. Points $\mathbf{r}'$ contributing to term (3.104) are thus located on the sphere of centre $\mathbf{r}$ and radius $R = c(t - t_0)$. At sufficiently long time, this sphere is completely outside of domain $\mathcal{D}$, as shown in Figure 3.4. So the initial

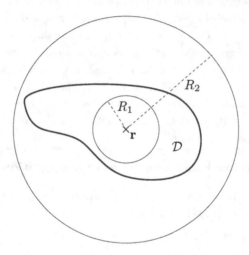

Fig. 3.4 Representation of the sphere centred in $\mathbf{r}$ and of radius $R = c(t - t_0)$ at two times $t_2 > t_1 > t_0$: at time $t = t_1$, the ball is inside domain $\mathcal{D}$ while it is outside $\mathcal{D}$ at time $t = t_2$, so that the term (3.104) vanishes.

condition term (3.104) vanishes identically inside the domain. Initial conditions are somehow forgotten, at least at an explicit level[12] in formula (3.103).

Finite propagation velocity and delay In formula (3.103) the term

$$\int_{\mathcal{D}} d\mathbf{r}' \, \frac{\rho(\mathbf{r}', t - |\mathbf{r} - \mathbf{r}'|/c)}{4\pi|\mathbf{r} - \mathbf{r}'|} \, \theta(t - t_0 - |\mathbf{r} - \mathbf{r}'|/c) \quad (3.105)$$

---

[12]Initial conditions are always implicitly present in the evolution of surface contributions to the total field (3.103).

describes the field resulting from the superposition at point $\mathbf{r}$ and time t of fields created by elementary sources distributed with density ρ. The elementary contribution of each source located in $\mathbf{r}'$ has been emitted at the delayed time $t_{\text{ret}} = t - |\mathbf{r} - \mathbf{r}'|/c$. It results from the propagation of the emitted signal at constant velocity c and its localised front-like structure. Density $\rho(\mathbf{r}', t')$ must be assessed at $t' = t_{\text{ret}}$.

Since it is only necessary to take into account the signals emitted between the instants t_0 and t, only delayed time $t_{\text{ret}} > t_0$ are acceptable, resulting in the presence of factor $\theta(t - t_0 - |\mathbf{r} - \mathbf{r}'|/c)$ in the integral (3.105). Thus, if the sources are located in a portion $\mathcal{D}_\rho$ of the entire domain as shown in Figure 3.5, then the superposition field (3.105) vanishes for all time t close to t_0 and for all sufficiently distant point, since no signal had enough time to reach such points. Note that there is no contradiction

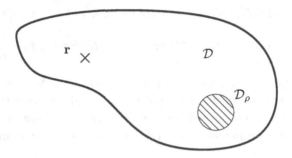

Fig. 3.5   In this figure, the sources are located within part $\mathcal{D}_\rho$ of domain $\mathcal{D}$. At the point $\mathbf{r}$, signal (3.105) from these sources is nonzero only for sufficiently large time t compared to t_0.

because the contributions of times $t < t_0$ are taken into account through the initial conditions. Note also that this property is an essential difference with diffusion. In the latter case indeed the superposition field has non-zero contributions for t infinitely close to t_0 even for points $\mathbf{r}$ far away from the source.

Multiple reflections in a finite domain   Surface terms

$$- \int_0^{t-t_0} d\tau \int_{\partial\mathcal{D}} d\Sigma' \; \phi(\mathbf{r}', t - \tau) \; \mathbf{n}' \cdot \boldsymbol{\nabla}_{\mathbf{r}'} \frac{\delta(\tau - |\mathbf{r} - \mathbf{r}'|/c)}{4\pi|\mathbf{r} - \mathbf{r}'|}$$

$$+ \int_0^{t-t_0} d\tau \int_{\partial\mathcal{D}} d\Sigma' \; \frac{\delta(\tau - |\mathbf{r} - \mathbf{r}'|/c)}{4\pi|\mathbf{r} - \mathbf{r}'|} \; \mathbf{n}' \cdot \boldsymbol{\nabla}_{\mathbf{r}'} \phi(\mathbf{r}', t - \tau), \quad (3.106)$$

represent the fields emitted by induced localised sources on the boundary $\partial\mathcal{D}$. Their surface distributions are similar to the superficial sources contributing to the electrostatic potential studied in Chapter 2. Emitted signals are again localised fronts propagating at the velocity c. Consequently, only delayed times $t_{\text{ret}} = t - |\mathbf{r} - \mathbf{r}'|/c$ contribute to temporal integrals, provided they are between t_0 and t of course.

Surface contributions can also be interpreted as resulting from reflection on the boundary $\partial\mathcal{D}$ of fields emitted by volume sources. This viewpoint is supported by the following argument. Consider the case of a flash source of density

$$\rho^{\text{pulse}}(\mathbf{r}, t) = \delta(\mathbf{r} - \mathbf{r}_0)\delta(t - t_0),$$

ignited at time t_0 while the field ϕ^{pulse} vanishes identically at earlier times. Assume that we have homogeneous Dirichlet boundary conditions. By definition, this field is nothing but the causal homogeneous Dirichlet Green's function,

$$\phi^{\text{pulse}}(\mathbf{r}, t) = G_{\text{HD}}^{+}(\mathbf{r}; \mathbf{r}_0; t - t_0).$$

Moreover, function $G_{\infty}^{+}(\mathbf{r} - \mathbf{r}_0; t - t_0)$ is also solution of the wave equation with ρ^{pulse}. For all times such that $c(t - t_0)$ is less than the minimum distance ℓ from $\mathbf{r}_0$ to the boundary, $G_{\infty}^{+}(\mathbf{r} - \mathbf{r}_0; t - t_0)$ satisfies also homogeneous Dirichlet condition on $\partial \mathcal{D}$. Since it also satisfies the same initial conditions as ϕ^{pulse}, uniqueness theorem implies that G_{∞}^{+} and G_{HD}^{+} coincide exactly for time t close enough to t_0,

$$G_{\text{HD}}^{+}(\mathbf{r}; \mathbf{r}_0; t - t_0) = G_{\infty}^{+}(\mathbf{r} - \mathbf{r}_0; t - t_0) \text{ for } t < t_0 + \frac{\ell}{c}. \qquad (3.107)$$

The identification (3.107) has a simple interpretation. As shown in Figure 3.6, the elementary field G_{∞}^{+} emitted by a pulse source did not have time to reach the boundary for $t < t_0 + \ell/c$ and there is no contribution from the reflected field! This interpretation is supported by the following analysis. Let us specify expression

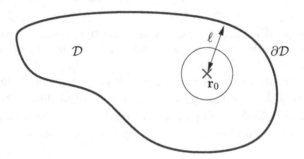

Fig. 3.6   In this figure, ℓ is the minimum distance between $\mathbf{r}_0$ and $\partial \mathcal{D}$. The circle represents a sphere of centre $\mathbf{r}_0$ and radius $R = c(t - t_0)$. For $t < t_0 + \ell/c$, this sphere has not yet reached the boundary $\partial \mathcal{D}$.

(3.103) to the present situation, where ϕ^{pulse} is the solution of the homogeneous wave equation for $t > t_0$, with initial conditions $\phi_0^{\text{pulse}} = 0$ and $\pi_0^{\text{pulse}} = c^2\delta(\mathbf{r} - \mathbf{r}_0)$, and with boundary condition $\phi^{\text{pulse}}(\mathbf{r}, t) = 0$ for $\mathbf{r} \in \partial \mathcal{D}$. Expression (3.103) then becomes

$$G_{\text{HD}}^{+}(\mathbf{r}; \mathbf{r}_0; t - t_0) = \frac{\delta(t - t_0 - |\mathbf{r} - \mathbf{r}_0|/c)}{4\pi|\mathbf{r} - \mathbf{r}_0|}$$
$$+ \int_0^{t - t_0} d\tau \int_{\partial \mathcal{D}} d\Sigma' \frac{\delta(\tau - |\mathbf{r} - \mathbf{r}'|/c)}{4\pi|\mathbf{r} - \mathbf{r}'|} \mathbf{n}' \cdot \boldsymbol{\nabla}_{\mathbf{r}'} G_{\text{HD}}^{+}(\mathbf{r}'; \mathbf{r}_0; t - t_0 - \tau). \qquad (3.108)$$

We get a perturbative expansion of the boundary effects by iterating this integral equation, where the p-th term describes the contribution coming from p reflections. For $t < t_0 + \ell/c$, each of these surface terms is zero, since $G_\infty^+(\mathbf{r}'; \mathbf{r}_0; t - t_0 - |\mathbf{r} - \mathbf{r}'|/c)$ vanishes on the boundary. This implies that G_{HD}^+ then reduces to G_∞^+, in agreement with the previous argument about uniqueness. In addition a surface term actually appears as soon as the front described by G_∞^+ reaches the boundary.

The above argument applies also to a pulse field with homogeneous Neumann boundary conditions. We then find $G_{\mathrm{HN}}^+ = G_\infty^+$ for $t < t_0 + \ell/c$. The surface term occurring in the analog of equation (3.108) only becomes active for later times $t > t_0 + \ell/c$. We ultimately bear in mind that surface contributions (3.106) include reflections on the boundary of the field emitted by volume sources. After a sufficiently long time, these reflections are obviously multiple.

Wave propagation Let us finally consider the case where boundaries are rejected to infinity so that the domain $\mathcal{D}$ is then the space $\mathbb{R}^3$. Let us also reject the initial conditions at $t_0 = -\infty$, and impose no initial field, $\phi_0(\mathbf{r}) = 0$ and $\pi_0(\mathbf{r}) = 0$ and that $\phi(\mathbf{r}, t)$ vanishes when $|\mathbf{r}| \to \infty$. Initial condition and surface terms then disappear in formula (3.103). We then recover the celebrated retarded potentials formula, namely

$$\phi(\mathbf{r}, t) = \int d\mathbf{r}' \, \frac{\rho(\mathbf{r}', t - |\mathbf{r} - \mathbf{r}'|/c)}{4\pi |\mathbf{r} - \mathbf{r}'|}. \tag{3.109}$$

Of the many possible applications of formula (3.109) let us mention here the determination of the field asymptotic form, far away from the sources, assuming the latter are localised in a finite portion of space. This is achieved by a quite standard calculation. This analysis appears for example in electromagnetism, in the proof of dipole radiation formulae. Let us consider a point $\mathbf{R}$ located far away from sources $r' \ll R$ for $\rho(\mathbf{r}', t') \neq 0$ and an observation point $\mathbf{r}$ close to $\mathbf{R}$, i.e. $|\mathbf{r} - \mathbf{R}| \ll R$. We then denote by x and x' the projections of $\mathbf{r} - \mathbf{R}$ and $\mathbf{r}'$ on the unit vector $\mathbf{n} = \mathbf{R}/|\mathbf{R}|$. The field then locally takes the form [13]

$$\phi(\mathbf{r}, t) \sim f(x - ct), \tag{3.110}$$

with

$$f(\xi) = \frac{1}{4\pi R} \int d\mathbf{r}' \, \rho\left(\mathbf{r}', \frac{x' - R - \xi}{c}\right). \tag{3.111}$$

So the field has locally the structure of a progressive plane wave, moving without deformation at speed c in the direction of $\mathbf{n}$.

---

[13] The form (3.110) is obtained for fixed $\mathbf{R}$. Function $f(\xi)$, defined by the formula (3.111) of course depends also on R and $\mathbf{n}$.

The progressive plane wave structure is naturally found in one-dimensional spatial geometries. A field $\phi(x,t)$ solution of the homogeneous wave equation therefore necessarily takes the well-known form

$$\phi(x,t) = f_1(x - ct) + f_2(x + ct).$$

It thus reduces to the superposition of two plane waves travelling in opposite directions. This superposition can result in the formation of a standing wave under suitable boundary conditions.

Conclusion   The wave equation exhibits a reversible nature owing to its invariance under the change $t \to -t$. It can thus lead in particular to standing waves just like Schrödinger equation. That said, symmetry of the wave equation with respect to spatial and temporal coordinates confers specific properties, such as wave propagation at finite speed, which have no equivalent in non-relativistic quantum mechanics.

3.2   Applications and examples

3.2.1   *Diffusion on a segment*

◊ Presentation ◊

Let us consider the one-dimensional diffusion problem confined to the segment $0 \le x \le L$. Through this example we will show how to implement the systematic construction of causal Green's functions from their spectral representations. Furthermore the obtained behaviours illustrate all general properties established in section 3.1.3.

We are interested here only in causal Green's functions $G_{\mathrm{HD}}^{+}(x;x';\tau)$ and $G_{\mathrm{HN}}^{+}(x;x';\tau)$ corresponding respectively to Dirichlet and Neumann homogeneous boundary conditions, i.e.

$$G_{\mathrm{HD}}^{+}(x;x';\tau) = 0 \qquad \text{for} \quad x = 0, L \qquad \forall \tau > 0,$$

$$\frac{\partial}{\partial x}G_{\mathrm{HN}}^{+}(x;x';\tau) = 0 \qquad \text{for} \quad x = 0, L \qquad \forall \tau > 0.$$

Remember that these boundary conditions respectively correspond to the presence of absorbing or reflecting walls. Furthermore the knowledge of these Green's functions enables one to determine explicitly the general solution of the diffusion equation, possibly in the presence of an arbitrary source, with Dirichlet and Neumann boundary conditions involving functions with an arbitrary time dependence. Among the many possible applications let us highlight, for example, diffusion of density with injection of particles at a point of the segment. One can also mention heat diffusion in a system confined between two parallel plates, in response to imposed variations of the plate temperatures. In this case the system symmetry leads to one-dimensional Green's function.

◇ **Analysis and solution** ◇

We showed on pp. 139-140 that the spectral representation of homogeneous Green's functions reads

$$G_H^+(x; x'; \tau) = \theta(\tau) \sum_n e^{-D\lambda_n \tau} \psi_n(x)\psi_n^*(x'),$$

where $\{\psi_n\}$ form an orthonormal basis of eigenfunctions of $-(\partial^2/\partial x^2)$, satisfying the same BC as G_H^+. We first determine the explicit form of these representations for G_{HD}^+ and G_{HN}^+, before studying the large time behaviour.

Explicit spectral formulas One easily computes in this particular case the eigenfunctions ψ_n and the corresponding eigenvalues λ_n, and eventually gets for homogeneous Dirichlet BC on the one hand

$$\psi_n(x) = \sqrt{\frac{2}{L}} \sin\left(\frac{n\pi x}{L}\right), \quad \lambda_n = \frac{n^2\pi^2}{L^2}, \quad n \in \mathbb{N}^*,$$

and for homogeneous Neumann BC on the other hand

$$\psi_0(x) = \sqrt{\frac{1}{L}}, \quad \lambda_0 = 0 \quad \text{and} \quad \psi_n(x) = \sqrt{\frac{2}{L}} \cos\left(\frac{n\pi x}{L}\right), \quad \lambda_n = \frac{n^2\pi^2}{L^2}, \quad n \in \mathbb{N}^*.$$

A simple resummation then provides

$$G_{HD}^+(x; x'; \tau) = \frac{\theta(\tau)}{2L} \left[\vartheta_3\left(\frac{\pi(x - x')}{2L}, e^{-\frac{\tau}{\tau_0}}\right) \right.$$

$$\left. -\vartheta_3\left(\frac{\pi(x + x')}{2L}, e^{-\frac{\tau}{\tau_0}}\right) \right], \quad (3.112)$$

and

$$G_{HN}^+(x; x'; \tau) = \frac{\theta(\tau)}{2L} \left[\vartheta_3\left(\frac{\pi(x - x')}{2L}, e^{-\frac{\tau}{\tau_0}}\right) \right.$$

$$\left. +\vartheta_3\left(\frac{\pi(x + x')}{2L}, e^{-\frac{\tau}{\tau_0}}\right) \right], \quad (3.113)$$

where we have introduced the characteristic time for diffusion $\tau_0 = L^2/(\pi^2 D)$, and the Theta function ϑ_3 defined by

$$\vartheta_3(u, q) = 1 + 2 \sum_{n=1}^{\infty} q^{n^2} \cos(2nu) .$$

Note that Laplace transforms of functions (3.112) and (3.113) are homogeneous Green's function of Helmholtz operator on the segment $[0, L]$, as shown in the general discussion, p. 141. These Green's functions were calculated in the segment $[-(L/2), (L/2)]$ p. 77. The reader is invited to exploit (2.52) and (2.53) for interesting sum formulas expressing series of trigonometric functions in terms of hyperbolic functions.

Long-time behaviours   Consider a time τ large compared to characteristic time τ_0, so that $q = e^{-\frac{\tau}{\tau_0}} \ll 1$. Using the asymptotic behaviour of the Theta function,

$$\lim_{q \to 0} \vartheta_3(u, q) = 1 + 2q \cos(2u) + O(q^4),$$

we get

$$G_{\mathrm{HD}}^+(x; x'; \tau) \simeq \frac{1}{L} \left[\cos \left(\frac{\pi(x - x')}{L} \right) - \cos \left(\frac{\pi(x + x')}{L} \right) \right] e^{-\frac{\tau}{\tau_0}} \qquad (3.114)$$

and

$$G_{\mathrm{HN}}^+(x; x'; \tau) - \frac{1}{L} \simeq \frac{1}{L} \left[\cos \left(\frac{\pi(x - x')}{L} \right) + \cos \left(\frac{\pi(x + x')}{L} \right) \right] e^{-\frac{\tau}{\tau_0}}, \qquad (3.115)$$

when $\tau \to \infty$. Asymptotic behaviours (3.114) and (3.115) show that $G_{\mathrm{HD}}^+(x; x'; \tau)$ vanishes exponentially fast, while $G_{\mathrm{HN}}^+(x; x'; \tau)$ converges just as quickly to $1/L$. We recover here the general properties established in section 3.1.3, respectively associated with absorbing and reflective natures of the edges of domain $[0, L]$. Time τ_0 can be interpreted as the time at which the system has forgotten the initial conditions. It also controls the exponential relaxation process towards the final steady state.

◊ Interpretation ◊

These Green's functions may represent the evolution of various physical quantities. In the following, we see them as probability densities of a particle performing Brownian motion. Their evolutions are compared and interpreted in this context.

Functions $G_{\mathrm{HD}}^+(x; x'; \tau)$ and $G_{\mathrm{HN}}^+(x; x'; \tau)$ are plotted on Figure 3.7 as functions of x, and for increasing times τ. The numerical calculation has been made with $L = 1$ and $x' = 0.5$, so that the starting point of the Brownian particle is in the middle of the segment. Functions G_{HD}^+ vanish at the segment edges, while its tangents become horizontal. Then, for the shortest time, these two homogeneous Green's functions are tightened around the particle starting point, and there is little difference between them. This is easily understood since boundary effects have little influence at short times. On the contrary G_{HD}^+ decreases and eventually vanishes completely when τ increases, while G_{HN}^+ becomes uniform and equal to 1.

Let us continue our observations by taking a starting point closer to an edge, for example $x' = 0.8$ again with $L = 1$. Corresponding Green's functions are plotted on Figure 3.8 for the same times τ as before. As expected, the presence of walls has a much faster effect. First, unlike the previous case, there is a notable difference between Green's functions G_{HD}^+ and G_{HN}^+ for the shortest time considered. Then, for an absorbing wall, the probability of presence of a particle decays faster than in the previous case. Conversely, for a reflecting wall, the particle remains on average longer in the vicinity of the wall, and therefore the probability of presence becomes a bit less rapidly homogeneous.

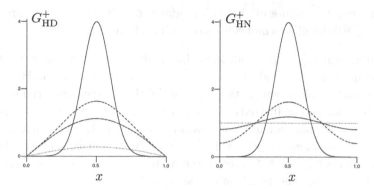

Fig. 3.7 Plots of functions $G_{HD}^{+}(x; x'; \tau)$ and $G_{HN}^{+}(x; x'; \tau)$ given by equations (3.112) and (3.113) as a function of x, for $L = 1$ and $x' = 0.5$, and for four increasing values of time τ.

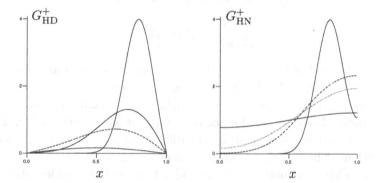

Fig. 3.8 Plot of functions $G_{HD}^{+}(x; x'; \tau)$ and $G_{HN}^{+}(x; x'; \tau)$ given by equations (3.112) and (3.113) as a function of x, with $L = 1$ and $x = 0.8$ and for the same times τ as on Figure 3.7. Axis scales are the same as those of Figure 3.7.

It is clear that this method also applies to diffusion problems in arbitrary dimension d, and in the hyper-rectangle $0 \le x_i \le L_i$, $i = 1, \ldots, d$. Such a generalisation is also possible for mixed boundary conditions, where some walls are reflective while others are absorbing. The final result can always be expressed in terms of the function ϑ.

3.2.2 *Fraunhofer diffraction*

◊ **Presentation** ◊

Fraunhofer diffraction is a fundamental application of Green's functions of d'Alembert operator to the theory of wave optics. Within the framework of a simple model, we will establish a mathematical formulation of Huygens-Fresnel principle, depending upon the nature of chosen boundary conditions. We then proceed to a discussion of Kirchhoff's formulas for small diffraction angles.

A fundamental approach of diffraction by an obstacle would require to start from Maxwell's equations since light is an electromagnetic wave. One might introduce a modelisation of the electromagnetic properties of the obstacle material by means of dielectric constant or conductivity. Such a program is in fact already tremendously complex. For example, boundary conditions on electric $\mathbf{E}$ and magnetic $\mathbf{B}$ fields are matching conditions instead of Dirichlet or Neumann conditions, due to the presence of waves transmitted through the material. Sharp corners of the obstacle may also induce point effects difficult to account for.

We will limit ourselves here to a much more modest approach! We first get rid of the tensorial nature of the electromagnetic field by adopting the so-called scalar model of light. The latter is then characterised by a scalar field ϕ called light vibration, which satisfies the wave equation in vacuum

$$[\frac{1}{c^2}\frac{\partial^2}{\partial t^2} - \Delta]\phi(\mathbf{r}, t) = 0, \tag{3.116}$$

where c is the speed of light. The ϕ field may of course be assimilated to any of the components of $\mathbf{E}$ or $\mathbf{B}$. Within the context of this minimalist description, it is reasonable to postulate homogeneous Dirichlet or Neumann boundary conditions on the obstacle. Their interest for realistic applications is certainly quite limited, but these assumptions allow to study the diffraction process without uncontrolled approximations. The resulting models combine simplicity with a rich physical content, so they play an important role in understanding the complex phenomena at play in realistic situations.

We consider the standard setting shown in Figure 3.9. A plate, supposed to be infinite, is placed at $z = 0$. This plate has a hole of arbitrary shape, and characterised by a length a. The coordinate origin is placed at a given point of this hole. A monochromatic light vibration reaches the plate from the $z < 0$ side. Under the so-called Fraunhofer conditions, we assume that this incident vibration is produced by a very distant source. In the vicinity of the hole, it reduces to a plane wave,

$$\phi(\mathbf{r}, t) = A_0 \, e^{i\mathbf{k}\cdot\mathbf{r} - i\omega t},$$

where $\mathbf{k}$ is the wave vector, $\omega = ck$ the frequency and A_0 a constant amplitude. We consider a normal incidence with respect to the plate, i.e. $\mathbf{k}$ is along the z-axis. We want to evaluate the vibration diffracted at a point located at a great distance compared to a in the $z > 0$ half-space. In practice, the corresponding light intensity measurement is performed using an optical device consisting of a photomultiplier or a photodiode located in the focal plane of a thin lens.

◊ Analysis and solution ◊

Here domain $\mathcal{D}$ corresponds to the $z > 0$ half-space. There is no source in this domain $\mathcal{D}$, and $\phi(\mathbf{r}, t)$ satisfies wave equation (3.116). Suppose that a source located in the other half-space $z < 0$ was switched on at t_0. Initial conditions (3.89), p. 159 then reduce in $\mathcal{D}$ to

$$\phi_0(\mathbf{r}) = 0 \quad \text{and} \quad \pi_0(\mathbf{r}) = 0 . \tag{3.117}$$

Domain boundary $\partial\mathcal{D}$ consists of the plane $z = 0$ and a fictitious surface S_∞ at infinity, defined for example as the large radius limit of a hemisphere centred at the origin. Given the initial conditions and the position of the source, the vibration $\phi(\mathbf{r}, t)$ and its spatial and temporal derivatives must vanish on S_∞. Integral equation (3.92) for $\phi(\mathbf{r}, t)$ involves any causal homogeneous Dirichlet or Neumann Green's functions G_{H}^+ of d'Alembert operator and takes the form

$$\phi(\mathbf{r}, t) = \int_0^\infty d\tau \int_{z'=0} d\Sigma' \, G_{\mathrm{H}}^+(\mathbf{r}; \mathbf{r}'; \tau) \, \mathbf{n}' \cdot \boldsymbol{\nabla}_{\mathbf{r}'} \phi(\mathbf{r}', t - \tau)$$

$$- \int_0^\infty d\tau \int_{z'=0} d\Sigma' \, \phi(\mathbf{r}', t - \tau) \, \mathbf{n}' \cdot \boldsymbol{\nabla}_{\mathbf{r}'} G_{\mathrm{H}}^+(\mathbf{r}; \mathbf{r}'; \tau), \tag{3.118}$$

obtained by sending t_0 to $-\infty$. On the $z = 0$ plane, we must distinguish part S_{p} corresponding to the plate, and part S_{h} corresponding to hole. The vibration will not be *a priori* imposed on S_{h} since it is a free surface. On S_{p} on the other hand, this vibration or its normal gradient will be determined by homogeneous Dirichlet or Neumann boundary conditions.

It is convenient to work with the best suited Green's function once the boundary conditions on S_{p} are chosen. It is clearly a Dirichlet or Neumann Green's function,

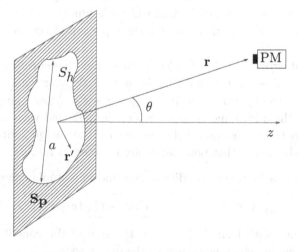

Fig. 3.9 Representation of the physical situation studied in this example: a plate S_{p} presents a hole S_{h}. The axes origin is chosen at a point on this hole. A photomultiplier (PM) measures the signal far away from the plate.

with homogeneous conditions over the whole boundary[14] $\partial \mathcal{D}$, i.e. on S_∞ and the plane $z = 0$. We first determine the Green's functions and we then deduce an expression for $\phi(\mathbf{r}, t)$ in terms of an integral over the surface hole S_h.

Dirichlet and Neumann Green's functions   Let us start with homogeneous Dirichlet conditions. In the static case of Laplace operator, we have already determined the Green's function for such a configuration in section 2.2.1 p. 92: it gives the electrostatic potential created by a point charge near a conducting plane. In the present situation, Green's function G_{HD}^+ may also be obtained by the method of images, with the result:

$$\boxed{G_{HD}^+(\mathbf{r}; \mathbf{r}'; \tau) = G_\infty^+(\mathbf{r} - \mathbf{r}'; \tau) - G_\infty^+(\mathbf{r} - \mathbf{r}'_{im}; \tau),} \tag{3.119}$$

where G_∞^+ is the Green's function of d'Alembert operator for the infinite system (3.101) and $\mathbf{r}'_{im}$ is symmetric of $\mathbf{r}'$ with respect to the plane $z = 0$. Let us emphasise again that the boundary conditions defining G_{HD}^+ are taken on the whole plane $z = 0$.

There are at least two ways to understand why the method of the images seen in the static case in Chapter 2 applies to the present situation. The first one is based on spectral representation (3.93) obtained on p. 162,

$$G_{HD}^+(\mathbf{r}; \mathbf{r}'; \tau) = \theta(\tau) \sum_n c \frac{\sin\left(c\sqrt{\lambda_n}\tau\right)}{\sqrt{\lambda_n}} \, \psi_n(\mathbf{r})\psi_n^*(\mathbf{r}').$$

Indeed, in this expression, $\psi_n(\mathbf{r})$ are eigenfunctions of Laplace operator with homogeneous Dirichlet BC. The argument developed on p. 92 therefore applies here as well. The second method consists in noticing that the Laplace transform over τ of $G_{HD}^+(\mathbf{r}; \mathbf{r}'; \tau)$ is the homogeneous Dirichlet Green's function of Helmholtz operator (see p. 159), for which the method of images presented on section 2.2.1 is also applicable.

Note also that expression (3.119) clearly illustrates the reflection mechanisms on the boundary described in section 3.1.6. For sufficiently short times τ, $G_{HD}^+(\mathbf{r}; \mathbf{r}'; \tau)$ effectively reduces to $G_\infty^+(\mathbf{r}; \mathbf{r}'; \tau)$ as $|\mathbf{r} - \mathbf{r}'_{im}| > |\mathbf{r} - \mathbf{r}'|$. After a single reflection on the plane $z = 0$, the additional contribution of the reflected wave appears in G_{HD}^+, corresponding to the elementary field created by the image source. No further reflection occurs because other boundaries are rejected at infinity.

For homogeneous Neumann conditions, we find in a similar manner

$$\boxed{G_{HN}^+(\mathbf{r}; \mathbf{r}'; \tau) = G_\infty^+(\mathbf{r} - \mathbf{r}'; \tau) + G_\infty^+(\mathbf{r} - \mathbf{r}'_{im}; \tau).} \tag{3.120}$$

The only difference with formula (3.119) is the sign of the contribution of the reflected wave or image, which now adds to the direct wave.

---

[14]Defining another Green's function with different conditions on S_p and S_h does not present any particular interest, given the difficulty to determine such function.

Light vibration with Dirichlet boundary conditions Suppose that $\phi(\mathbf{r}, t)$ satisfies to homogeneous Dirichlet boundary conditions $\phi(\mathbf{r}, t) = 0$ on S_p. Given the form of these boundary conditions as well as linearity of the wave equation, we can say that the time dependence of $\phi(\mathbf{r}, t)$ at a given point takes the form

$$\phi(\mathbf{r}, t) = A_D(\mathbf{r})e^{-i\omega t},$$

where $A_D(\mathbf{r})$ is a complex amplitude. We then obtain, using G_{HD}^+ in[15] equation (3.118),

$$A_D(\mathbf{r}) = \int_0^\infty d\tau \int_{S_\mathrm{h}} d\Sigma'\, A_D(\mathbf{r}')e^{i\omega\tau}\frac{\partial}{\partial z'}G_{\mathrm{HD}}^+(\mathbf{r}; \mathbf{r}'; \tau). \tag{3.121}$$

The amplitude of the light vibration at any point of $\mathcal{D}$ is therefore fully determined by its value over the hole S_h. The latter is not itself easily accessible as it would require full resolution of the problem in the whole space.

Integral expression (3.121) can be simplified as follows by exploiting the properties of G_∞^+. Starting from the formula (3.119), we easily find

$$\frac{\partial}{\partial z'}G_{\mathrm{HD}}^+(\mathbf{r}; \mathbf{r}'; \tau) = -2\frac{\partial}{\partial z}G_\infty^+(\mathbf{r} - \mathbf{r}'; \tau) \quad \text{at} \quad z' = 0.$$

This allows us to write integral (3.121) as

$$A_D(\mathbf{r}) = -2\frac{\partial}{\partial z}\int_0^\infty d\tau \int_{S_\mathrm{h}} d\Sigma'\, A_D(\mathbf{r}')\, e^{i\omega\tau}\, \frac{c}{4\pi|\mathbf{r} - \mathbf{r}'|}\, \delta\left[c\tau - |\mathbf{r} - \mathbf{r}'|\right],$$

where we used formula (3.101), p. 166, for G_∞^+. Integration over τ is immediate and gives the exact result

$$\boxed{A_D(\mathbf{r}) = -\frac{1}{2\pi}\int_{S_\mathrm{h}} d\Sigma'\, A_D(\mathbf{r}')\frac{\partial}{\partial z}\left[\frac{e^{ik|\mathbf{r} - \mathbf{r}'|}}{|\mathbf{r} - \mathbf{r}'|}\right].} \tag{3.122}$$

Light vibration with Neumann conditions Homogeneous Neumann boundary conditions stipulate that $\partial\phi/\partial z = 0$ on S_p. As in Dirichlet case, light vibration reads

$$\phi(\mathbf{r}, t) = A_N(\mathbf{r})e^{-i\omega t}.$$

As proposed in exercise 3.4, the reader can follow the same approach as above to show that amplitude $A_N(\mathbf{r})$ at any point of $\mathcal{D}$ is simply given by the exact formula

$$\boxed{A_N(\mathbf{r}) = -\frac{1}{2\pi}\int_{S_\mathrm{h}} d\Sigma'\, \frac{\partial A_N}{\partial z'}(\mathbf{r}')\frac{e^{ik|\mathbf{r} - \mathbf{r}'|}}{|\mathbf{r} - \mathbf{r}'|}.} \tag{3.123}$$

◊ **Interpretation** ◊

The content of formulas (3.122) and (3.123) is discussed in relation to Huygens-Fresnel principle. We then deduce the respective interference patterns at infinity. We finally conclude with a brief comment on other approaches.

---

[15]Let us recall that the vector $\mathbf{n}'$ is oriented toward the outside of $\mathcal{D}$ and therefore towards $z < 0$.

Huygens-Fresnel principle   This principle states that each point of a wave surface behaves as a secondary source emitting a spherical wave. It played a key role historically, and it is still often used today in wave optics as a starting point for diffraction calculations, as it proves sufficient to accurately describe many practical situations. It however appears like a recipe from a more fundamental point of view based on wave equation, even if it is very tasteful one!

In the case of Neumann conditions on the plate, formula (3.123) seems at first sight to be a mathematical translation of Huygens-Fresnel principle. Indeed, each point of the hole S_h emits a spherical wave of amplitude $e^{ik|\mathbf{r}-\mathbf{r}'|}/|\mathbf{r} - \mathbf{r}'|$. However, these points do not necessarily belong to a wave surface, because the incident plane wave is certainly distorted inside the hole. For Dirichlet conditions over S_p, formula (3.122) shows even more severe deviations from the predictions of Huygens-Fresnel principle, since the wave emitted by each point of hole S_h is no longer isotropic. But we shall see below that these distortions become negligible at infinity in directions close to the Oz axis along which the incident wave propagates.

Interference pattern at infinity   Let us now work at much larger distances from the hole than its characteristic size a. Then in each of integrals (3.122) and (3.123), we can replace the envelope factor $1/|\mathbf{r} - \mathbf{r}'|$ by $1/r$ for all points $\mathbf{r}'$ belonging to S_h and use asymptotic expansion

$$|\mathbf{r} - \mathbf{r}'| = r - \frac{\mathbf{r} \cdot \mathbf{r}'}{r} + O(r'/r) \tag{3.124}$$

in the phase factor $e^{ik|\mathbf{r}-\mathbf{r}'|}$. By noting θ the angle of $\mathbf{r}$ with Oz axis (see Figure 3.9) and $\mathbf{k}_\theta$ the wave number along the unit vector $\mathbf{r}/r$ with modulus k, we finally find

$$\boxed{A_D(\mathbf{r}) \simeq -\frac{ik\cos\theta\ e^{ikr}}{2\pi r} \int_{S_h} d\Sigma'\ A_D(\mathbf{r}')\ e^{-i\mathbf{k}_\theta \cdot \mathbf{r}'}} \tag{3.125}$$

and

$$\boxed{A_N(\mathbf{r}) \simeq -\frac{e^{ikr}}{2\pi r} \int_{S_h} d\Sigma'\ \frac{\partial A_N}{\partial z'}(\mathbf{r}')\ e^{-i\mathbf{k}_\theta \cdot \mathbf{r}'}} \tag{3.126}$$

when $r \to \infty$.

The interference pattern is obtained by computing the light intensity I proportional to $|A_D(\mathbf{r})|^2$ or $|A_N(\mathbf{r})|^2$. The main difference between these two expressions comes from the anisotropic factor $\cos^2\theta$ appearing in I_D. The nature of the boundary conditions is therefore less important for small diffraction angles θ.

Comparison with other approaches   It is not unusual to find in the literature applications of Huygens-Fresnel principle, where one imposes both Dirichlet *and* Neumann boundary conditions on the light vibration. Note that this choice is usually mathematically inconsistent, because in general there is no solution of the wave equation satisfying these two conditions simultaneously. In fact, it is not necessary to invoke these boundary conditions in order to derive the so-called Kirchhoff approximation. In the case studied here, it follows from the steps presented below.

Suppose that the hole size a is large compared with the wavelength $\lambda = 2\pi/k$ of the incident wave. It is then legitimate to assume that $\phi(\mathbf{r}, t)$ is extremely small in the immediate vicinity of plate S_p in domain $\mathcal{D}$. Diffraction should indeed be weaker in directions forming a large angle with the direction Oz of the incident wave. This is indeed observed experimentally, provided that the plate is opaque, and regardless of the precise form of boundary conditions. At the same level of approximation, it is also legitimate to assume that the wave surface is undisturbed near the hole except at a distance of order λ from its contour. In integral equation (3.92), p. 161 for $\phi(\mathbf{r}, t)$ with G_∞^+, this amounts to keeping only the contribution from S_h, replacing $\phi(\mathbf{r}', t)$ by its incident form $A_0\, e^{i\mathbf{k}\cdot\mathbf{r}' - i\omega t}$ on S_h. One then gets Kirchhoff formula for the diffracted amplitude in directions close to Oz:

$$A_K(\mathbf{r}) \simeq -\frac{ikA_0\, e^{ikr}}{2\pi r} \int_{S_h} d\Sigma'\, e^{-i\mathbf{k}_\theta \cdot \mathbf{r}'}.$$

This formula can also be established starting from expressions (3.125) and (3.126) for θ small: one only has to replace $A_D(\mathbf{r}')$ and $A_N(\mathbf{r}')$ by $A_0\, e^{i\mathbf{k}\cdot\mathbf{r}'}$ in the vicinity of S_h. This approximation is consistent with the assumption of small deformation of the incident wavefront in such a neighbourhood.

3.2.3 *Emission of sound waves*

◊ **Presentation** ◊

Consider a body oscillating in a perfect fluid. These oscillations may involve both its time-dependent volume V_c and the position of its centre of mass. In both cases, which may actually be simultaneously present, we assume that the oscillation amplitude is of the order of the body characteristic size ℓ. These oscillations induce local variations of the fluid density which will naturally propagate. The body thus emits sound waves, which we will characterise at large distances.

Except in the immediate vicinity of the body, we assume that the fluid density variation $\delta\rho$ remains small compared with the homogeneous density of the fluid at rest ρ_0, $\delta\rho \ll \rho_0$. We also assume that local transformations are adiabatic and reversible. This allows us to relate the pressure variation δP to $\delta\rho$ via $\delta P = c^2 \delta\rho$ and c is the speed of sound given by $c = [(\partial P/\partial\rho)_S]^{1/2}$, where the partial derivative is taken at constant entropy. It is then possible to show from the Euler equation and the mass conservation equation that the induced flow derives from a potential. In other words there is a potential $\phi(\mathbf{r}, t)$ such that the velocity field $\mathbf{u}(\mathbf{r}, t)$ in the fluid reads $\mathbf{u}(\mathbf{r}, t) = \boldsymbol{\nabla}\phi(\mathbf{r}, t)$. Moreover ϕ satisfies the PDE

$$\left[\frac{1}{c^2}\frac{\partial^2}{\partial t^2} - \Delta\right]\phi(\mathbf{r}, t) = 0. \tag{3.127}$$

Equation (3.127) describing sound propagation is a wave equation of type (3.85) without source. It is valid at any point of the fluid which is not reached by the oscillating body. Note that inside the region where the body surface is moving, fluid density varies sharply from 0 to ρ_0 so that the approximation $\delta\rho \ll \rho_0$ is not valid. In addition, Neumann boundary conditions must be written on the moving body surface which makes them more difficult to handle. Similarly to the study of Fraunhofer diffraction, it is far more convenient to work in a domain outside a fixed closed surface Σ_f, which contains the body at any time. Causal Green's function G_∞^+ of d'Alembert operator then allows to determine the behaviour of the quantities of interest at large distances.

We consider the regime $\ell \ll cT$, where T is the characteristic time scale of variations of the body shape and/or position, which is achieved for a sufficiently small object or slow enough oscillations. Noting $\lambda = cT$ the characteristic wavelength of emitted sound waves, we have $\ell \ll \lambda$. Note that this condition implies that the velocity field induced in the fluid, which is at most of order ℓ/T, is very small compared to the speed of sound c. The corresponding Mach number of order u/c is very small compared to 1, and condition $\delta\rho \ll \rho_0$ is achieved away from the immediate vicinity of the body.

◇ Analysis and solution ◇

Let us introduce the domain $\mathcal{D}$ between a closed surface Σ_f and a fictitious surface Σ_∞ at infinity. Assume that the body starts moving at time t_0, at which the fluid was at rest. The velocity potential and all its time derivatives are thus identically zero at any point of $\mathcal{D}$ and time t_0. At a later time t, we assume that transmitted waves did not have time to reach the boundaries of the system under consideration, so that the velocity potential and all its derivatives are identically zero on Σ_∞ (see Figure 3.10). We first show that $\phi(\mathbf{r}, t)$ is simply expressed in terms of $\phi(\mathbf{r}', t)$ and $\boldsymbol{\nabla}\phi(\mathbf{r}', t)$ with $\mathbf{r} \in \Sigma_f$. We then infer its asymptotic form at large distance.

Expression of potential as a surface integral Velocity potential ϕ satisfies the wave equation without source (3.127) throughout domain $\mathcal{D}$. Thus it satisfies an integral equation of the form (3.92), p. 161 with any causal Green's function of d'Alembert operator. It is more convenient here to use Green's function G_∞^+ for an infinite system, rather than Dirichlet or Neumann Green's function in $\mathcal{D}$. Indeed, since Σ_f is a free surface, values of $\phi(\mathbf{r}, t)$ or $\boldsymbol{\nabla}\phi(\mathbf{r}, t)$ are not imposed a *priori*. Moreover these functions remain to be determined.

We therefore apply formula (3.103), p. 169. Here, no source is present in $\mathcal{D}$, while initial conditions (3.89), p. 159 are $\phi_0(\mathbf{r}') = 0$ and $\pi_0(\mathbf{r}') = 0$. Since the

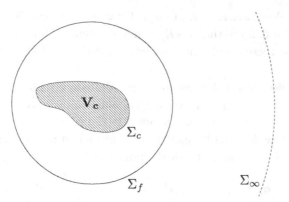

Fig. 3.10 Body of volume V_c moving in a fluid. Σ_c is the body surface, Σ_f is a fixed surface containing the body at any time during its motion, and Σ_∞ is a fictitious wall at infinity.

surface Σ_∞ does not provide any contribution, only the contribution coming from Σ_f remains, that is to say

$$\phi(\mathbf{r}, t) = \int_{\Sigma_f} d\Sigma' \mathbf{n}' \cdot \boldsymbol{\nabla}_{\mathbf{r}'} \phi(\mathbf{r}', t - |\mathbf{r} - \mathbf{r}'|/c) \frac{1}{4\pi |\mathbf{r} - \mathbf{r}'|}$$

$$+ \boldsymbol{\nabla}_{\mathbf{r}} \cdot \left[\int_{\Sigma_f} d\Sigma' \mathbf{n}' \phi(\mathbf{r}', t - |\mathbf{r} - \mathbf{r}'|/c) \frac{1}{4\pi |\mathbf{r} - \mathbf{r}'|} \right]. \quad (3.128)$$

In order to establish this expression, we used $\boldsymbol{\nabla}_{\mathbf{r}'} G_\infty^+(\mathbf{r} - \mathbf{r}', t - t') = -\boldsymbol{\nabla}_{\mathbf{r}} G_\infty^+(\mathbf{r} - \mathbf{r}', t - t')$. We then considered a time t large enough so that the sound wave reaches the point $\mathbf{r}$, i.e. $t > t_0 + |\mathbf{r} - \mathbf{r}'|/c$ for all $\mathbf{r}'$ in Σ_f. The first term is a monopolar contribution decreasing at large distances as $1/r$. The second term includes a dipolar part in $1/r^2$, as well as a monopolar part resulting from the action of operator $\boldsymbol{\nabla}_{\mathbf{r}}$ on $\phi(\mathbf{r}', t - |\mathbf{r} - \mathbf{r}'|/c)$ inside the integral. As in the study of Fraunhofer diffraction in section 3.2.2, we see that each point of Σ_f behaves as a secondary source emitting monopolar and dipolar spherical waves which propagate at velocity c.

Large distance expansion Surface Σ_f is completely arbitrary. We choose it fairly close to the moving body, so that $|\mathbf{r}'|$ on Σ_f is of order ℓ. Let us then introduce $R = |\mathbf{r}|$ with $R \gg \ell$. We can use the same type of asymptotic expansion as in p. 182 for Fraunhofer diffraction, i.e.

$$\boldsymbol{\nabla}_{\mathbf{r}} \left[\frac{\phi(\mathbf{r}', t - |\mathbf{r} - \mathbf{r}'|/c)}{|\mathbf{r} - \mathbf{r}'|} \right] = -\frac{\mathbf{n}}{Rc} \frac{\partial \phi}{\partial t}(\mathbf{r}', t - R/c + \mathbf{n} \cdot \mathbf{r}'/c) + O\left(\frac{1}{R^2}\right),$$

with the unit vector $\mathbf{n} = \mathbf{r}/r$. The field $\phi(\mathbf{r}, t)$ then becomes, up to terms of order $O(1/R^2)$,

$$\phi(\mathbf{r}, t) = \frac{1}{4\pi R} \oint_{\Sigma_f} d\Sigma' \mathbf{n}' \cdot \boldsymbol{\nabla}_{\mathbf{r}'} \phi(\mathbf{r}', t - R/c + \mathbf{n} \cdot \mathbf{r}'/c)$$

$$- \frac{\mathbf{n}}{4\pi c R} \cdot \oint_{\Sigma_f} d\Sigma' \mathbf{n}' \frac{\partial \phi}{\partial t}(\mathbf{r}', t - R/c + \mathbf{n} \cdot \mathbf{r}'/c). \quad (3.129)$$

Field ϕ therefore decreases as $1/R$ at large distances, with coefficients depending on its value over Σ_f at delayed time $t - R/c$, in agreement with propagation at velocity c. Expressions for these coefficients are determined in the following paragraph.

Amplitude of dominant terms at large distance The time $(\mathbf{n} \cdot \mathbf{r}'/c)$ is of order ℓ/c. Under the assumption $\ell \ll \lambda$, it is small compared to T, the characteristic variation time scale of ϕ. Functions involved in the two surface integrals appearing in asymptotic expression (3.129) can thus be developed in the vicinity of the same delayed time $t - R/c$, common to all the points of Σ_f,

$$\nabla_{\mathbf{r}'}\phi(\mathbf{r}', t - R/c + \mathbf{n} \cdot \mathbf{r}'/c) = \nabla_{\mathbf{r}'}\phi(\mathbf{r}', t - R/c) + \frac{\mathbf{n} \cdot \mathbf{r}'}{c} \frac{\partial}{\partial t} [\nabla_{\mathbf{r}'}\phi(\mathbf{r}', t - R/c)] + \cdots$$

$$\frac{\partial \phi}{\partial t}(\mathbf{r}', t - R/c + \mathbf{n} \cdot \mathbf{r}'/c) = \frac{\partial \phi}{\partial t}(\mathbf{r}', t - R/c) + \cdots.$$

It provides an expansion of each corresponding surface integral in powers of ℓ/λ. The zeroth-order term reduces to the flux of the velocity field through Σ_f evaluated at time $t - R/c$. However the fluid behaves as an incompressible fluid in the region between the body surface Σ_c and Σ_f thanks to the condition $\ell \ll \lambda$, so that

$$\oint_{\Sigma_f} d\Sigma' \mathbf{n}' \cdot \mathbf{u}(\mathbf{r}', t - R/c) = -\dot{V}_c(t - R/c) + \cdots.$$

A simple order of magnitude estimate of $\delta\rho/\rho_0$ near the body[16] shows that the neglected terms are at least of order $(\ell/\lambda)^2$. We finally find, by replacing the surface integrals in the asymptotic expression (3.129) by their expansions in powers of ℓ/λ,

$$\phi(\mathbf{r}, t) = -\frac{1}{4\pi R}[\dot{V}_c(t - R/c) + \dot{\mathbf{A}}(t - R/c) \cdot \mathbf{n}/c + O(\frac{\ell^2}{\lambda^2})] + O(\frac{1}{R^2}), \qquad (3.130)$$

with

$$\mathbf{A}(t - R/c) = -\oint_{\Sigma_f} d\Sigma' [\mathbf{n}' \cdot \mathbf{u}(\mathbf{r}', t - R/c)] \mathbf{r}' + \oint_{\Sigma_f} d\Sigma' \phi(\mathbf{r}', t - R/c)\mathbf{n}'. \quad (3.131)$$

Note that the term involving $\dot{\mathbf{A}}(t - R/c)/c$ is smaller than term $\dot{V}_c(t - R/c)$ by a factor of order ℓ/λ.

◊ Interpretation ◊

Using Green's function G_∞^+ for the infinite system allowed us to determine the asymptotic behaviour of the velocity potential at large distances, which decays as $1/R$. If body oscillations induce volume changes caused by contractions or dilations, as shown in Figure 3.11 for example, then $\phi(\mathbf{r}, t)$ is isotropic when $R \to \infty$, with

---

[16]Since $\mathbf{u}$ varies on a time scale T, while the pressure δP varies over a spatial scale ℓ in the body vicinity, Euler equation provides a relation between the orders of magnitude of $\mathbf{u}$ and δP. Furthermore, using the adiabatic relation $\delta P = c^2 \delta\rho$ and the estimate $|\mathbf{u}| \propto \ell/T$ in the vicinity of the body, we eventually find that $\delta\rho/\rho_0$ is of order ℓ^2/λ^2.

an amplitude $\dot{V_c}$. Discarding terms of order ℓ/λ, the corresponding radial velocity field is isotropic:

$$\mathbf{u}(\mathbf{r}, t) \sim \frac{\ddot{V_c}(t - R/c)}{4\pi c R} \mathbf{n} \quad \text{when} \quad R \to \infty,$$

discarding terms of order (l/λ) smaller.

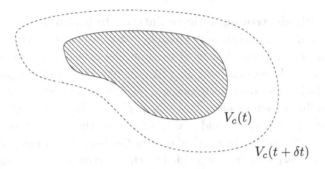

Fig. 3.11   Body expansion or contraction.

If there is no change in volume, i.e. $\dot{V_c} = 0$, the leading contribution comes from the anisotropic term $\dot{\mathbf{A}}$. This situation appears for a rigid body oscillating around its equilibrium position, as shown in Figure 3.12. The corresponding radial velocity

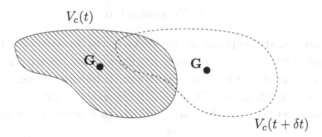

Fig. 3.12   Rigid body oscillating around its equilibrium position.

field is anisotropic,

$$\mathbf{u}(\mathbf{r}, t) \sim \frac{\mathbf{n} \cdot \ddot{\mathbf{A}}(t - R/c)}{4\pi c^2 R} \mathbf{n} \quad \text{when} \quad R \to \infty.$$

Its amplitude vanishes in the direction perpendicular to $\ddot{\mathbf{A}}$ and is maximal in the direction of $\ddot{\mathbf{A}}$. Note that a complete knowledge of $\mathbf{A}$ is tricky, because it would require the computation of the velocity field near the body. If we assume an instantaneous response of the velocity field in the fluid, we necessarily have $|\mathbf{A}| \propto V_c |\mathbf{V}|$ where $\mathbf{V}$ is the body speed. For example, if the body is a spherical ball performing oscillations along the Ox axis, the above assumption leads us to use, in the vicinity of the ball, the velocity field (2.94) computed in section 2.2.2 on p. 97 for a uniform motion in an incompressible fluid! We then find

$$\mathbf{A} = \frac{3}{2} V_c \mathbf{V}.$$

Notice that $\dot{\mathbf{A}}/c$ is indeed of order ℓ/λ as estimated after equation (3.131).

Analogy with electromagnetic radiation In light of the previous results, the sound wave radiation is then only possible if the body or parts of it are accelerated. So the situation is quite similar to the one observed in electromagnetism, where only accelerated charges radiate. Just like the electric $\mathbf{E}$ and magnetic $\mathbf{B}$ fields, the radiated velocity field only decreases as $1/R$. Thus the energy flow through a large sphere of radius R remains constant when $R \to \infty$, in agreement with the absence of dissipation in a perfect fluid. Note of course that $\mathbf{u}$ does not have the same symmetries as $\mathbf{E}$ and $\mathbf{B}$. In particular, a radial isotropic velocity field in $1/R$ may appear, while this geometry is excluded by charge conservation in electromagnetism. In absence of body volume variation, $\mathbf{u}$ is relatively similar to dipolar radiation fields $\mathbf{E}$ and $\mathbf{B}$. However, the emission is then maximal in the direction of body acceleration in hydrodynamics, while it is orthogonal to the charges acceleration in electromagnetism.

3.2.4   *Wavefront in supersonic regime*

◊ **Presentation** ◊

It is well known that a supersonic aircraft gives rise to a shock wave resulting in a violent sound. The study of supersonic motion in a fluid is a complex problem that involves various aspects of fluid mechanics and turbulence. Here, we do not consider these aspects in all their complexity. We start instead with a very simple model, presenting however the required characteristics to describe the wavefront structure.

Let us assume that a point-like object moves in a homogeneous fluid at a constant velocity v larger than sound speed c in this environment. The huge perturbation caused within the fluid in the vicinity of the object propagates, inducing pressure and density changes. These distortions are decaying far enough from the object. They can then be reasonably assimilated to a low amplitude sound wave whose linearised evolution is governed by a wave equation without source. Strictly speaking, the precise shape of this wave should be determined by a matching procedure involving the behaviour of quantities of interest in the vicinity of the object, similarly to the method used in the preceding example for a body emits a sound wave in the subsonic regime. Such a procedure is extremely difficult to implement here. We limit ourselves to a much more rudimentary approach, consisting in introducing a phenomenological source term in the propagation equation. We admit that the field

$\phi(\mathbf{r}, t)$, representing for example the pressure, satisfies the PDE

$$\left[\frac{1}{c^2}\frac{\partial^2}{\partial t^2} - \Delta\right]\phi(\mathbf{r}, t) = \rho_p(\mathbf{r}, t), \qquad (3.132)$$

where source $\rho_p(\mathbf{r}, t)$ models the complex process emission induced by the point object. At this level of modelisation, it is natural to identify $\rho_p(\mathbf{r}, t)$ with the density associated to the point-like object, i.e.

$$\rho_p(\mathbf{r}, t) = \rho_0 \, \delta(x - vt)\delta(y)\delta(z), \qquad (3.133)$$

where the Ox axis coincides with the rectilinear trajectory of the particle. In heuristic form (3.133) the arbitrary constant ρ_0 has the dimension of a mass times a frequency squared. In the following we study the shape of the wave front generated by the supersonic object via wave equation (3.132).

◊ Analysis and solution ◊

Domain $\mathcal{D}$ under consideration is the whole space, and boundary $\partial\mathcal{D}$ is thus sent at infinity. Pressure $\phi(\mathbf{r}, t)$ is assumed to vanish on $\partial\mathcal{D}$, so that the fluid is at rest infinitely far away from the object. It is also at rest at the initial time t_0 with a homogeneous pressure and an identically zero velocity field. Corresponding initial conditions (3.89) are $\phi_0(\mathbf{r}) = 0$ and $\pi_0(\mathbf{r}) = 0$ in the whole domain. General expression for the solution of the wave equation (3.92), p. 161, reduces here to the source term. Using the causal Green's function G_∞^+ for an infinite system given by (3.101), and sending t_0 to $-\infty$, we find

$$\phi(\mathbf{r}, t) = \int_{-\infty}^{t} dt' \int d\mathbf{r}' \, \frac{c}{4\pi|\mathbf{r} - \mathbf{r}'|}\delta\left[c(t - t') - |\mathbf{r} - \mathbf{r}'|\right] \rho_p(\mathbf{r}', t'). \qquad (3.134)$$

Integration of the expression (3.134) on $\mathbf{r}$ is straightforward, thanks to the Dirac distributions in ρ_p, leading to

$$\phi(\mathbf{r}, t) = \frac{\rho_0 c}{4\pi} \int_{-\infty}^{t} dt' \, \frac{\delta[c(t - t') - \sqrt{y^2 + z^2 + (x - vt')^2}]}{\sqrt{y^2 + z^2 + (x - vt')^2}}. \qquad (3.135)$$

To carry out the final integration over t', we must first determine the support of the Dirac distribution. We then deduce an explicit form for $\phi(\mathbf{r}, t)$.

Support of Dirac distribution It is useful to make the change of variable $t'' = t' - x/v$ in integral (3.135), to get

$$\phi(\mathbf{r}, t) = \frac{\rho_0 c}{4\pi} \int_{-\infty}^{t-x/v} dt'' \, \frac{\delta[c(t - x/v - t'') - \sqrt{d^2 + v^2 t''^2}]}{\sqrt{d^2 + v^2 t''^2}}, \qquad (3.136)$$

where $d = \sqrt{y^2 + z^2}$ is the distance from point $\mathbf{r}$ to Ox axis. The support of Dirac distribution in integral (3.136) is the set of times t'' such that

$$c(t - x/v - t'') = \sqrt{d^2 + v^2 t''^2}. \qquad (3.137)$$

After squaring this equation, we find a quadratic equation for t'' which has real solutions if and only if its discriminant is positive, or

$$(x - vt)^2 \geq (v^2/c^2 - 1)d^2. \tag{3.138}$$

The two solutions of equation (3.137) are then

$$t''_\pm = \frac{1}{v(v^2/c^2 - 1)} \left[x - vt \pm (v/c)\sqrt{(x - vt)^2 - (v^2/c^2 - 1)d^2} \right]. \tag{3.139}$$

Support on the real axis of the Dirac distribution of interest thus reduces to two points $t''_\pm$ if condition (3.138) is satisfied, and the empty set otherwise.

Contributions from the two roots $t''_\pm$ For roots $t''_\pm$ to contribute to integral (3.136), they must naturally be within the integration interval, i.e. they must verify $t''_\pm \leq t - x/v$. If $t - x/v < 0$, $x - vt$ is positive and one gets $t''_\pm > 0$ by simple inspection of the formula (3.139). In this case roots are outside the integration interval, and integral (3.135) is identically zero. In addition, this integral is obviously also zero if condition (3.138) is not satisfied. One finally obtains

$$\phi(\mathbf{r}, t) = 0 \qquad \text{for} \quad x > vt - d\sqrt{v^2/c^2 - 1}.$$

If $x - vt$ is negative, it is clear that expressions (3.139) are negative, so that both roots $t''_\pm$ are in the integration interval, provided that existence condition (3.138) is also satisfied. This is achieved for

$$x < vt - d\sqrt{v^2/c^2 - 1}.$$

Then one just has to apply identity (A.2), p. 265. Let us point out that defining

$$f(t'') = c(t - x/v - t'') - \sqrt{d^2 + v^2 t''^2} \qquad \text{and} \qquad g(t'') = \frac{1}{\sqrt{d^2 + v^2 t''^2}}$$

an intermediate result is

$$\frac{g(t''_\pm)}{|f'(t''_\pm)|} = \frac{1}{\left| c\sqrt{d^2 + v^2 t''^2_\pm} + v^2 t''_\pm \right|} = \frac{1}{c\sqrt{(vt - x)^2 - d^2(v^2/c^2 - 1)}}.$$

We thus obtain the final expression for $\phi(\mathbf{r}, t)$,

$$\phi(\mathbf{r}, t) = \frac{\rho_0}{2\pi} \frac{\theta(vt - x - d\sqrt{v^2/c^2 - 1})}{\sqrt{(vt - x)^2 - d^2(v^2/c^2 - 1)}}. \tag{3.140}$$

◊ **Interpretation** ◊

Formula (3.140) for $\phi(\mathbf{r}, t)$ highlights a remarkable structure of the wave front, which we shall now interpret while discussing its physical relevance. We conclude with an analogy with Cherenkov effect.

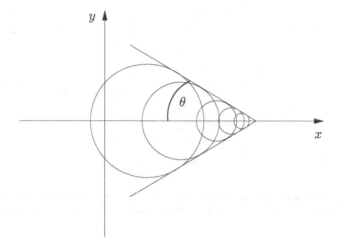

Fig. 3.13   Shock wave caused by an object moving at a speed $v > c$.

Wavefront structure   We start by taking advantage of the Heaviside function appearing in the final result (3.140). It actually gives the wavefront equation at time t,

$$x = vt - d\sqrt{v^2/c^2 - 1} \quad \text{with} \quad d = \sqrt{y^2 + z^2}. \tag{3.141}$$

The half-angle θ of the cone (see Figure 3.13) is easily obtained from (3.141). It is given by the relation

$$\sin\theta = \frac{d}{\sqrt{(x - vt)^2 + d^2}} = \frac{c}{v}. \tag{3.142}$$

This angle therefore depends only on the ratio between the speeds of sound and of the object. It decreases as the object speed increases, as one would intuitively think. This result can also be easily interpreted in terms of sound propagation, as shown in Figure 3.14. In this figure a sound emitted by the object at time t_1 at the point A is found at time t at point B of the wavefront. At the same time t, the object is in M. Equation (3.142) means that the sound and the object have respectively travelled distances $c\Delta t$ and $v\Delta t$ during time interval $\Delta t = t - t_1$.

Despite its simplicity the model introduced here correctly captures the conical shape of the real wavefront. Indeed the underlying mechanism is somehow purely kinematic, relying on the source moving at a speed greater than c. This mechanism does not depend on the model under consideration, and is based on the reasonable assumption that pressure fluctuations also move at speed c outside of the linear regime of standard acoustic. Thus each point simultaneously receives the contributions of two signals propagating respectively to the front and the rear of the object (see Figure 3.15), and emitted at the two times (3.139) obtained by a purely geometric argument.

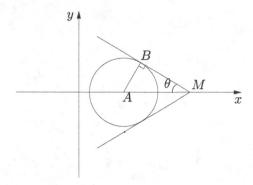

Fig. 3.14   At time t_1, the object is in A and emits a sound. At a subsequent time t, the sound is in B and the object in M. The angle θ then satisfies $\sin\theta = (c\Delta t/v\Delta t) = (c/v)$ with $\Delta t = t - t_1$.

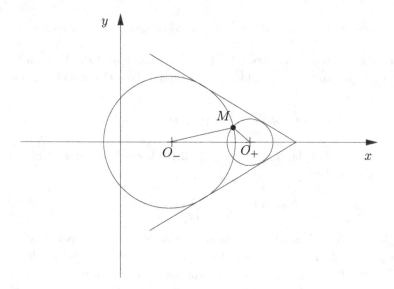

Fig. 3.15   Each point M inside the cone is at the intersection of two circles O_+ and O_- centred on the $0x$ axis. The centres of these circles are on either side of the considered point, and correspond to the points where the two phonons received in M at time t were emitted. These phonons were respectively emitted at times $t'_\pm = t''_\pm + (x/v)$.

Singularity on the wavefront   Field (3.140) becomes singular on the conical wavefront, which is the signature of a shock wave. The simple result (3.140) suggests the occurrence of such a shock wave, it cannot be described quantitatively since it was obtained assuming that pressure fluctuations become small away from the object! This inconsistency of the model could be removed by a more realistic approach, taking into account the finite size of the moving object in particular. Note finally that within the shock wave cone, the field value decreases asymptotically as $1/t$.

Cherenkov effect The phenomenon highlighted by this simple model appears in other areas of physics. For example if we now consider the electromagnetic field produced by a charged particle in motion, we also need to study a wave equation with a source term like ρ_p. Note that this equation is fundamental, while PDE (3.132) does not have the same status of course. Imagine now that the particle moves at speed v in a dielectric medium. The velocity of light in the medium c depends on the radiation frequency, and is related to the speed of light in vacuum c_0 by the relation $c = c_0/n$, where n is the refractive index of the dielectric material. It is then possible to reach speeds $v > c$, typically in high energy physics experiments where particle velocities are large. The resulting electromagnetic radiation has similar characteristics to those of the previous shock wave: It is the Cherenkov effect.

Regardless of the field value within the cone formed by the wavefront, it is the shape of the cone itself, and in particular the angle θ, which is of great utility in particle detectors. The wave vector corresponding to Cherenkov radiation is indeed perpendicular to the wavefront, and it therefore makes an angle $\pi/2 - \theta$ with Ox axis. Measuring this angle gives back the particle speed. Let us give an order of magnitude. To detect a speed v such that $v/c_0 \simeq 0.99$, and for a material of index $n = 1.03$ used in a detector of the *Large Hadron Collider* we find $\pi/2 - \theta \simeq 11°$.

3.2.5 *On the instantaneity of heat propagation*

◇ **Presentation** ◇

Diffusion equation presents an instantaneous character that is not consistent with Einstein's theory of relativity. Green's function $G_\infty^+(\mathbf{r} - \mathbf{r}'; t - t')$ associated with this equation, given by formula (3.41), p. 141, is non-zero for all times t greater than the initial time t'. Consider for example a temperature diffusion problem. The previous property of Green's function G_∞^+ implies that the temperature is changed at arbitrarily large distances from the source point, at any later time. Yet the theory of relativity imposes that no object or signal can propagate faster than the speed of light in vacuum. So diffusion equation does not meet one of the fundamental principles of physics. It is still very useful in non-relativistic conditions!

We will consider here a more elaborate version of heat equation that takes into account the effects of finite speed propagation. Let us consider a material in solid phase. From a microscopic point of view, thermal agitation propagates by phonons travelling at material sound speed. A natural modification of heat equation then leads to Cattaneo's equation including an additional term involving the sound propagation speed c. This equation is somehow the combination of diffusion and wave equation and reads

$$\left(-\Delta + a^2\frac{\partial}{\partial t} + \frac{1}{c^2}\frac{\partial^2}{\partial t^2}\right)\phi(\mathbf{r}, t) = \rho(\mathbf{r}, t), \tag{3.143}$$

where a^2 plays the role of the inverse of a diffusion coefficient, ϕ is the temperature and ρ a source term. In the limit $a \to 0$ we find the wave equation, whereas in the limit $c \to \infty$ we recover the diffusion equation.

We will compute the causal Green's function G_∞^+ associated with Cattaneo's equation (3.143), in the infinite one-dimensional system. We thus highlight the impossibility of a propagation speed greater than c. The three-dimensional case is studied in exercise 3.11, p. 211.

◇ Analysis and solution ◇

As Cattaneo's operator

$$\mathcal{O} = -\frac{\partial^2}{\partial x^2} + a^2\frac{\partial}{\partial t} + \frac{1}{c^2}\frac{\partial^2}{\partial t^2}$$

is invariant by translations in space and time, the homogeneous causal Green's function of an infinite system has the form $G_\infty^+(x, x'; t, t') = G_\infty^+(x - x'; t - t')$. It is therefore sufficient to determine G_∞^+ for $x' = 0$ and $t' = 0$. The corresponding PDE reads

$$\left(-\frac{\partial^2}{\partial x^2} + a^2\frac{\partial}{\partial t} + \frac{1}{c^2}\frac{\partial^2}{\partial t^2}\right)G_\infty^+(x; t) = \delta(x)\delta(t) \tag{3.144}$$

with causality condition $G_\infty^+(x; t) = 0$ for $t < 0$.

In order to make equation (3.144) more symmetrical, it is useful to set

$$G_\infty^+(x; t) = e^{-\frac{c^2a^2t}{2}} g_\infty^+(x; t).$$

The PDE for g_∞^+ then becomes

$$\left(-\frac{\partial^2}{\partial x^2} - m^2 + \frac{1}{c^2}\frac{\partial^2}{\partial t^2}\right)g_\infty^+(x; t) = \delta(x)\delta(t) \tag{3.145}$$

where $m^2 = c^2a^4/4$. This equation has a form familiar to particle physicists: if we change the sign of m^2, it becomes the Klein-Gordon equation, well known in field theory. We also propose to adapt the method below to obtain the Green's function of Klein-Gordon operator in exercise 3.12, p. 212. This method consists in taking the spatio-temporal Fourier transform of PDE (3.145) as we did for d'Alembert operator on p. 166.

Fourier transform Let us introduce

$$\widehat{g}_\infty^+(k;z) = \int_{-\infty}^{\infty} dt \int_{-\infty}^{\infty} dx \; e^{izt-ikx} g_\infty^+(x;t) = \int_0^{\infty} dt \int_{-\infty}^{\infty} dx \; e^{izt-ikx} g_\infty^+(x;t),$$

where we used causality condition on g_∞^+. As function g_∞^+ describes the propagation-diffusion of a signal, it is not assured that it decreases at large time for all values of x. It is then suitable to work with a complex frequency $z = \omega + i\gamma$, whose imaginary part γ is strictly positive. This ensures the convergence of the time integral when $t \to +\infty$.

Let us take a term by term Fourier transform of PDE (3.145). Partial derivatives with respect to space and time become simple multiplications by powers of iz and $-ik$. Furthermore, as the transform of the product of delta functions obviously reduces to 1, we eventually get

$$(k^2 - m^2 - z^2/c^2)\widehat{g}_\infty^+(k;z) = 1.$$

By choosing γ large enough, i.e. $\gamma > mc$, this equation is inversible for all values of k and ω with the result

$$\widehat{g}_\infty^+(k;z) = \frac{1}{k^2 - m^2 - z^2/c^2},$$

and function $g_\infty^+(x;t)$ is thus given by the inverse transformation

$$g_\infty^+(x;t) = \frac{1}{(2\pi)^2} \int_{-\infty+i\gamma}^{+\infty+i\gamma} dz \int_{-\infty}^{\infty} dk \; \frac{e^{-izt+ikx}}{k^2 - m^2 - z^2/c^2}. \tag{3.146}$$

At this stage it only remains to perform the integrals over k and then z.

Integration over k To perform the integration over k we write

$$\frac{1}{k^2 - m^2 - z^2/c^2} = \frac{1}{\left(k - (m^2 + z^2/c^2)^{1/2}\right)\left(k + (m^2 + z^2/c^2)^{1/2}\right)}$$

to explicitly identify the poles of the integrand in k. The determination of the function $(m^2 + z^2/c^2)^{1/2}$ is chosen so that its imaginary part is positive for $\gamma = \operatorname{Im} z > 0$. We will specify this choice of determination as well as the domain of analyticity of this function in z in the next paragraph. The integral over k can be calculated by combining Jordan's Lemma with the residue theorem. For $x > 0$, it is useful to complete the real axis of integration on k by a large semi-circle in the upper complex plane. Indeed, the integral over the semicircle goes to zero in the infinite radius limit by Jordan's lemma. The integral on the real axis is then simply given by the residue of pole $k = (m^2 + z^2/c^2)^{1/2}$ by Cauchy's Theorem. For $x < 0$ the appropriate closed contour is obtained by adding to the real axis a half-circle in the lower half complex plane, and then only the opposite pole $k = -(m^2 + z^2/c^2)^{1/2}$ contributes. The final result for all x is given by the synthetic formula

$$g_\infty^+(x;t) = \frac{ic}{4\pi} \int_{-\infty+i\gamma}^{+\infty+i\gamma} dz \; \frac{\exp(-izt + i(m^2c^2 + z^2)^{1/2}|x|/c)}{(m^2c^2 + z^2)^{1/2}}. \tag{3.147}$$

Focus on the propagation   We should now specify the analytical properties of the function $(m^2 + z^2/c^2)^{1/2}$. It has two branch points in $z = \pm imc$. Each of them must be the starting point of a branch cut, which can be chosen arbitrarily. For simplicity it is convenient here to take a single cut joining these two points, namely the segment $[-imc, imc]$. Function $(m^2 + z^2/c^2)^{1/2}$ is then analytic in the whole complex plane except for the segment $[-imc, imc]$. In addition we choose the determination such that we have $(m^2 + z^2/c^2)^{1/2} = \sqrt{m^2 + \omega^2/c^2}$ for $z = \omega$ with $\omega > 0$ real. This choice ensures $\mathrm{Im}(m^2 + z^2/c^2)^{1/2} > 0$ in the upper half complex plane in z. This positivity is therefore ensured on the integration axis $]-\infty + i\gamma, \infty + i\gamma[$, passing above the cut $[-imc, imc]$ by virtue of the condition $\gamma > mc$ (see Figure 3.16). Finally, note that $(m^2 + z^2/c^2)^{1/2} \sim z/c$ when $|z| \to \infty$.

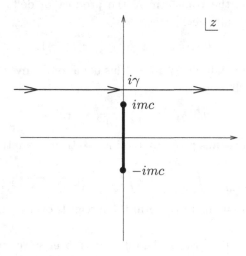

Fig. 3.16   Integration contour for integral (3.147). There is a cut between the two branch points.

Before performing the integral over z, we show that the signal described by $g_\infty^+(x; t)$ cannot travel faster than the speed c. This feature is a direct consequence of the above analyticity properties[17]. Thus if $|x| > ct$, it is useful to complement the integration axis $]-\infty + i\gamma, \infty + i\gamma[$ with a large semicircle in the upper half complex plane. The integral over the semicircle tends to zero as the radius goes to infinity by Jordan's lemma. The integrand is analytic in the whole interior of the closed contour thus constructed, so that integral (3.147) vanishes! Thus, Green's function $G_\infty^+(x; t) = 0$ vanishes for $|x| > ct$. For $t < 0$ we recover the causal condition. For $t > 0$ we see that there is indeed propagation at finite speed c, unlike in the pure diffusion case.

Integration over z   Let us consider the other case, i.e. $|x| < ct$. It is now more convenient to complete $]-\infty + i\gamma, \infty + i\gamma[$ with a large semicircle in the

---

[17]The discussion below is similar to that of exercise 1.9, Chapter 1 p. 51.

lower half complex plane, so that this closed integration contour encloses the cut $[-imc, imc]$. Again Jordan's lemma helps to get rid of integral over the semicircle. We can moreover deform this contour by virtue of Cauchy theorem to surround infinitely closely the cut $[-imc, imc]$, as shown in Figure 3.17. The contribution

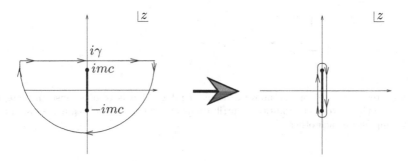

Fig. 3.17   Deformation of the integration contour around the cut between $z = imc$ and $z = -imc$.

of semicircles surrounding the branching points tends to 0 as the square root of the semicircles radius, and leaves only the two integrals on each side of the cut between $+imc$ and $-imc$. For $z = \pm\delta + imcw$ with $\delta \to 0^+$ and $w \in [-1, 1]$, we get $(m^2 + z^2/c^2)^{1/2} = \pm m\sqrt{1 - w^2}$ according to the chosen determination. It finally leads to

$$g_\infty^+(x;t) = \frac{c}{4\pi} \int_{-1}^1 dw \left[\frac{e^{mc(tw + i\sqrt{1-w^2}|x|/c)}}{\sqrt{1 - w^2}} + \frac{e^{mc(tw - i\sqrt{1-w^2}|x|/c)}}{\sqrt{1 - w^2}} \right].$$

Previous expression can be transformed with successive changes of variable $w \to \beta$ with $w = \sin\beta$ and then $\beta \to \beta + i\alpha$ with α real and defined by parametrisation

$$t = \sqrt{t^2 - x^2/c^2} \, \cos(i\alpha) \quad \text{and} \quad \frac{|x|}{c} = -i\sqrt{t^2 - x^2/c^2} \, \sin(i\alpha)$$

valid for $ct > |x|$. We find

$$g_\infty^+(x;t) = \frac{c}{4\pi} \int_{-\pi+i\alpha}^{\pi+i\alpha} d\beta \, e^{mc\sqrt{t^2 - x^2/c^2} \, \sin\beta}. \tag{3.148}$$

This last integral on a segment of the upper half complex plane can be reduced to an integral over the real axis. Integral

$$\oint d\beta \, e^{mc\sqrt{t^2 - x^2/c^2} \, \sin\beta}, \tag{3.149}$$

on the closed contour of Figure 3.18 vanishes due to analyticity of integrand inside this contour. Furthermore the contributions of each vertical segments are opposite. Expression (3.148) thus reduces to an integral over the real segment $[-\pi, \pi]$, which we can relate to the definition of modified Bessel function

$$I_0(\rho) = \frac{1}{2\pi} \int_{-\pi}^{\pi} d\theta \, e^{\rho \, \cos\theta}. \tag{3.150}$$

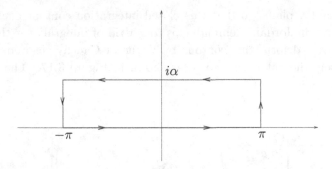

Fig. 3.18 Integral (3.149) on the above contour of integration is zero because the integrand is analytic inside the contour. In addition, contributions of the two vertical segments are of opposite sign and compensate each other.

Taking into account the prefactor $e^{-a^2c^2t/2}$, and the cancellation of $g_\infty^+(x;t)$ for $|x| > ct$ with a Heaviside function $\theta(ct - |x|)$, we finally derive the causal Green's function of Cattaneo operator in one dimension

$$G_\infty^+(x;t) = \theta(ct - |x|)\frac{c}{2}e^{-\frac{a^2c^2t}{2}}I_0\left(\frac{a^2c^2}{2}\sqrt{t^2 - x^2/c^2}\right). \tag{3.151}$$

◇ Interpretation ◇

Beyond the particular form of this Green's function, we have again highlighted the fundamental importance of analyticity properties. It is also possible to check that we find the Green's function of diffusion equation in the limit $c \to \infty$. This limit is analysed using the saddle point method presented in the next chapter for integral (3.150), as proposed in exercise 4.6, p. 258. Moreover, one can take the limit $a \to 0$ directly in formula (3.151) to recover the expression for the causal Green's function of d'Alembert operator in one dimension, determined in exercise 3.6, p. 208.

It is instructive to compare the evolution of Cattaneo Green's function to that of d'Alembert operator. There is little difference at short times. At intermediate time Cattaneo Green's function spreads when $|x|$ increases from 0 to ct as a result of the diffusion process, while the Green's function for d'Alembert operator remains constant. At large time, Cattaneo Green's function is close to the form predicted by the diffusion equation, except when $|x|$ exceeds ct where the propagating nature prevails again and strictly cancels it.

Note finally that the second derivative term can lead to a wave-like behaviour for temperature, similar to pressure or density in the presence of a sound wave. This is called second sound in condensed matter physics because temperature behaves similarly to density in acoustics.

3.2.6 *Polarisability of the hydrogen atom*

◊ Presentation ◊

Consider a hydrogen atom submitted to a homogeneous and time-dependent external electric field $\mathbf{E}_{\text{ext}}(t)$. In the non-relativistic limit, the system can be studied in the context of ordinary quantum mechanics. The atom being neutral, its overall translational motion is not affected by a spatially uniform electric field. It is of course sufficient in practice for the electric field to be uniform at the scale of the atom. This condition is achieved for example in the case of an atom submitted to an electromagnetic wave in the visible range. Note that the action of magnetic field is negligible because the inner electron speed is small compared to the speed of light. To determine the atom response, it is therefore sufficient to look at the dynamics of its internal state, described by the relative particle of mass $m = m_e M_p/(m_e + M_p)$, where M_p and m_e are respectively the proton and electron masses. In the absence of an external field, the Hamiltonian reduces to

$$H_0 = -\frac{\hbar^2}{2m}\Delta - \frac{e^2}{4\pi\epsilon_0 r}, \tag{3.152}$$

which includes Coulomb attractive potential $-e^2/(4\pi\epsilon_0)r$ between the proton of charge e and the electron of charge $-e$. In the presence of $\mathbf{E}_{\text{ext}}(t)$ the Hamiltonian of the relative particle becomes

$$H(t) = H_0 + e\mathbf{E}_{\text{ext}}(t) \cdot \mathbf{r}. \tag{3.153}$$

Let us study the polarisation of the atom under the influence of a weak field $\mathbf{E}_{\text{ext}}(t)$ connected adiabatically at $t = -\infty$. We assume that the atom is initially in its ground state $|\psi_0\rangle$ with energy E_0. Using the general framework and notations from linear response presented in Chapter 1, steady state[18] $\mathcal{E}_0$ is here identified with $|\psi_0\rangle$, and $A(t)$ with the polarisation

$$\mathbf{p}(t) = -e\int d\mathbf{r} \, [\phi(\mathbf{r},t)]^* \, \mathbf{r} \, \phi(\mathbf{r},t), \tag{3.154}$$

where $\phi(\mathbf{r}, t)$ is the wave function solution of Schrödinger equation

$$i\hbar\frac{\partial\phi}{\partial t}(\mathbf{r},t) = H(t)\phi(\mathbf{r},t)$$

with the initial condition $\phi(\mathbf{r}, -\infty) = \psi_0(\mathbf{r})$. Note that the polarisation in the ground state is zero, since $\psi_0(\mathbf{r}) = \psi_0(r)$ by rotation invariance. External perturbation $F(t)$ is here set equal to the field $\mathbf{E}_{\text{ext}}(t)$, so that the corresponding susceptibility $\chi(\omega)$ is nothing but the polarisability of the atom in its ground state.

---

[18]In the evolution generated by H_0, the wave function $|\psi_0\rangle$ oscillates at frequency $E_0/\hbar$, but the corresponding mean value of any observable A is constant.

◇ **Analysis and solution** ◇

The calculation of the susceptibility falls within the standard time-dependent perturbation theory in quantum mechanics, that has been formulated using Green's functions in section 3.1.4. We establish here the expression of $\chi(\omega)$ in terms of such functions, and then deduce its spectral representation involving eigenstates of Hamiltonian H_0. It allows to prove some interesting analytical properties.

Perturbative expansion of polarisation In this perturbative calculation, it is convenient to consider first a finite initial time t_0 and then to take the limit $t_0 \to -\infty$. Moreover the causal Green's function should be defined in the whole space with boundary conditions at infinity. It can be written as

$$G^{+(0)}(\mathbf{r}; \mathbf{r}_1; \tau) = \langle \mathbf{r} | e^{-iH_0\tau/\hbar} | \mathbf{r}_1 \rangle,$$

while the perturbation is

$$W(\mathbf{r}, t) = e\mathbf{E}_{\text{ext}}(t) \cdot \mathbf{r}.$$

Wave function $\phi(\mathbf{r}, t)$ is given by applying formula (3.73), p. 154 with $\phi_0(\mathbf{r}) = \psi_0(\mathbf{r})$, leading to

$$\phi(\mathbf{r}, t) = e^{-iE_0(t-t_0)/\hbar}\, \psi_0(\mathbf{r})$$
$$-\frac{i}{\hbar}\, e \int d\mathbf{r}_1\, \psi_0(\mathbf{r}_1) \int_0^{t-t_0} d\tau_1\, G^{+(0)}(\mathbf{r}; \mathbf{r}_1; \tau_1)$$
$$\times e^{-iE_0(t-t_0-\tau_1)/\hbar}\, \mathbf{E}_{\text{ext}}(t - \tau_1) \cdot \mathbf{r}_1 + \cdots, \quad (3.155)$$

where we used

$$\int d\mathbf{r}'\, G^{+(0)}(\mathbf{r}; \mathbf{r}'; \tau)\, \psi_0(\mathbf{r}') = e^{-iE_0(t-t_0)/\hbar}\, \psi_0(\mathbf{r}).$$

Then after inserting this perturbative expansion in the expression (3.154) of polarisation and taking the limit $t_0 \to -\infty$, we find

$$\mathbf{p}(t) = 2\frac{e^2}{\hbar} \int d\mathbf{r}\, \psi_0(\mathbf{r})\, \mathbf{r} \int d\mathbf{r}_1\, \psi_0(\mathbf{r}_1)$$
$$\text{Re}\left(i \int_0^{+\infty} d\tau_1 G^{+(0)}(\mathbf{r}; \mathbf{r}_1; \tau_1)\, e^{iE_0\tau_1/\hbar}\, \mathbf{E}_{ext}(t - \tau_1) \cdot \mathbf{r}_1 \right) + \ldots. \quad (3.156)$$

This first order expression in the perturbation takes general form (1.4), seen in Chapter 1, p. 4. Moreover the response function K_0 involves $G^{+(0)}$, in agreement with the general interpretation of Green's functions presented in section 3.1.1.

Expression of $\chi(\omega)$ in terms of Green's functions For a monochromatic field of complex frequency z,

$$\mathbf{E}_{\text{ext}}(t) = \text{Re}\big[\mathbf{E}_z \exp(-izt)\big],$$

connected adiabatically at $t \to -\infty$ ($\text{Im}(z) > 0$), the integral over τ_1 in (3.156) then appears as the Laplace transform of Green's function $G^{+(0)}$ for different values of the parameter of the form $s = \pm iE_0/\hbar - iz$ or $s = \pm iE_0/\hbar + iz^*$. We have shown that the corresponding Laplace transforms are proportional to Green's function $G_\lambda(\mathbf{r}; \mathbf{r}_1)$ associated with the resolvent $[\lambda + H_0]^{-1}$,

$$G_\lambda(\mathbf{r}; \mathbf{r}_1) = \langle \mathbf{r}| \frac{1}{\lambda + H_0} |\mathbf{r}_1\rangle, \tag{3.157}$$

for $\lambda = \hbar z - E_0 = \lambda_+$ or $\lambda = -\hbar z - E_0 = \lambda_-$. The final result indeed takes the form expected from the general analysis, namely

$$\mathbf{p}(t) = \text{Re}\big(\chi(z)\mathbf{E}_z \exp(-izt)\big),$$

with the susceptibility

$$\chi(z) = e^2 \int d\mathbf{r} \int d\mathbf{r}_1 \, \psi_0(\mathbf{r}) \, \psi_0(\mathbf{r}_1) \, x \, x_1 \, [G_{\lambda_+}(\mathbf{r}; \mathbf{r}_1) + G_{\lambda_-}(\mathbf{r}; \mathbf{r}_1)]. \tag{3.158}$$

As a result of the ground state rotational invariance and Green's functions symmetry properties[✠], polarisation is isotropic and collinear with the applied electric field, so that $\chi(z)$ is a pure scalar. Note that formula (3.158) is *a priori* valid only in the upper complex plane ($\text{Im } z > 0$).

✠ **Comment:** Green's functions are symmetrical with respect to any plane passing through the origin. They satisfy for example the symmetry relation $G_{\lambda_\pm}(x, y, z; x_1, y_1, z_1) = G_{\lambda_\pm}(x, -y, z; x_1, -y_1, z_1)$. Thus spatial integrals over rectangle terms of the form xy_1 in formula (3.156) are zero, since $\psi_0(\mathbf{r}) = \psi_0(|\mathbf{r}|)$. Note that these terms are no longer zero for another non-rotationally invariant eigenstate, and the susceptibility becomes a tensor.

Spectral representation The spectral representation of $\chi(z)$ in terms of eigenfunctions ψ_n of H_0 is obtained from integral expression (3.158), by using formula (2.77), p. 88 for each Green's function $G_{\lambda_\pm}$, with the result

$$\chi(z) = 2e^2 \sum_n |\langle \psi_0|x|\psi_n\rangle|^2 \frac{(E_n - E_0)}{[(E_n - E_0)^2 - \hbar^2 z^2]}. \tag{3.159}$$

The sum in (3.159) is divided into a discrete part associated with bound states of negative energy, and an integral part associated with ionised states of positive energy. In the following we explicitly give the corresponding eigenfunctions, dropping out the generic label n to index individual functions. Bound states are characterised by three integers n, l, m. The principal quantum number n ($n \geq 1$) determines the energy $E_{n-1} = -e^2/(n^2 a_B)$, where $a_B = 4\pi\epsilon_0 \hbar^2/(me^2)$ is Bohr radius. The azimuthal quantum number l sets the eigenvalue $l(l+1)\hbar^2$ of the square of the orbital angular momentum: for a given n, it can only take n values, $0 \leq l \leq n-1$. Magnetic quantum number m sets the eigenvalue $m\hbar$ of a component of the angular momentum along an arbitrary direction: for a given l, it can only take $(2l + 1)$ values, $-l \leq m \leq l$. Bound states will henceforth be denoted $|\psi_{nlm}\rangle$, so that the ground state is now $|\psi_{100}\rangle$. Ionised states, also known as scattering states, are themselves characterised by a real positive wave number k, and two integers l and m with the same physical meaning as for bound states. Wave number k determines the energy $E_k = \hbar^2 k^2/(2m)$, while l can take any integer value with $-l \leq m \leq l$. From now on we shall note these states $|\psi_{klm}\rangle$.

Taking into account the previous detailed form of H_0 spectrum, formula (3.159) becomes

$$\chi(z) = 2e^2 \left(\sum_{n\geq 2}^{\infty} |\langle \psi_{100}|x|\psi_{n10}\rangle|^2 \frac{(E_{n-1} - E_0)}{[(E_{n-1} - E_0)^2 - \hbar^2 z^2]} \right.$$
$$\left. + \int_0^{\infty} \mathrm{d}k |\langle \psi_{100}|x|\psi_{k10}\rangle|^2 \frac{(E_k - E_0)}{[(E_k - E_0)^2 - \hbar^2 z^2]} \right). \qquad (3.160)$$

Indeed, it turns out that only states with $l = 1$ and $m = 0$ contribute to the spectral representation of $\chi(z)$. The reader can see it by calculating the matrix elements in representation (3.158), with the expressions of eigenfunctions $\psi_{nlm}(r, \theta, \varphi)$ and $\psi_{klm}(r, \theta, \varphi)$ in spherical coordinates[19].

Analytical properties of the polarisability Using the explicit expression of wave functions, we show that $|\langle \psi_{100}|x|\psi_{n10}\rangle|^2$ decreases as $1/n^3$ when $n \to \infty$, while $|\langle \psi_{100}|x|\psi_{k10}\rangle|^2$ decreases as $1/k^8$ when $k \to \infty$. The series on n and the integral over k in spectral representation (3.160) are absolutely convergent. Therefore, as the latter is valid for $\mathrm{Im}\, z > 0$, $\chi(z)$ is analytic in the upper half complex plane.

The behaviour of $\chi(z)$ on the real axis is obtained by taking the limit of representation (3.160) when $\mathrm{Im}\, z \to 0^+$. Real points $\pm\omega_{n-1}$ with $\omega_{n-1} = (E_{n-1} - E_0)/\hbar$ are simple poles of $\chi(z)$. These infinitely many poles accumulate in the vicinity of ionisation frequency $\pm\omega_i$ with $\omega_i = |E_0|/\hbar$. Moreover, for $\omega \neq \pm\omega_{n-1}$, $|\omega| < \omega_i$, $\chi(\omega)$ is purely real.

For $|\omega| > \omega_i$, the presence of a singularity in the integrand in $k = k(\omega) = \sqrt{2m(|\omega| - \omega_i)}/\hbar$ induces a non-zero imaginary part in $\chi(\omega)$ by virtue of identity (A.1), i.e.

$$\chi''(\omega) = \frac{\pi m e^2 |\omega|}{\hbar^2 \omega k(\omega)} |\langle \psi_{100}|x|\psi_{k(\omega)10}\rangle|^2. \qquad (3.161)$$

---

[19]Angular parts are spherical harmonics $Y_{lm}(\theta, \varphi)$ described in Appendix G. Radial parts are confluent hypergeometric functions, as detailed in [50].

Imaginary part $\chi''(\omega)$ remains finite when $\omega \to \omega_i^+$, or $\omega \to -\omega_i^-$, since $|\langle\psi_{100}|x|\psi_{k(\omega)10}\rangle|^2$ then vanishes as $k(\omega)$. Actually the two symmetrical points $\pm\omega_i$ are branch points for $\chi(z)$. More precisely, in the vicinity of $z = \pm\omega_i$, $\chi(z)$ has a logarithmic singularity $\ln(z \mp \omega_i)$: the appearance of a finite imaginary part when crossing the singularity is a simple consequence of identity $\ln Z = \ln|Z| + i \arg Z$. A branch cut necessarily starts from each branch point $\pm\omega_i$. One may require that these branch cuts constitute only one finite curve finishing at $\pm\omega_i$ and contained in the lower half complex plane. The corresponding analytical structure of $\chi(z)$ in the whole complex plane is summarised in Figure 3.19.

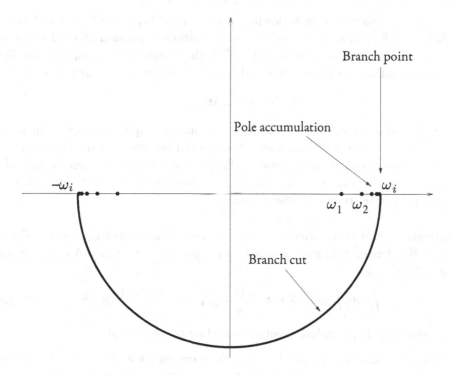

Fig. 3.19   Analytic structure of $\chi(z)$.

Real part $\chi'(\omega)$ and imaginary part $\chi''(\omega)$ are respectively even and odd functions. At zero frequency, $\chi(0) = \chi'(0)$ is purely real and reduces to static polarisability $\chi(0) = 36\pi\epsilon_0 a_B^3/2$. At large frequencies $\chi(\omega)$ behaves like

$$\chi(\omega) \simeq \frac{-2e^2}{\hbar^2\omega^2} \langle\psi_{100}|x(H_0 - E_0)x|\psi_{100}\rangle$$

or, using the expression of $\psi_{100}(r)$ which is simply proportional to e^{-r/a_B},

$$\chi(\omega) \simeq -\frac{e^2}{m\omega^2} \quad \text{when } |\omega| \to \infty. \tag{3.162}$$

The dominant term in the asymptotic expansion of $\chi(\omega)$ is purely real. It is identical to the polarisability of a free classical particle with a charge $-e$ that one immediately derives from Newton's laws of motion

$$m\frac{d^2\mathbf{r}}{dt^2} = -e\mathbf{E}_{\text{ext}}(t).$$

In other words, quantum effects as well as contributions of interactions between proton and electron become negligible in the high frequency limit. This limit is governed by pure inertia effects, which can moreover be treated classically. This result is very general and applies to other physical systems.

The first corrective term to leading behaviour (3.162) is still real and varies as $1/\omega^4$. The first imaginary term in the asymptotic expansion of $\chi(\omega)$ is not an integer power of $1/\omega$, but behaves as $1/|\omega|^{9/2}$: this indicates that response function $K_0(\tau)$ is not infinitely differentiable and has a singularity in $\tau^{7/2}$ at the origin.

<h2 style="text-align:center">◇ Interpretation ◇</h2>

The analytical structure of $\chi(z)$ is very rich, and exemplifies general considerations of section 1.1. We first interpret the essential features and then comment on various analogies with classical systems. Finally, we conclude by mentioning spontaneous emission, which is a fundamental relativistic quantum effect, and by briefly discussing the limitations of linear response.

Singularities and resonances   Imaginary part $\chi''(\omega)$ controls the power $\overline{\mathcal{P}}$ provided by the driving field and averaged over a period $T = 2\pi/|\omega|$. A simple calculation indeed leads to

$$\overline{\mathcal{P}} = \frac{1}{T}\int_0^T dt \int d\mathbf{r}\, [\phi(\mathbf{r},t)]^* \, \frac{\partial H(t)}{\partial t} \, \phi(\mathbf{r},t) = \frac{\omega\chi''(\omega)}{2}\,|\mathbf{E}_\omega|^2. \tag{3.163}$$

The product $\omega\chi''(\omega)$ is indeed positive according to formula (3.161).

Presence of singularities on the real axis is consistent with the absence of dissipation in the considered system, which is a consequence of the conservative nature of H_0. Simple poles in $z = \pm\omega_{n-1}$ originate from a resonance phenomenon. For these particular frequencies, the exciting field induces transitions between the ground state and excited states. It implies the divergence of $\chi(\pm\omega_{n-1})$, because $\phi(\mathbf{r},t)$ does not remain close to $\psi_0(r)$ as assumed *a priori* in the perturbative approach. The imaginary nature of $\chi(\omega)$ in the vicinity of $\pm\omega_{n-1}$, obtained by setting $z = \pm\omega_{n-1} + i\epsilon$ with $\epsilon \to 0^+$, means that as expected, the exciting field then provides energy to the atom. The accumulation of poles in the vicinity of $\pm\omega_i$ is due to the existence of Rydberg states with an energy arbitrarily close to the ionised states onset threshold $E = 0$.

For $\omega \neq \pm\omega_{n-1}$ with $|\omega| < \omega_i$, $\chi(\omega)$ is purely real and dipole $\mathbf{p}(t)$ oscillates in phase with the exciting field. The latter does not have to provide energy to maintain this oscillation, since there is no dissipative processes in the system. Appearance of a non-zero imaginary part in $\chi(\omega)$ for $|\omega| > \omega_i$ stems from the resonant coupling between the ground state and an ionised state of wave number $k(\omega)$, coupling which is induced by the exciting field. Corresponding divergences, similar to those previously observed for $\omega = \pm\omega_{n-1}$, are now smoothed out by the integration over all k. In other words the continuous nature of ionised states spectrum prevents $\chi(\omega)$ to diverge.

Analogies The analytical form of $\chi(\omega)$ has similarities with the ones obtained in other situations in physics that it is useful to clarify and comment. First $\chi(\omega)$ behaves as

$$\chi(\omega) \simeq \frac{2e^2\omega_{n-1}|\langle\psi_{1,0,0}|x|\psi_{n,1,0}\rangle|^2}{\hbar(\omega_{n-1}^2 - \omega^2)} \tag{3.164}$$

in the vicinity of $\pm\omega_{n-1}$, just like the so called Thomson's model without dissipation, presented in an example of section 1.2.2 (see p. 24). This very simple phenomenological model thus perfectly reproduces the singularity of $\chi(\omega)$ at a given resonance for $\pm\omega_{n-1}$, in the framework of classical dynamics. Adjustments of the restoring frequency and the effective mass require however knowing the spectrum of H_0. Note that they are specific to each resonance.

The mechanism inducing a finite imaginary part in $\chi(\omega)$ for $|\omega| > \omega_i$ also appears for the dielectric constant of an electron gas at equilibrium, as shown in section 1.2.4. In the latter case, it is the integration over all possible electron velocities, distributed according to Maxwell-Boltzmann statistics, that smoothes singularities coming from the surfing phenomenon on the applied external wave. These surfing electron pump part of the exciting wave energy. As described above in the atom case, this energy absorption is not induced by dissipation in the system, but results from a resonant effect.

Spontaneous emission Spontaneous emission is a fundamental relativistic effect whose description requires the introduction of quantum electrodynamics. In this context, the coupling between the atom and the fluctuations of the electromagnetic field induces an instability of the atom excited states. The latter have a finite lifetime: They spontaneously relax by emitting a photon. At a phenomenological level, this process can be addressed by assigning an imaginary part $-i\hbar/\tau_{\text{life}}$ to the excited state energy. Its evolution then exhibits a damping factor $\exp(-t/\tau_{\text{life}})$, so that the polarisability has now simple poles in the lower half complex plane. In other words, the polarisability becomes analytic on the real axis, in agreement with the emergence of dissipative processes resulting from spontaneous emission, the energy provided by the exciting field being carried away by emitted photons. Thus the introduction of a finite speed of light c moves the singularities in the lower half complex plane! Note finally that the behaviour of polarisability in the vicinity of each pole is then exactly reproduced by Thomson's model presented in section 1.2.2.

Beyond linear response Predictions from linear response must be understood in an asymptotic sense, where the exciting field amplitude tends to zero. It is natural to ask whether these predictions are relevant to small but finite applied field. The above analysis already provides answers. For example, if the exciting field frequency corresponds to a resonance, the atom undergoes a transition to an excited state. In general one speculates that the ionisation always occurs at sufficiently long times for any frequency, due to non-linear effects. Note that each term of the perturbation series remains finite at all times at zero frequency. The final state is however still ionised because of a tunnel effect. Ionisation then occurs on a time scale growing exponentially fast when the applied field tends to zero. This mechanism, whose dependency on the applied field is an essential singularity, cannot be reproduced by the perturbative expansion. We finally note that for a sufficiently low exciting field, the atom induced polarisability is given for human time scales by the linear approach.

3.3   Exercises

■ **Exercise 3.1. Uniqueness of solutions for diffusion and wave equations**

Using the first Green's formula, show that diffusion equation (3.26), p. 136, has a unique solution if we impose the IC (3.27) and either Dirichlet or Neumann BC (see equations (3.28) and (3.29)).

1. Same question for d'Alembert operator with initial conditions (3.89), p. 159 and either Dirichlet or Neumann boundary conditions.

Solution p. 316.

■ **Exercise 3.2. Reciprocity relations**

Show, using the second Green's formula that causal Green's functions of d'Alembert operator with homogeneous boundary conditions are symmetric, that is to say, $G_{\mathrm{H}}^{+}(\mathbf{r}; \mathbf{r}'; t - t') = G_{\mathrm{H}}^{+}(\mathbf{r}'; \mathbf{r}; t - t')$.

Solution p. 317.

■ Exercise 3.3. Equation for long cables

The PDE given below has been established by Lord Kelvin to study the transmission of an electrical signal in long submarine cables. It has since been used to describe many other phenomena such as signal transmission along dendrites in neural networks. The equation is given by

$$\lambda^2 \frac{\partial^2 V(x,t)}{\partial x^2} - \tau_0 \frac{\partial V(x,t)}{\partial t} - V(x,t) + V_0 = \rho(x,t). \tag{3.165}$$

Here $V(x,t)$ is the potential at a time t and a point x of the transmission line (cable or dendrite), V_0 is a constant external potential and $\rho(x,t)$ is proportional to the current injected at the point x at time t. Give the expression for $V(x,t)$ in terms of $\rho(x,t)$ using as boundary conditions $V(x,t) \to V_0$ for $x \to \pm\infty$ and $t \to -\infty$.

Solution p. 318.

■ Exercise 3.4. Neumann conditions in diffraction theory

Derive expression (3.123), p. 181, for the amplitude $A_N(\mathbf{r})$, corresponding to Fraunhofer diffraction with homogeneous Neumann BC on the plate S_p.

Solution p. 318.

■ Exercise 3.5. Green's function for d'Alembert operator in $2+1$ dimensions

The purpose of this exercise is to determine in $2+1$ dimension the causal Green's function of d'Alembert operator. More precisely, this function denoted G_2^+ is the solution of PDE

$$\left(\frac{1}{c^2} \partial_t^2 - \partial_x^2 - \partial_y^2 \right) G_2^+(x - x_0, y - y_0, t - t_0) = \delta(x - x_0)\delta(y - y_0)\delta(t - t_0)$$

vanishing when $r \to \infty$. We write in a similar way G_3^+, the Green's function in $3+1$ dimensions given by equation (3.101), p. 166.

1. Show that

$$G_2^+(x - x_0, y - y_0, t - t_0) = \int_{-\infty}^{+\infty} dz \, G_3^+(x - x_0, y - y_0, z, t - t_0). \tag{3.166}$$

2. Use this result to compute G_2^+.

3. Comment on this result.

4. Consider a set of sources in $3+1$ dimensions, located at time t_0 on the straight line parallel to the z-axis and passing through the point of coordinates $(x_0, y_0, 0)$, so that

$$\rho(x', y', z') = \lambda\delta(x' - x_0)\delta(y' - y_0)\delta(t' - t_0)$$

where λ is a constant. By considering the wave equation in $3 + 1$ dimensions associated with this source, discuss the results of questions **1.** and **3.** from the point of view of $3 + 1$ dimensions.

Solution p. 319.

■ **Exercise 3.6. Green's function of d'Alembert operator in $1 + 1$ dimensions**

The delayed Green's function of d'Alembert operator in $1 + 1$ dimension is the function G_1^+ solution of PDE

$$\left(\frac{1}{c^2}\partial_t^2 - \partial_x^2\right)G_1^+(x - x_0, t - t_0) = \delta(x - x_0)\delta(t - t_0), \qquad (3.167)$$

with BC $G_1^+(x - x_0, t - t_0) \to 0$ when $|x| \to \infty$. Compute G_1^+ from its spectral representation (3.93), p. 162.

Solution p. 320.

■ **Exercise 3.7. Laplacian Green's function G_∞ in dimension $d \geq 3$**

We propose in this exercise to calculate the Green's function G_∞ of the Laplace operator in dimension $d \geq 3$ from Green's function G_∞^+ for diffusion in the same space $\mathbb{R}^d$.

1. Show that

$$G_\infty(\mathbf{r} - \mathbf{r}') = D \int_0^\infty \mathrm{d}t\, G_\infty^+(\mathbf{r} - \mathbf{r}'; t)$$

where D is the diffusion coefficient.

2. Find then the result of exercise 2.2, p. 111, by performing a suitable change of variable in the above integral, and using the definition of Euler Γ function,

$$\Gamma(\alpha + 1) = \int_0^\infty \mathrm{d}u\, u^\alpha\, e^{-u}.$$

Solution p. 321.

■ Exercise 3.8. Heat diffusion in a ball

Consider a ball of radius R at an initial uniform temperature T_0. It is immersed at time $t = 0$ in a bath at temperature T_b, which immediately sets the temperature of the ball boundary at $T = T_b$ (see Figure 3.20). Derive an expression for temperature

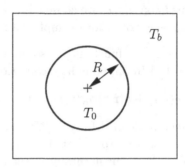

Fig. 3.20   Ball of radius R initially at uniform temperature T_0, and then immersed in a bath at temperature T_b.

$T(r, t)$ at any point of the ball and at any time. To this aim, given the problem spherical symmetry, we will work in the set of functions depending only on $r = |\mathbf{r}| \in [0, R]$ and vanishing at $r = R$. A complete basis of this set is given by functions $\{\psi_n\}$ defined by

$$\psi_n(r) = \sqrt{\frac{2}{R}} \frac{1}{r} \sin(\frac{n\pi r}{R}), \qquad n \in \mathbb{N}^*.$$

These functions are orthonormal, i.e.

$$\int_0^R dr \, r^2 \psi_n(r) \psi_m(r) = \delta_{nm}.$$

Express $T(r, t)$ directly in this basis.

Solution p. 322.

■ Exercise 3.9. From Dirichlet to Robin boundary conditions

1. *Preliminary question:* In this question we work with the set of functions defined on $\mathbb{R}^+$ (origin included) and consider two functions b and ϕ in this set such that $b(0) = 0$ and

$$b = -\frac{d\phi}{dx} + h\phi, \tag{3.168}$$

with $\phi(x) \to 0$ when $x \to +\infty$ and h, a positive constant. Determine ϕ by the method of variation of constants and show that

$$\phi(x) = \int_0^{+\infty} du \, e^{-hu} b(x + u). \tag{3.169}$$

2. Consider differential equation

$$(\mathcal{O}_\mathbf{r} + \mathcal{O}_t)\,\phi(\mathbf{r}, t) = \rho \qquad\qquad (3.170)$$

in a volume $\mathcal{D}$ of boundary $\partial\mathcal{D}$. Here $\rho(\mathbf{r}, t)$ is a source while $\mathcal{O}_\mathbf{r}$ and $\mathcal{O}_t$ are one of the differential operators corresponding to diffusion equation or wave equation. Initial conditions at a time t_0 on $\phi(\mathbf{r}, t)$ are chosen adequately according to the type of operator to ensure uniqueness of ϕ, for example with Dirichlet BC on $\partial\mathcal{D}$.

Imagine that one imposes on a function ϕ_R, solution of PDE (3.170), the same initial conditions but so-called Robin boundary conditions,

$$\mathbf{n} \cdot \boldsymbol{\nabla}\phi_R(\mathbf{r}, t) + h(\mathbf{r}, t)\phi_R(\mathbf{r}, t) = 0 \qquad \forall \mathbf{r} \in \partial\mathcal{D}, \qquad (3.171)$$

with $h(\mathbf{r}, t)$ a positive function defined for $\mathbf{r} \in \partial\mathcal{D}$ and for any time t. Show that ϕ_R is unique following exercise 3.1. Note that the purpose of the following exercise is to show that these Robin BC occur naturally in certain physical situations.

3. We now assume that $\mathcal{D}$ consists of half-line $[0, \infty[$ and h is a positive constant. More precisely $\phi_R(x, t)$ satisfies Robin conditions (3.171) at $x = 0$ and tends to zero as $x \to +\infty$. We assume that initial conditions are $\phi_R(x, t_0) = 0$ for the diffusion, completed by $\partial_t\phi_R(x, t_0) = 0$ for the wave equation.

Let us consider the function

$$b_D(x, t) = -\frac{\partial}{\partial x}\phi_R(x, t) + h\phi_R(x, t). \qquad\qquad (3.172)$$

Show that b_D is solution of a PDE of type (3.170) and identify the corresponding source. What are the BC satisfied by $b_D(x, t)$?

4. Let $G_{\mathrm{HD}}^+(x; x'; t - t')$ be the causal Green's function with homogeneous Dirichlet BC of operator $(\mathcal{O}_x + \mathcal{O}_t)$. Express $b_D(x, t)$ as a function of G_{HD}^+, ρ, and derivatives of ρ. Give an expression for $\phi_R(x, t)$.

Solution p. 322.

■ **Exercise 3.10. Robin conditions for heat equation**

There are physical situations for which the most realistic boundary conditions to impose are neither Dirichlet nor Neumann. This exercise illustrates this property in the case of heat equation. Imagine that a body of volume $\mathcal{D}$ is at temperature $T(\mathbf{r}, t)$ and its edge $\partial\mathcal{D}$ is in contact with a thermostat at a constant and uniform temperature T_0.

1. We recall Fourier's law connecting heat flow $\mathbf{j}_Q$ to the temperature gradient,

$$\mathbf{j}_Q = -\lambda\boldsymbol{\nabla}T,$$

where λ is the thermal conductivity. Let us also recall the Stefan-Boltzmann law, which states that the amount of heat received by a body per unit of time and through a surface element $d\Sigma$ of its boundary is given by

$$\frac{dQ}{dt} = -a(T^4 - T_0^4)d\Sigma.$$

In this expression, positive constant a is Stefan constant and T, the temperature of the considered surface element. In the case where $|T - T_0| \ll T_0$ show that the temperature difference $\phi = T - T_0$ satisfies Robin boundary conditions

$$[\mathbf{n} \cdot \boldsymbol{\nabla}\phi(\mathbf{r}, t) + h\phi(\mathbf{r}, t)]\Big|_{\partial\mathcal{D}} = 0,$$

where h is a positive constant to be determined. Function $\phi(\mathbf{r}, t)$ is a homogeneous solution of diffusion equation with Robin BC.

2. We now consider a semi-infinite bar whose temperature variations in the transverse direction are negligible. We can therefore restrict ourselves to a one-dimensional problem defined on the half-line $x \geq 0$. Let $G_R^+(x; x'; \tau)$ be the causal Green's function of diffusion equation with Robin BC in $x = 0$ vanishing as $x \to +\infty$. Using the results of the previous exercise show that

$$G_R^+(x; x'; \tau) = \int_0^{+\infty} du \, e^{-hu} \left(\frac{\partial}{\partial x'} G_{HD}^+(x + u; x'; \tau) + h G_{HD}^+(x + u; x'; \tau) \right).$$

3. Show then that

$$G_R^+(x; x'; \tau) = G_{HN}^+(x; x'; \tau) - 2h \int_0^\infty du \, e^{-hu} G_\infty^+(x + u; -x'; \tau), \qquad (3.173)$$

where G_∞^+ is the causal Green's function of diffusion on $\mathbb{R}$ and G_{HN}^+, the causal Green's function on $[0, \infty[$ with homogeneous Neumann BC in $x = 0$.

Solution p. 323.

■ Exercise 3.11. Cattaneo's equation in 3D

This exercise proposes to determine the causal Green's function G_∞^+, with BC at infinity, for Cattaneo equation in 3D using the results from paragraph 3.2.5 in 1D. Remember that Cattaneo equation in 3D reads:

$$\left(-\Delta + a^2 \frac{\partial}{\partial t} + \frac{1}{c^2} \frac{\partial^2}{\partial t^2} \right) \phi(\mathbf{r}, t) = \rho(\mathbf{r}, t).$$

1. Following the same steps as in the 1D case, show that

$$G_\infty^+(\mathbf{r}; t) = e^{-\frac{c^2 a^2 t}{2}} g_\infty^+(\mathbf{r}; t)$$

with

$$g_\infty^+(\mathbf{r}; t) = \frac{1}{(2\pi)^4} \int_{-\infty+i\gamma}^{+\infty+i\gamma} dz \int dk \, \frac{e^{-izt+i\mathbf{k}\cdot\mathbf{r}}}{k^2 - m^2 - z^2/c^2},$$

where γ is large enough real number.

2. Then integrate on the angles to get

$$g_\infty^+(\mathbf{r};t) = -\frac{i}{(2\pi)^3 r} \int_{-\infty+i\gamma}^{+\infty+i\gamma} dz \int_{-\infty}^{+\infty} dk \frac{ke^{-izt+ikr}}{k^2 - m^2 - z^2/c^2}.$$

3. Conclude noting that the above expression can be written as

$$g_\infty^+(\mathbf{r};t) = -\frac{1}{(2\pi)^3 r} \frac{\partial}{\partial r} \left(\int_{-\infty+i\gamma}^{+\infty+i\gamma} dz \int_{-\infty}^{+\infty} dk \frac{e^{-izt+ikr}}{k^2 - m^2 - z^2/c^2} \right)$$

and using result (3.151), p. 198 obtained for the one-dimensional case.

Solution p. 324.

■ **Exercise 3.12. Klein–Gordon equation**

This exercise proposes to calculate causal Green's functions G_1^+ and G_3^+ for Klein–Gordon equation

$$\left(\frac{1}{c^2} \partial_t^2 - \Delta + m^2 \right) G^+(\mathbf{r}, t) = \delta(\mathbf{r})\delta(t),$$

respectively in one and three dimensions, with BC at infinity.

1. Show first that in one dimension

$$G_1^+(x;t) = \frac{1}{(2\pi)^2} \int_{-\infty+i\gamma}^{+\infty+i\gamma} dz \int_{-\infty}^{+\infty} dk \frac{e^{-izt+ikx}}{\left(k - (z^2/c^2 - m^2)^{1/2}\right)\left(k + (z^2/c^2 - m^2)^{1/2}\right)},$$

where γ is a large enough real number and with a suitable choice of determination for $(z^2/c^2 - m^2)^{1/2}$. Perform the integral over k and show that $G_1^+(x;t) = 0$ if $|x| > ct$. Then conclude the calculation of G_1^+ along the lines of the calculation led at section 3.2.5 for Cattaneo equation.

2. Compute G_3^+ following the method proposed in the previous exercise for Cattaneo equation. We recall the definitions of the Bessel functions:

$$J_0(\rho) = \frac{1}{2\pi} \int_{-\pi}^{\pi} d\theta e^{i\rho\cos\theta} \quad \text{and} \quad J_1(\rho) = -\frac{dJ_0}{d\rho}.$$

Solution p. 324.

Chapter 4

Saddle-point method

A physicist often has to estimate quantities within a certain limit. This limit may correspond for example to a low temperature, a small coupling constant, a large system, or to the classical limit $\hbar \to 0$ etc. Moreover these physical quantities are often expressed as an integral. This chapter is devoted to the presentation of a very general method, called the saddle-point method, to estimate these integrals in some asymptotic limits. More precisely, their common feature is the emergence of a very pronounced maximum of the integrand. Schematically, the saddle-point method then consists in replacing the integrand by its asymptotic form near its maximum. The integral thus takes a Gaussian form incorporating the dominant contribution and that of the region near the saddle. The saddle-point method thus provides a unifying framework, relating a *priori* very different physical mechanisms. Another major advantage of this method is the systematic inclusion of contributions from the region near the saddle-point, which are not always easily accessible by other approaches.

As shown in the second part of this chapter, applications of the saddle-point method involve any type of integral, ranging from the simplest integral on a single real variable, to the most delicate on a field, going through the case of a discrete number of integration variables. We first address the case of a single integral of the form

$$I(\lambda) = \int_{\mathcal{D}} \mathrm{d}x \; e^{-f(x;\lambda)},$$

where $f(x; \lambda)$ is a real function depending on a parameter λ and a real variable x inside a domain $\mathcal{D} \subset \mathbb{R}$. This canonical case contains the essence of the general method. We discuss the asymptotic validity of the saddle-point estimate by evaluating the corresponding corrections. We then present the stationary phase method, applicable to the case where $f(x; \lambda)$ is a pure phase $f(x; \lambda) = i\varphi(x; \lambda)$ with φ real. This variant of the saddle-point method has specific difficulties arising from the oscillating nature of the integrand whose modulus is constant. In fact these two variants can be combined in a unified framework, where x becomes a complex variable z. The function $f(z, \lambda)$ is supposed to be analytic in z, while the domain of integration $\mathcal{D}$ is replaced by a path γ in the complex plane. Assuming sufficient analyticity properties of f, the introduction of a complex variable allows to deform γ into another path γ_s passing through a saddle point z_c. The integrand modulus $|e^{-f(z\lambda)}|$, seen as a function of two variables $\mathrm{Re}\, z$ and $\mathrm{Im}\, z$, has a saddle structure in the vicinity of z_s. This property is the origin of the generic name of this method, also called steepest descent method.

We then present the case of a multiple integral over several real variables in a domain $\mathcal{D}$ of $\mathbb{R}^d$. The corresponding Gaussian integrals are easily expressed in terms of the eigenvalues of a quadratic form. We finally consider a functional integral over a field $\phi(\zeta)$ indexed by a continuous variable ζ. It is defined as the continuous limit of a multiple integral over a discrete number of variables. The field at the saddle point is given by a variational principle. The dominant asymptotic form again reduces to a Gaussian integral which can be expressed, at least formally, in terms of the eigenvalues of a quadratic functional.

The second part of this chapter is devoted to the presentation of different examples. The first one consists in obtaining Stirling's formula for $N!$ with N large. The control parameter $\lambda = N$ can describe for instance the number of particles of an ideal gas at thermodynamic equilibrium, and the resulting asymptotic behaviour ensures the extensivity of its free energy. Another simple application to statistical mechanics is the proof of the equivalence of micro-canonical and canonical ensembles in the thermodynamic limit. Again the control parameter λ identifies with N, while the integration variable is the system energy in the canonical ensemble. We then study the partition function of a classical system, which provides an example of an integral over more than one variable. This system is assumed to be a crystal at a sufficiently low temperature. In this regime the partition function can be estimated by the saddle-point method where λ is the inverse temperature, while the saddle point is made of the sites of the crystal lattice formed at zero temperature. An example of functional integral comes from the Hubbard-Stratanovitch representation of the Ising model. The leading term at the saddle point corresponds here to the usual mean field theory. Finally, we establish the representation of the thermal propagator for a quantum system in terms of a path integral, and we derive the corresponding semi-classical approximation.

4.1 General properties

4.1.1 *Simple integral*

We consider here the simple integral

$$I(\lambda) = \int_{\mathcal{D}} \mathrm{d}x \; e^{-f(x;\lambda)}, \qquad (4.1)$$

on a domain $\mathcal{D} \subset \mathbb{R}$ where $f(x;\lambda)$ is a function taking real values. First we point out the preliminary assumptions for the application of the saddle-point method when the control parameter λ goes to infinity. This method is then implemented through the derivation of the so-called saddle point formula. We then discuss the nature of asymptotic approximate form thus obtained. We conclude this section by extending the method to the case where $f(x;\lambda)$ is a pure phase, i.e. a function taking pure imaginary values.

$\diamond$ Context of application and essence of the method $\diamond$

Let us assume the function $f(x;\lambda)$ admits a minimum at $x = x_\mathrm{s}(\lambda)$ inside integration domain $\mathcal{D}$ and that this minimum becomes increasingly narrower as $\lambda \to \infty$. Consequently the integrand $e^{-f(x;\lambda)}$ presents a very pronounced narrow peak in $x = x_\mathrm{s}(\lambda)$ for λ large enough. This situation is illustrated in Figure 4.1, where several curves representing the integrand $e^{-f(x,\lambda)}$ with increasing values of λ are presented, for a typical example of function $f(x;\lambda)$. It will be quantitatively specified later on.

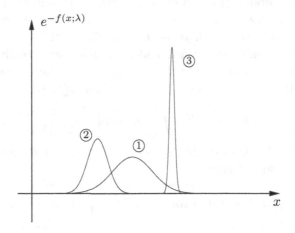

Fig. 4.1   This figure illustrates how the integrand $e^{-f(x;\lambda)}$ becomes narrower when $\lambda \to \infty$. Curves ①, ② and ③ correspond to $\lambda_1 < \lambda_2 < \lambda_3$.

In order to estimate $I(\lambda)$ in the limit $\lambda \to \infty$, we note that leading contributions should come from the region where x remains close to $x_\mathrm{s}(\lambda)$. Following this idea, it seems legitimate to replace $f(x;\lambda)$ by its Taylor expansion to second order in $(x - x_\mathrm{s}(\lambda))$: This is the essence of the saddle-point method. The corresponding estimate of $I(\lambda)$ is detailed in the following paragraph.

◇ **Saddle-point formula** ◇

Since $x_s(\lambda)$ is a minimum of $f(x; \lambda)$, we necessarily have[1]

$$\frac{\partial f}{\partial x}(x_s(\lambda); \lambda) = 0 \quad \text{and} \quad \frac{\partial^2 f}{\partial x^2}(x_s(\lambda); \lambda) > 0.$$

Let us introduce

$$\boxed{F_c(\lambda) = f(x_s(\lambda); \lambda) \quad \text{and} \quad C(\lambda) = \frac{\partial^2 f}{\partial x^2}(x_s(\lambda); \lambda).}$$

Taylor expansion of $f(x; \lambda)$ near the minimum $x_s(\lambda)$ reads

$$f(x; \lambda) = F_c(\lambda) + \frac{C(\lambda)}{2}(x - x_s(\lambda))^2 + O((x - x_s(\lambda))^3),$$

under the usual assumptions of differentiability. Using this expansion truncated to second order in the integrand $e^{-f(x;\lambda)}$, we get an approximate expression of $I(\lambda)$ written as

$$I(\lambda) \simeq e^{-F_c(\lambda)} \int_{\mathcal{D}} dx \, \exp\left[-\frac{C(\lambda)}{2}(x - x_s(\lambda))^2\right]. \tag{4.2}$$

Domain $\mathcal{D}$ is often only a part of $\mathbb{R}$. Since the integrand quickly vanishes around $x_s(\lambda)$ on a scale $\sigma(\lambda) = 1/\sqrt{C(\lambda)}$, we can extend the integration domain to $\mathbb{R}$ provided that the whole peak is included in $\mathcal{D}$. In the typical case where $\mathcal{D} = [L_1, L_2]$, this will only be justified under the necessary condition that distances from the minimum to the border $|L_i - x_s(\lambda)|$ are very large compared to the peak width $\sigma(\lambda)$. Approximate expression (4.2) is then replaced by

$$I(\lambda) \simeq e^{-F_c(\lambda)} \int_{-\infty}^{\infty} dx \, \exp\left[-\frac{(x - x_s(\lambda))^2}{2(\sigma(\lambda))^2}\right].$$

Elementary calculation of the full Gaussian integral over $\mathbb{R}$ eventually leads to the so-called saddle-point formula:

$$\boxed{\begin{array}{l}
\textbf{Saddle-point formula for} \quad : \quad I(\lambda) = \displaystyle\int_{\mathcal{D}} dx \, e^{-f(x;\lambda)}, \\[4mm]
\qquad\qquad I_{\text{saddle}}(\lambda) \;=\; \sqrt{2\pi}\,\sigma(\lambda)\,e^{-F_c(\lambda)}, \qquad\qquad (4.3) \\[4mm]
\text{with } F_c(\lambda) = f(x_s(\lambda); \lambda) \text{ and } \sigma(\lambda) = \left[\dfrac{\partial^2 f}{\partial x^2}(x_s(\lambda); \lambda)\right]^{-1/2}.
\end{array}}$$

The interpretation of approximation (4.3) is straightforward. The term $e^{-F_c(\lambda)}$ is simply the peak height. It can be seen as the contribution from $x_s(\lambda)$ itself.

---

[1]We will not consider the case where the first three partial derivatives of $f(x; \lambda)$ with respect to x vanish at $x = x_s(\lambda)$, while the fourth derivative is strictly positive. This situation is far more unusual in practice.

It is called dominant[✠] since the exponential factor ensures that it dominates all other contributions in general. The term

$$\sqrt{2\pi}\,\sigma(\lambda) \qquad (4.4)$$

is proportional to the peak width. It takes into account contributions of points close to $x_{\mathrm{s}}(\lambda)$. Returning to a terminology coming from statistical physics, this contribution is said to come from fluctuations of the variable x in the vicinity of $x_{\mathrm{s}}(\lambda)$. Note that approximation (4.3) simply amounts to multiply the peak height by its width, as expected from a rough argument!

> ✠ **Comment:** There are however cases where this is not the case. Consider for example a quantum particle in one dimension, confined by a potential presenting an absolute minimum at the origin, with $V(0) = 0$. One can calculate the corresponding thermal propagator in the semi-classical approximation, which is actually an application of the saddle-point method to path integral as will be discussed in section 4.2.5. Then at low temperature, the leading behaviour for the propagator near the origin takes the form $e^{-\beta\hbar\omega/2}$ where $\omega = \sqrt{V''(0)/m}$ is the vibration frequency in the locally harmonic potential. This behaviour comes from the contribution from the width of the saddle, that is to say from the determinant in formula (4.83)!

◇ Asymptotic nature of the saddle point formula ◇

Expression (4.3) only constitutes the asymptotic form of $I(\lambda)$ when $\lambda \to \infty$ if *a priori* neglected contributions are indeed small compared to $I_{\mathrm{saddle}}(\lambda)$. The corresponding corrections come from the non strictly Gaussian nature of integrand near $x_{\mathrm{s}}(\lambda)$, from the existence of other peaks, and finally from boundary effects. These three contributions are analysed successively in the following. As we shall see, a complete separation of the various contributions is not easy to formulate rigorously. Therefore the estimates should rather be seen as recipes, which nevertheless predict the correct behaviour!

Contribution of non-quadratic terms in the peak vicinity Let us estimate the contribution of terms with order higher than two in the Taylor expansion of f in the neighbourhood of $x_{\mathrm{s}}(\lambda)$. To this end we set

$$f(x;\lambda) = F_c(\lambda) + \frac{C(\lambda)}{2}(x - x_{\mathrm{s}}(\lambda))^2 + R(x;\lambda),$$

where the remainder $R(x;\lambda)$ is given by the series

$$R(x;\lambda) = \sum_{p \geq 3} \frac{\partial^p f}{\partial x^p}(x_{\mathrm{s}}(\lambda);\lambda)\, \frac{(x - x_{\mathrm{s}}(\lambda))^p}{p!}.$$

Since the remaining contribution to integral $I(\lambda)$ is *a priori* small, a perturbative expansion in powers of the remainder is legitimate. It consists in expanding e^{-R}

in powers of R and then replacing R by the previous Taylor expansion. We then carry out a term by term integration of the resulting expansion in powers of $(x - x_\mathrm{s}(\lambda))$. Integration domain $\mathcal{D}$ is extended to the whole $\mathbb{R}$, as in the calculation of approximate form (4.3), boundary effects being treated separately. This leads to moments of the Gaussian $\exp[-(x - x_\mathrm{s}(\lambda))^2/2(\sigma(\lambda))^2]$ at arbitrary order, which are all well defined. Odd moments are zero by symmetry, and even moments of order $2p$ are proportional to $\sigma(\lambda)^{2p}$. We thus obtain a perturbative expansion of the overall peak contribution,

$$\sqrt{2\pi}\,\sigma(\lambda)\,e^{-F_c(\lambda)}\,[1 + \sum_{p=2}^{\infty} c_p(\lambda)\,(\sigma(\lambda))^{2p}], \tag{4.5}$$

where each coefficient $c_p(\lambda)$ is a linear combination of products of derivatives of f in $x_\mathrm{s}(\lambda)$,

$$\prod_i \left(\frac{\partial^{p_i} f}{\partial x^{p_i}}(x_\mathrm{s}(\lambda); \lambda)\right)^{q_i} \quad \text{with} \quad \sum_i p_i q_i = 2p, \tag{4.6}$$

affected with purely numerical factors. For example, we have

$$c_2(\lambda) = -\frac{1}{8}\frac{\partial^4 f}{\partial x^4}(x_\mathrm{s}(\lambda); \lambda),$$

$$c_3(\lambda) = -\frac{1}{48}\frac{\partial^6 f}{\partial x^6}(x_\mathrm{s}(\lambda); \lambda) + \frac{5}{24}[\frac{\partial^3 f}{\partial x^3}(x_\mathrm{s}(\lambda); \lambda)]^2. \tag{4.7}$$

In order for approximate form (4.3) to be asymptotically exact, it is necessary that each correction of order p in the perturbative series (4.5) tends to zero as $\lambda \to \infty$. The structure of coefficients $c_p(\lambda)$ then impose the following condition:

$$\boxed{\begin{array}{l} \textbf{Validity of saddle-point formula (4.3):} \\[2mm] \forall q \geq 3, \quad [\sigma(\lambda)]^q\,\frac{\partial^q f}{\partial x^q}(x_\mathrm{s}(\lambda); \lambda) \to 0 \text{ when } \lambda \to \infty. \end{array}} \tag{4.8}$$

In practice these limit behaviours are almost always sufficient for the sum of all corrections in series (4.5) to be effectively negligible at large λ. Strictly speaking, we must also check the convergence of series (4.5), which is not necessarily assured. In case of asymptotic convergence only, saddle-point formula (4.3) may nevertheless provide the exact leading behaviour of the overall peak contribution when $\lambda \to \infty$.

Contribution from secondary peaks   In addition of $x_\mathrm{s}(\lambda)$, function $f(x; \lambda)$ may admit other local minima at points $x_i(\lambda)$ belonging to domain $\mathcal{D}$ with $f(x_i(\lambda); \lambda) > F_c(\lambda)$. In addition to the main peak in $x_\mathrm{s}(\lambda)$, integrand $e^{-f(x;\lambda)}$ then has secondary peaks at $x_i(\lambda)$, also assumed to be very narrow.

If the peaks are well separated and have a small overlap, then the contribution of each peak may be determined independently. Let us admit that for each of them the corresponding saddle-point formula

$$\sqrt{2\pi}\,\sigma_i(\lambda)\,e^{-F_i(\lambda)}, \tag{4.9}$$

is asymptotically valid when $\lambda \to \infty$. Then the dominant contribution to the integral $I(\lambda)$ is the largest of the expressions (4.3) and (4.9). Note that this is not necessarily the contribution from the main peak, since a larger secondary peak can actually dominate. For large λ however, the heights vary generally exponentially faster than the widths so that the main peak indeed gives the dominant contribution. Note that the prerequisite of small overlap of peaks imposes that their relative distances are large compared to their widths, i.e. $|x_i(\lambda) - x_j(\lambda)| \gg \sigma_i(\lambda), \sigma_j(\lambda)$.

Boundary contribution and asymptotic formula Consider the case where the domain reduces to a segment $[L_1, L_2]$, and assume that there is a single peak at $x_s(\lambda)$, whose contribution is asymptotically given by saddle-point formula (4.3). So a rough order of magnitude estimate of the contribution of each edge L_1 and L_2 is then simply given respectively by $(L_2 - L_1)e^{-f(L_1;\lambda)}$ and $(L_2 - L_1)e^{-f(L_2;\lambda)}$. These contributions are generally negligible compared with the saddle-point estimate (4.3), provided that the peak width $\sigma(\lambda)$ is not exponentially small when λ diverges.

If all of the above conditions are met for each analysed contributions, then the saddle-point formula indeed provides the asymptotic behaviour of $I(\lambda)$ for large λ.

Case of a multiplicative control parameter It often happens in practice that the control parameter λ appears multiplicatively in function f, which then takes the form $f(x; \lambda) = \lambda g(x)$. Suppose that $g(x)$ has an absolute minimum at x_0, this point lying strictly inside domain $\mathcal{D}$. The main peak position $x_s(\lambda)$ is then independent of λ and obviously reduces to this point x_0. We leave to the reader to check that all the conditions leading to asymptotic formula (4.3) are then fulfilled so that

---

Saddle point formula for $: I(\lambda) = \int_{\mathcal{D}} \mathrm{d}x \; e^{-\lambda g(x)}$

$$I_{\text{saddle}}(\lambda) = \sqrt{\frac{2\pi}{\lambda g''(x_0)}} \; e^{-\lambda g(x_0)} \qquad (4.10)$$

x_0 : Absolute minimum of $g(x)$.

The corrections to dominant asymptotic behaviour (4.10) are easily obtained from perturbation expansion (4.5), generated by local deviations from the Gaussian shape in the vicinity of x_0. Width $\sigma(\lambda)$ is inversely proportional to $\lambda^{1/2}$, and each derivative of f of order q is of course proportional to λ itself. Each term of order p,

$$c_p(\lambda) \, (\sigma(\lambda))^{2p} \qquad (4.11)$$

is thus a polynomial in $1/\lambda$. For example $c_2 \, (\sigma(\lambda))^4$ is proportional to $1/\lambda$, while $c_3 \, (\sigma(\lambda))^6$ is a second order polynomial in $1/\lambda$ with a null constant term. As a result, the series (4.5) can be reorganised as a perturbative expansion in powers of

$1/\lambda$: each term at a given order comes from a finite number of terms (4.11). All other corrections to the saddle point formula, coming for instance from secondary peaks or boundaries, are in general exponentially smaller.

Counterexample   The saddle-point method is not applicable in general when function $f(x;\lambda)$ takes the form $f(x;\lambda) = g(\lambda x)$. Indeed, due to the underlying scale invariance of this form, integrand $e^{-f(x;\lambda)}$ does not locally reduce to a Gaussian in the limit $\lambda \to \infty$. For example, if we consider an ideal Fermi gas at very low temperatures, thermodynamic quantities can be rewritten as integrals over energy E involving derivative of the Fermi-Dirac distribution. This distribution, which is a function of βE with $\beta = 1/(k_B T)$, has a very sharp peak near the Fermi energy when $\beta \to \infty$. However, it is not possible to approximate the peak by a Gaussian. It is in fact necessary in this case to make the change of variable $\beta E \to x$.

$$\Diamond \ \textbf{Extension to the case of a pure phase} \ \Diamond$$

Consider a function $f(x;\lambda)$ which is a pure phase, $f(x;\lambda) = i\varphi(x;\lambda)$ with φ real. Integral $I(\lambda)$ is now of the type

$$I(\lambda) = \int_{\mathcal{D}} \mathrm{d}x \, e^{-i\varphi(x;\lambda)} \tag{4.12}$$

and we propose once again to find its asymptotic form when λ diverges, up to some hypotheses usually satisfied by φ. Unlike the previous case where f was real, here the integrand keeps a constant modulus equal to 1 and in general oscillates. Estimations immediately become much more difficult, because of cancellations between positive and negative contributions. We first argue that the neighbourhood of an arbitrary integration point contributes little as a result of destructive interferences. We then show that the neighbourhood of a point where φ is extremal should give the dominant contributions.

Destructive interference in the vicinity of an arbitrary point   Consider an arbitrary point x_a strictly inside domain $\mathcal{D}$. We then perform a Taylor series expansion of phase $\varphi(x,\lambda)$ near point x_a,

$$\varphi(x;\lambda) = \varphi(x_a;\lambda) + k_a(\lambda)(x - x_a) + \frac{1}{2}\frac{\partial^2 \varphi}{\partial x^2}(x_a;\lambda)(x - x_a)^2 + \cdots, \tag{4.13}$$

with $k_a(\lambda) = (\partial \varphi/\partial x)(x_a;\lambda)$. Local wave number k_a controls the oscillations wavelength in the vicinity of x_a. Suppose that k_a diverges when $\lambda \to \infty$. As λ increases, $e^{-i\varphi(x;\lambda)}$ oscillates more and more rapidly in the vicinity of x_a. Destructive interferences appear when integrating this phase factor on such a neighbourhood, resulting in a vanishing contribution in the limit $\lambda \to \infty$. The contributions of this neighbourhood to $I(\lambda)$ are then expected to be small.

Contribution from the vicinity of phase extremum Wave number k_a vanishes if the point x_a coincides with an extremum of φ, $x_a = x_s(\lambda)$. The phase φ is then stationary in a neighbourhood of $x_s(\lambda)$. In other words integrand $e^{-i\varphi(x;\lambda)}$ now remains essentially constant over a length scale larger than in the vicinity of an arbitrary point with $k_a \neq 0$. The interferences are therefore *a priori* much less destructive than previously. To estimate the corresponding contribution of the neighbourhood of $x_s(\lambda)$, we substitute $\varphi(x, \lambda)$ by its local quadratic form $\varphi(x_s(\lambda); \lambda) + \frac{1}{2}\frac{\partial^2 \varphi}{\partial x^2}(x_s(\lambda); \lambda) \, (x - x_s(\lambda))^2$, obtained by truncating the Taylor series. Extending the domain $\mathcal{D}$ to $\mathbb{R}$, we get a Gaussian integral of pure imaginary covariance. The calculation is presented in detail in Appendix E, pp. 280-281. It leads to the so-called stationary phase formula for the contribution from vicinity of the extremum under consideration to $I(\lambda)$:

$$
\textbf{Stationary phase} \;:\; I(\lambda) = \int_{\mathcal{D}} \mathrm{d}x \; e^{-i\varphi(x;\lambda)}
$$

$$
I_{\text{sta}}(\lambda) = \sqrt{2\pi} \left[i\frac{\partial^2 \varphi}{\partial x^2}(x_s(\lambda); \lambda) \right]^{-1/2} e^{-i\varphi(x_s(\lambda); \lambda)} \quad (4.14)
$$

$$
x_s(\lambda) \quad \text{such that} \;:\; \frac{\partial \varphi}{\partial x}(x_s(\lambda); \lambda) = 0.
$$

In the above formula the choice of determination for $Z^{1/2}$ is $\sqrt{|Z|}e^{-i\mathrm{Arg}(Z)/2}$. A direct analysis of other contributions to $I(\lambda)$ is much more difficult than in the previous case where the integrand was real and with constant sign. Note in particular that a perturbative estimation of deviations from the local quadratic form of φ, similar to series (4.5) brings up ill-defined moments. Boundary contributions are also more difficult to evaluate, while all the extrema *a priori* give contributions of the same order as expression (4.14)! Integrals (4.1) and (4.12) can in fact be studied by a standard method by passing in the complex plane, as we shall see. In this unified framework, we find again the stationary phase formula (4.14). It gives the asymptotic behaviour of $I(\lambda)$ for large λ under conditions eventually similar to those for saddle point formula (4.3).

4.1.2 *Integral on a path in the complex plane*

The saddle point and stationary phase methods are actually special cases of a more general method for integrals in the complex plane of the type

$$
I(\lambda) = \int_{\gamma} \mathrm{d}z \; e^{-f(z;\lambda)}, \quad (4.15)
$$

where γ is a path on $\mathbb{C}$ connecting two points. Here function $f(z; \lambda)$ is analytic with respect to complex variable z, in a region including the integration contour.

Real integral (4.1) is a particular form of expression (4.15), for a path γ reducing to segment $[L_1, L_2]$, and a function f whose restriction to this segment is purely real. Integral (4.12) on a pure phase factor is another particular form, with the same path γ and a function f becoming purely imaginary on $[L_1, L_2]$.

The goal is always to assess $I(\lambda)$ when $\lambda \to \infty$, assuming that function $f(z, \lambda)$ then presents large variations when z travels along the integration path γ. In the following we make a clear distinction between the demonstration of the saddle point formula and its practical use. For the latter we make an analogy with the residue theorem. Finally we end this section by checking that expressions (4.3) and (4.14) are indeed special forms of the complex saddle-point method.

◇ **Proof of the saddle point formula** ◇

Deformation of the original path γ   Before we ask what is a saddle point[2] z_s in the complex case, we start with a simple observation. Suppose the function f to integrate has such a saddle point z_s but that z_s is not on the integration path γ. Suppose then that the analyticity domain of f is large enough so as to include not only the path γ but also the point z_s. Then application of Cauchy's theorem allows one to rewrite $I(\lambda)$ as

$$I(\lambda) = \int_{\gamma_s} dz \, e^{-f(z;\lambda)}, \tag{4.16}$$

where γ_s is a path contained in the analyticity domain and obtained by deformation of γ, passing through the point z_s under consideration (see Figure 4.2). We discuss below the characteristics of the saddle point z_s, and clarify below how to choose this path.

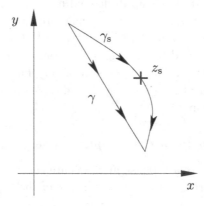

Fig. 4.2   The original integration contour γ is deformed into γ_s to pass through the saddle point z_s. The specific choice of γ_s is specified below.

---

[2]We shall temporarily write z_s instead of $z_s(\lambda)$.

Saddle point Compared to the case where the integration involves a real variable, it is natural to ask what the analogue of the saddle point x_s is. To answer this question, let us call $P(x, y; \lambda)$ and $Q(x, y; \lambda)$ respectively the real and imaginary parts of $f(z; \lambda)$, i.e.

$$f(z; \lambda) = P(x, y; \lambda) + iQ(x, y; \lambda) \qquad \text{with } z = x + iy.$$

According to previous analysis, it seems natural to say that the analogue to x_s is a point $z_s = x_s + iy_c$ where P has a minimum while Q is extremal. It imposes in particular that the first partial derivatives of these functions are zero:

$$\boxed{\frac{\partial f}{\partial z}(z_s; \lambda) = 0.}$$

However function $f(z; \lambda)$ is analytic in a region including the contour of integration. As mentioned in Appendix A, p. 263, P and Q are then harmonic functions. Function P has therefore no local minimum, since its Laplacian vanishes:

$$\frac{\partial^2 P}{\partial x^2} + \frac{\partial^2 P}{\partial y^2} = 0.$$

To fix ideas and go ahead with the argument, let us assume that $(\partial^2 P/\partial x^2) < 0$ at z_s. We then have $(\partial^2 P/\partial y^2) > 0$ at z_s. In other words, if z_s corresponds to a maximum of P in the x direction, then it is a minimum of P in the y direction. In fact we are just recalling here that harmonic function P has the shape of a saddle around z_s. We show on Figure 4.3 the corresponding shape for e^{-P}.

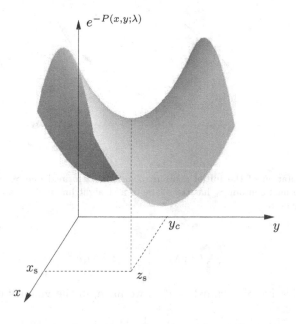

Fig. 4.3 The function $e^{-P(x, y; \lambda)}$ has the shape of a saddle around point z_s.

Path of steepest descent   At this level a natural idea is to ensure that deformed path γ_s corresponds, in the vicinity of z_s, to the path of steepest descent on the saddle corresponding to e^{-P}. Note that the terminology "descent" refers here to the situation as seen from point z_s. Such a choice ensures indeed that the variations of e^{-P} around saddle point z_s are the strongest and therefore the dominant contribution to the integral indeed comes from z_s. However we have not considered so far the part of e^{-f} behaving as e^{-iQ}, which could undermine this argument by producing destructive interferences around the point z_s. This is actually not the case thanks to analyticity properties of $f(z; \lambda)$. Instead, the steepest descent path corresponds actually to one of the two paths where imaginary part Q varies the least in the vicinity of z_s! To be convinced of this property, let us take the line parametrised by the equation

$$z - z_s = \pm \rho e^{i\alpha} \tag{4.17}$$

for a path γ_s in the vicinity of z_s, where α is the angle of this line with the x-axis (see Figure 4.4) and where the two signs correspond to two portions of the path on both sides of z_s. Let us also write

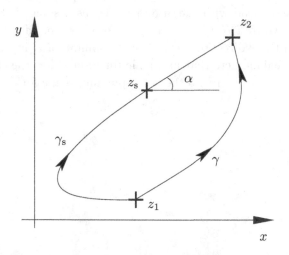

Fig. 4.4   Representation of the initial contour γ and the deformed one γ_s passing through the saddle point. Deformed contour γ_s has the shape of a straight line making an angle α with the x axis in the vicinity of z_s.

$$\frac{\partial^2 f}{\partial z^2}(z_s; \lambda) = \left| \frac{\partial^2 f}{\partial z^2}(z_s; \lambda) \right| e^{i\theta}. \tag{4.18}$$

As the first derivative of f vanishes at z_s we have, in the vicinity of z_s:

$$f(z; \lambda) - f(z_s; \lambda) \simeq \frac{1}{2} \frac{\partial^2 f}{\partial z^2}(z_s; \lambda)(z - z_s)^2. \tag{4.19}$$

Remember that in the limit $\lambda \to \infty$, $\left|\frac{\partial^2 f}{\partial z^2}(z_s; \lambda)\right|$ is supposed to be large. Expansion (4.19) for f together with parametrisation (4.17) gives for P and Q:

$$P(x, y; \lambda) - P(x_s, y_c; \lambda) \simeq \frac{1}{2}\rho^2 \left|\frac{\partial^2 f}{\partial z^2}(z_s; \lambda)\right| \cos(\theta + 2\alpha)$$

and

$$Q(x, y; \lambda) - Q(x_s, y_c; \lambda) \simeq \frac{1}{2}\rho^2 \left|\frac{\partial^2 f}{\partial z^2}(z_s; \lambda)\right| \sin(\theta + 2\alpha). \tag{4.20}$$

The path of steepest descent corresponds to take the largest positive value for $\cos(\theta + 2\alpha)$, namely $\theta + 2\alpha = 2\pi n$ with n integer. But in this case, $\sin(\theta + 2\alpha)$ vanishes and equation (4.20) then shows that the changes of the imaginary part Q in the vicinity of z_s are very small, of the order of $O(\rho^3)$. In summary, the path of steepest descent corresponds to the choice

$$\alpha = -\frac{\theta}{2} + n\pi, \tag{4.21}$$

or to $e^{i\alpha} = \pm e^{-i\theta/2}$. The two alternative signs correspond of course to portions of the path on both sides of z_s. The situation is summarised in Figure 4.5.

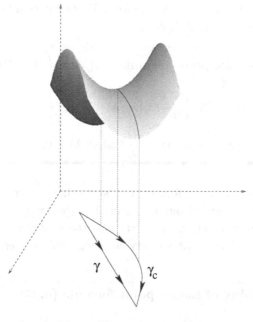

Fig. 4.5   Representation of $\exp[-P(x, y : \lambda)]$ as a function of x and y. In the vicinity of z_s, path γ_s corresponds to the path of steepest descent.

Complex saddle point formula   Since

$$\mathrm{d}z = \mathrm{d}\rho e^{i\alpha} = \pm \mathrm{d}\rho e^{-i\frac{\theta}{2}}$$

we obtain from the expansion (4.18) and the choice (4.21), for each part of the path:

$$I(\lambda) \simeq 2e^{-f(z_s(\lambda);\lambda)}e^{-i\frac{\theta}{2}} \int_0^\infty \mathrm{d}\rho \, \exp\left[-\frac{1}{2}\rho^2 \left|\frac{\partial^2 f}{\partial z^2}(z_s(\lambda);\lambda)\right|\right] \qquad (4.22)$$

$$\simeq e^{-f(z_s(\lambda);\lambda)}e^{-i\frac{\theta}{2}}\sqrt{2\pi}\left|\frac{\partial^2 f}{\partial z^2}(z_s(\lambda);\lambda)\right|^{-1/2}, \qquad (4.23)$$

where we extended the integration to the entire line (4.17), just like in the case of the real integral on p. 216. Note that expression (4.22) is a simple way to understand the choice of path of steepest descent since with such a choice, the argument of the exponential is purely real. The same strategy is recalled in Appendix E for the calculation of some Gaussian integrals. We can express the prefactor in formula (4.23) as a function of the second derivative of f at $z_s(\lambda)$ only. To this end, it is necessary to introduce the analytic function $Z^{-1/2}$ of the complex variable Z. To define it unambiguously, we choose the determination

$$Z^{-1/2} = e^{-i \arg Z/2}/\sqrt{|Z|}, \qquad (4.24)$$

ensuring that for Z real and positive, $Z^{-1/2}$ is real positive. Of course, it is also necessary to introduce a branch cut starting from $Z = 0$ which is a singular point. We choose it here along the negative real axis. Using relation (4.18) and this choice of determination, formula (4.23) becomes:

Complex saddle point formula: $I(\lambda) = \displaystyle\int_\gamma \mathrm{d}z \; e^{-f(z;\lambda)}$

$$I_{\mathrm{saddle}}(\lambda) = \sqrt{2\pi}\left[\frac{\partial^2 f}{\partial z^2}(z_s(\lambda);\lambda)\right]^{-1/2} e^{-f(z_s(\lambda);\lambda)} \qquad (4.25)$$

$$z_s(\lambda) \quad \text{such that} \quad \frac{\partial f}{\partial z}(z_s(\lambda);\lambda) = 0.$$

Validity of the asymptotic formula   Estimating corrections to saddle point formula (4.25) may be carried out by extending the analysis introduced for real integral (4.1) in previous section. In particular the contributions of non-Gaussian fluctuations in the saddle point vicinity are negligible under similar conditions, namely

---

Validity of saddle point formula (4.25):

$$\forall p \geq 3, \qquad \frac{\partial^p f}{\partial z^p}(z_s(\lambda);\lambda)[\frac{\partial^2 f}{\partial z^2}(z_s(\lambda);\lambda)]^{-p/2} \to 0 \text{ when } \lambda \to \infty. \qquad (4.26)$$

These conditions are necessary to ensure that formula (4.25) is indeed the asymptotic form of $I(\lambda)$ when $\lambda \to \infty$. Let us stress that in addition to these conditions one must check that the contributions of the other regions along the deformed path γ_s remain negligible compared to that of the saddle point. It may indeed happen that path γ_s passes through regions where the integrand takes large values that may overcome the saddle contribution. For instance, if we consider the integral

$$I(\lambda) = \int_1^2 dx e^{-\lambda x^2},$$

the saddle point is located in $z_s = 0$. However the asymptotic behaviour of $I(\lambda)$ when $\lambda \to \infty$ is $e^{-\lambda}/2\lambda$, while the saddle point contribution goes as $1/\sqrt{\lambda}$. A similar mechanism might happen in the presence of several saddle points.

◇ Practical use and extensions ◇

The complex saddle point formula is used in practice somehow similarly to the residue theorem, with the following difference. For the residue theorem, the original path γ is deformed to enclose a simple pole while in the case of complex saddle point formula, the path γ is deformed to pass through the saddle point and to avoid all singularities! In practice, it is sufficient to ensure that this deformation of path γ is possible, and then to simply apply formula (4.25).

Saddle point formula extends to integrals of the form

$$I(\lambda) = \int_\gamma dz\, \ell(z) e^{-f(z;\lambda)} \tag{4.27}$$

where $\ell(z)$ is a slowly varying function. In this case we get, under assumptions (4.26),

$$I(\lambda) \simeq \ell(z_s)\sqrt{2\pi} \left[\frac{\partial^2 f}{\partial z^2}(z_s(\lambda);\lambda)\right]^{-1/2} e^{-f(z_s(\lambda);\lambda)}. \tag{4.28}$$

◇ Back on the real axis ◇

Let us return to the case of simple integral (4.1) originally defined by a path γ reduced to a segment of the real axis. In this case, the imaginary part Q of f is identically zero on γ. One of complex saddle points z_s is then simply the real point $z_s(\lambda) = x_s(\lambda)$ minimising f on the real axis. In addition, the path of steepest descent in the saddle structure of e^{-P} is nothing but the real axis itself. The path γ_s then coincides with γ, and since $(\partial^2 f/\partial z^2)(z_s(\lambda);\lambda) = (\partial^2 f/\partial x^2)(x_s(\lambda);\lambda)$, complex formula (4.25) coincides with real formula (4.3), as expected!

Finally, in the case of integral (4.12), p. 220, involving a pure phase factor, it is again possible to choose the point $z_s(\lambda)$ on the real axis, corresponding to the point $x_s(\lambda)$ maximising the phase. However the line of steepest slope is no longer on the real axis. Indeed, as

$$(\partial^2 f/\partial z^2)(z_s(\lambda);\lambda) = i(\partial^2 \varphi/\partial x^2)(x_s(\lambda);\lambda),$$

Equations (4.18) and (4.21) show that this line makes an angle of $\pi/4$ with the real axis, as shown in Figure 4.6. Thus path γ_s is here different from γ. Complex saddle point formula (4.25) then gives back stationary phase formula (4.14), p. 221. Note that the method of path deformation in the complex plane allows to estimate corrections to this formula, and so to convincingly justify its asymptotic character.

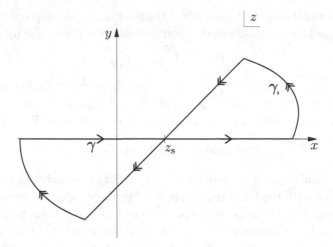

Fig. 4.6   For integral (4.12), the path of steepest descent makes an angle of $\pi/4$ with the original path which is the real axis. Note that it is exactly this deformation contour that is used to compute the Gaussian case, presented in Appendix E, p. 280.

4.1.3   *Case of multiple integral*

In this section, we first consider an integral over d real variables with an integration domain $\mathcal{D}$ included in $\mathbb{R}^d$. We then turn to the case of a functional integral involving an infinite number of integration variables. We show that the saddle point formula established for a simple integral is easily generalised to these two situations, at least formally. Indeed its implementation is more difficult than in the one-dimensional case, because it requires the diagonalisation of a $d \times d$ matrix in finite dimension, or the calculation of the spectrum of an operator for a functional integral. Similarly considerations about its asymptotic character become much more difficult and we shall not enter into a general discussion about this problem here.

◇ Finite number of variables ◇

Let $\boldsymbol{x} = (x_1, .., x_d)$ denote d real variables and $f(\boldsymbol{x}; \lambda)$ a real function of these variables depending on a control parameter λ. We consider the integral

$$I(\lambda) = \int_{\mathcal{D}} d\boldsymbol{x} \, e^{-f(\boldsymbol{x};\lambda)}, \tag{4.29}$$

with the notation $d\boldsymbol{x} = \prod_{i=1}^{d} dx_i$. Suppose that $f(\boldsymbol{x}; \lambda)$ has an absolute minimum[3] at the saddle point $\boldsymbol{x} = \boldsymbol{x}_s(\lambda)$, located inside of domain $\mathcal{D}$. If this minimum becomes narrower and more pronounced when $\lambda \to \infty$, then it is reasonable to replace the integrand $e^{-f(\boldsymbol{x};\lambda)}$ by its local Gaussian approximation in the vicinity of saddle point $\boldsymbol{x}_s(\lambda)$ while extending the integration domain to the whole space $\mathbb{R}^d$. This

---

[3]The function $f(\boldsymbol{x}; \lambda)$ may have other secondary minima. This will generally be more and more frequent as dimension d increases.

approximation is the obvious extension to d dimensions of the saddle-point method for a simple integral. Its specific form is set forth below.

Local Gaussian approximation We perform a Taylor expansion of $f(x; \lambda)$ in the neighbourhood of $x_s(\lambda)$. It leads to

$$f(x; \lambda) = f(x_s(\lambda); \lambda) + \frac{1}{2}(x - x_s(\lambda))^T \cdot C(\lambda) \cdot (x - x_s(\lambda)) + \cdots , \qquad (4.30)$$

where $C(\lambda)$ is the symmetric $d \times d$ matrix of elements

$$C_{ij}(\lambda) = \frac{\partial^2 f}{\partial x_i \partial x_j}(x_s(\lambda); \lambda).$$

We used in expression (4.30) the generic notation

$$\mathbf{u}^T \cdot M \cdot \mathbf{v} = \sum_{ij} u_i M_{ij} v_j, \qquad (4.31)$$

for two vectors $\mathbf{u}$ and $\mathbf{v}$ of $\mathbb{R}^d$, M a $d \times d$ matrix, and where superscript ^{T} denotes the transposition. Keeping only the quadratic terms in Taylor series (4.30) and replacing $\mathcal{D}$ by $\mathbb{R}^d$, we obtain the Gaussian approximation of integral $I(\lambda)$

$$I_{\text{saddle}}(\lambda) = e^{-F_c(\lambda)} \int_{\mathbb{R}^d} dx \, \exp\left[-\frac{1}{2}(x - x_s(\lambda))^T \cdot C(\lambda) \cdot (x - x_s(\lambda))\right], \quad (4.32)$$

with $F_c(\lambda) = f(x_s(\lambda); \lambda)$.

It thus remains to compute a multiple Gaussian integral. This calculation is presented in the general case in Appendix E, p. 282. The result reads

$$\int_{\mathbb{R}^d} dx \, e^{-\frac{1}{2} x^T \cdot C(\lambda) \cdot x}$$
$$= \sqrt{\frac{(2\pi)^d}{\text{Det}(C(\lambda))}}. \qquad (4.33)$$

We refer the reader to Appendix E for details ✠.

✠ **Comment:** Let us however briefly indicate here that the occurrence of the determinant of $C(\lambda)$ is easily understood in the case where matrix $C(\lambda)$ is diagonal,

$$C(\lambda) = \begin{pmatrix} \mu_1(\lambda) & & \\ & \mu_2(\lambda) & \\ & & \ddots \\ & & & \mu_d(\lambda) \end{pmatrix}.$$

The multiple integral factorises in this case. The integral over component x_i of $\mathbf{x}$ gives $\sqrt{2\pi/\mu_i(\lambda)}$. The product of these terms leads to results (4.33).

Note that $\text{Det}\, C(\lambda) > 0$. Indeed the d eigenvalues $\mu_1(\lambda), ..., \mu_d(\lambda)$ are necessarily real and strictly positive since $C(\lambda)$ is a real symmetric matrix and $x_s(\lambda)$ is a minimum of f. We therefore find the result:

$$\textbf{Saddle point formula for} \quad : \quad I(\lambda) = \int_{\mathcal{D}} d\boldsymbol{x}\; e^{-f(\boldsymbol{x};\lambda)},$$

$$I_{\text{saddle}}(\lambda) \;=\; e^{-F_c(\lambda)} \sqrt{\frac{(2\pi)^d}{\text{Det}(C(\lambda))}} \qquad (4.34)$$

$$F_c(\lambda) = f(\boldsymbol{x}_{\text{s}}(\lambda);\lambda) \text{ and } C_{ij}(\lambda) = \frac{\partial^2 f}{\partial x_i \partial x_j}(\boldsymbol{x}_{\text{s}}(\lambda);\lambda)$$

$$\text{and } \frac{\partial f}{\partial x_i}(\boldsymbol{x}_{\text{s}}(\lambda);\lambda) = 0.$$

Regarding the asymptotic nature of formula (4.34), one must check that the terms of order higher than two in Taylor series (4.30) give contributions tending to 0 when $\lambda \to \infty$. As in the case of a single real variable, this condition is fulfilled if

$$\textbf{Validity of saddle point formula (4.34):}$$

$$\forall q \geq 3 \qquad \text{with} \quad q = \sum_{i=1}^{p} q_i, \qquad (4.35)$$

$$\frac{1}{\sqrt{\mu_{i_1}^{q_1}(\lambda).....\mu_{i_p}^{q_p}(\lambda)}} \; \frac{\partial^q f}{\partial x_{i_1}^{q_1}...\partial x_{i_p}^{q_p}}(\boldsymbol{x}_{\text{s}}(\lambda);\lambda) \to 0 \quad \text{when} \quad \lambda \to \infty.$$

Under this condition, $\mu_{i_1}(\lambda), ..., \mu_{i_p}(\lambda)$ corresponds to an arbitrary set of p eigenvalues of matrix $C(\lambda)$. It is analogous to condition (4.8), p. 218. As in the simple integral case, condition (4.35) is automatically satisfied for f of the form $f(\boldsymbol{x};\lambda) = \lambda g(\boldsymbol{x})$.

$\Diamond$ Infinite number of variables and transition to the functional integral $\Diamond$

Under certain conditions, the multiple integral becomes a functional integral when the number of integration variables becomes infinite. We first describe briefly how this integral is constructed from a suitable limit. We then establish the corresponding saddle point formula.

Construction of the functional integral   The limit process at play takes the following generic form. Start from a multiple integral such as (4.29) in a space of $d = N$ dimensions. From now on let us write ϕ_i the i-th variable while i belongs to the set $[1, ..., N]$, assuming ϕ_i real to fix ideas. In most physical situations, the index i is associated with a parameter $\zeta(i, N)$ depending also on the total number of variables N. For example this parameter may be a position, as in the Hubbard-Stratanovitch representation of the Ising model, or a pseudo-time as in the path integral representation of the quantum thermal propagator. These representations are examples described in the second part of this chapter.

At this level variable ϕ_i can be seen as a field $\phi(\zeta)$, depending on a parameter ζ taking discrete values for finite N. If the set of values $\zeta(i, N)$ becomes dense in a given domain Ω when $N \to \infty$, then a configuration $\{\phi_i, i = 1, ..., N\}$ leads to a continuous field denoted $\phi(\cdot) = \phi(\zeta)$ where ζ belongs to Ω. The elementary volume in the initial integration space for finite N then defines the measure $d[\phi(\cdot)]$ of a functional integral on the domain of fields $\phi(\cdot)$, i.e.

$$\lim_{N \to \infty} \prod_{i=1}^{N} d\phi_i = d[\phi(\cdot)]. \tag{4.36}$$

In this limit $N \to \infty$, the function of N variables $f(\phi_1, \phi_2, ..., \phi_N; \lambda)$ defines a functional, often denoted[4] $S_\lambda[\phi(\cdot)]$. This limit process from an initial multiple integral thus leads to introduce the functional integral

$$\boxed{I(\lambda) = \int d[\phi(\cdot)] \, \exp\{-S_\lambda[\phi(\cdot)]\}.} \tag{4.37}$$

In the above presentation, we deliberately omitted all mathematical difficulties arising in the limit process $N \to \infty$. In some cases, like that of Feynman-Kac representation of the density matrix in statistical mechanics, this process can be controlled rigorously[5]. In the following we assume that functional integral (4.37) is well defined.

Functional saddle-point formula As in the discrete case, a very narrow peak in the integrand of (4.37) may appear when $\lambda \to \infty$. This peak is reached for a field $\phi_c(\cdot)$ corresponding to an extremum of $S_\lambda[\phi(\cdot)]$, so that the functional derivative[6] of $S_\lambda[\phi(\cdot)]$ vanishes at $\phi_c(\cdot)$, i.e.

$$\frac{\delta S_\lambda[\phi(\cdot)]}{\delta \phi(\zeta)}[\phi_c(\cdot)] = 0 \qquad \forall \zeta \in \Omega. \tag{4.38}$$

It is then reasonable to expand $S_\lambda[\phi(\cdot)]$ up to second order in $(\phi - \phi_c)$ to obtain the following Gaussian approximation

$$I_{\text{saddle}}(\lambda) = e^{-S_\lambda[\phi_c(\cdot)]} \int d[\phi(\cdot)] \, \exp\left\{-\frac{1}{2} \int_{\Omega^2} d\zeta_1 d\zeta_2 \, [\phi(\zeta_1) - \phi_c(\zeta_1)]\right.$$
$$\left. \times C_\lambda(\zeta_1, \zeta_2) \, [\phi(\zeta_2) - \phi_c(\zeta_2)]\right\}. \tag{4.39}$$

Covariance $C_\lambda(\zeta_1, \zeta_2)$ appearing in Gaussian functional integral (4.39) is defined as

$$C_\lambda(\zeta_1, \zeta_2) = \frac{\delta^2 S_\lambda}{\delta \phi(\zeta_1) \delta \phi(\zeta_2)}[\phi_c(\cdot)]. \tag{4.40}$$

---

[4]The notation S comes from field theory, where this quantity then represents the action.
[5]See the book [85] for a detailed presentation of this demonstration.
[6]Appendix H contains reminders about functional derivatives.

This Gaussian integral is formally expressed in terms of the eigenvalues $\mu_\lambda(\nu)$ of operator C_λ, where ν is a spectral parameter[✠] characterising the eigenvalue under consideration. These eigenvalues are real, since C_λ is Hermitian. In addition the condition that S presents a minimum at the saddle point also implies that they are all positive. Then introducing the density of eigenfunctions $g_\lambda(\nu)$ and generalising result (E.11), p. 283, valid for multiple Gaussian integrals, saddle point formula (4.39) for $I(\lambda)$ then reduces to

---

Saddle point formula for $I(\lambda) = \displaystyle\int d[\phi(\cdot)] \; \exp\bigl\{-S_\lambda[\phi(\cdot)]\bigr\}$

$$\tag{4.41}$$

$$I_{\text{saddle}}(\lambda) = e^{-S_\lambda[\phi_c(\cdot)]} \; \exp\left\{\frac{1}{2}\int d\nu \; g_\lambda(\nu) \; \ln\bigl[2\pi/\mu_\lambda(\nu)\bigr]\right\}.$$

[✠] **Comment:** Spectral parameter ν allows labelling the eigenvalues in the continuous part of the spectrum of a given operator. It is somehow equivalent to index i for discrete eigenvalues μ_i of matrix $C(\lambda)$ in finite dimension d. For example eigenvalues of operator $-\lambda\Delta$ in a segment of length L with homogeneous Dirichlet boundary conditions take the form $\mu_\lambda(k) = \lambda k^2$ with $k = p\pi/L$ and $p \in \mathbb{N}^*$. In the limit $L \to \infty$, wave-number k can be seen as a spectral parameter taking continuous real and positive values. Setting here $\nu = k$, the corresponding density of eigenfunctions becomes $g_\lambda(\nu)/L = 1/\pi$, and the domain of integration over ν is $[0, +\infty]$.

Just as with single or multiple integrals, the contributions of the saddle point itself to formula (4.41), $\exp(-S_\lambda[\phi_c(\cdot)])$, is multiplied by the contribution of its neighbourhood. In practice the determination of eigenvalues $\mu_\lambda(\nu)$ is of course a cumbersome problem in general, so that expression (4.41) often remains difficult to use. In some cases however eigenvalues can be simply calculated, or their product is expressed in a compact and practical way[7], in terms of quantities specific to the saddle field ϕ_c. The analysis of corrections to the saddle point approximation is equally difficult. With this in mind, note that the vast extent of integration domain, typically the set of all continuous functions $\phi(\zeta)$, leads to a complex landscape representing $\exp(-S_\lambda[\phi(\cdot)])$. In particular many other local minima may exist, so that one needs to evaluate their contributions when λ tends to infinity.

---

[7]Such simplification appears in the semi-classical approximation for the thermal propagator for instance, as we shall see in the next section of this chapter.

4.2 Applications and examples

4.2.1 *Stirling's formula and indistinguishability*

◇ Presentation ◇

The notion of indistinguishability plays a crucial role in statistical mechanics in solving the Gibbs paradox, or through spectacular effects related to the bosonic or fermionic nature of particles. Its importance was for instance demonstrated in the simple case of an ideal classical gas with Maxwell-Boltzmann statistics. The canonical partition function of this gas reads

$$Z_{\text{IG}}(N, V, T) = \frac{1}{N!} \int_{V^N \otimes \mathbb{R}^{3N}} \prod_{i=1}^{N} \left[\frac{\mathrm{d}\mathbf{r}_i \mathrm{d}\mathbf{p}_i}{(2\pi\hbar)^3}\right] \exp\left[-\beta \sum_{i=1}^{N} \frac{\mathbf{p}_i^2}{2m}\right], \qquad (4.42)$$

for N identical particles of mass m, contained in a volume V at temperature T where $\beta = 1/(k_B T)$ and k_B is the Boltzmann constant. Prefactor $1/N!$ in this expression comes from the indistinguishability of particles: It allows counting only once all microscopic configurations differing only by a permutation of particle coordinates. The integration on particle positions $\mathbf{r}_i$ is immediate, whereas the purely Gaussian integration on their impulsions $\mathbf{p}_i$ is easily achieved by applying result (E.1), p. 279, of Appendix E. We then find the well-known formula

$$Z_{\text{IG}}(N, V, T) = \frac{1}{N!} \left(\frac{V}{\lambda_D^3}\right)^N, \qquad (4.43)$$

where $\lambda_D = (h/\sqrt{2\pi m k_B T})$ is the thermal de Broglie wavelength. The issue of the extensivity of free energy

$$F_{\text{IG}}(N, V, T) = -k_B T \ln Z_{\text{IG}}(N, V, T)$$

then reduces to the study of the asymptotic behaviour of $N!$ when $N \to \infty$.

Let us introduce Euler function Γ, such that $\Gamma(N + 1) = N!$. The desired asymptotic behaviour can simply be obtained starting from integral representation

$$\boxed{\forall \lambda \geq 0, \ \Gamma(\lambda + 1) = \int_0^\infty \mathrm{d}x \ e^{-x} \ x^\lambda,} \qquad (4.44)$$

which is of the generic type (4.1), p. 214, with

$$I(\lambda) = \Gamma(\lambda + 1), \qquad \mathcal{D} = [0, \infty[\qquad \text{and} \quad f(x; \lambda) = x - \lambda \ln x.$$

The saddle-point method applied to integral (4.44) will then provide the required asymptotic behaviour as we shall now see.

◇ Analysis and solution ◇

We first determine the saddle point and its characteristics, leading to the celebrated Stirling formula. We then calculate the first corrections to this formula.

Stirling's formula Function f presents a unique minimum at $x_c(\lambda) = \lambda$ since

$$\frac{\partial f}{\partial x} = 1 - \frac{\lambda}{x} \quad \text{and} \quad \frac{\partial^2 f}{\partial x^2}(x_c(\lambda); \lambda) = \frac{1}{\lambda} > 0.$$

This minimum belongs to integration domain $\mathcal{D}$. General saddle point formula (4.3), p. 216, then reduces here to

$$\boxed{\Gamma_{\text{saddle}}(\lambda + 1) = \lambda^\lambda e^{-\lambda} \sqrt{2\pi\lambda}} \tag{4.45}$$

which is nothing but Stirling's formula!

Let us now examine the conditions of validity (4.8). We then find

$$\sigma(\lambda) = \sqrt{\lambda} \quad \text{and} \quad \frac{\partial^p f}{\partial x^p}(x_c(\lambda); \lambda) = \frac{(-1)^p (p-1)!}{\lambda^{p-1}}$$

for $p \geq 2$, leading to

$$[\sigma(\lambda)]^p \frac{\partial^p f}{\partial x^p}(x_c(\lambda); \lambda) = O(\frac{1}{\lambda^{p/2-1}}).$$

Near the border $x = 0$ of integration domain $\mathcal{D}$, $\exp(-f(x; \lambda))$ is also exponentially small compared to $\exp(-f(x_c(\lambda); \lambda))$. These considerations justify the asymptotic validity of Stirling's formula (4.45) when $\lambda \to \infty$.

Table 4.1   Stirling's formula and corrections.

N	$N!$	Stirling Equation (4.45)	Stirling + corrections Equation (4.46)
4	2	1.919	1.999
3	6	5.84	5.99
4	24	23.51	23.99
5	120	118.02	119.99

Before examining the corrections to Stirling's formula, we stop a moment to point out that the peak width $\sigma(\lambda)$ diverges when $\lambda \to \infty$. This may seem surprising at first sight and the reader may wonder whether the peak is actually narrow. In fact, as $x_c(\lambda) = \lambda$ diverges too, the peak width should be compared to the value of $x_c(\lambda)$, as shown in Figure 4.7. The intuitive picture of the narrow peak coincides with the conditions of validity checked above.

Corrections to Stirling's formula From the previous remark on the contribution of boundary at $x = 0$, the main corrections to Stirling's formula come from deviations from the Gaussian. The first correction is thus computed following the method presented on p. 218 leading to

$$\Gamma(\lambda + 1) = \lambda^\lambda e^{-\lambda} \sqrt{2\pi\lambda} \left[1 + \frac{1}{12\lambda} + O(\frac{1}{\lambda^2}) \right]. \tag{4.46}$$

If the obtained formulas are asymptotically valid on a mathematical level only in the limit of very large λ, it turns out they are already remarkably accurate for values less than 5 as shown in Table 4.1.

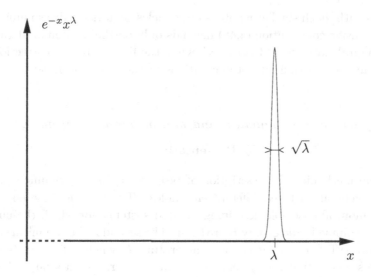

Fig. 4.7 The width $\sigma(\lambda) = \sqrt{\lambda}$ of the peak of function $e^{-x}x^{\lambda}$ diverges when $\lambda \to \infty$. However, this width remains small compared to $x_c(\lambda) = \lambda$, and therefore the peak around the maximum gets increasingly narrower when rescaling in units of λ in the limit $\lambda \to \infty$.

◇ Interpretation ◇

The above analysis allows to study the free energy behaviour in the thermodynamic limit (TL) defined by $N, V \to \infty$ with fixed density $\rho = N/V$ and fixed temperature T. Substituting $N!$ by asymptotic expansion (4.46) with $\lambda = N$, we get:

$$\beta \frac{F_{\text{IG}}}{N} = \ln(\rho\lambda_D^3) - 1 + \frac{\ln(2\pi N)}{2N} + \frac{1}{12N^2} + O(1/N^3) \qquad (4.47)$$

in the thermodynamic limit. The free energy per particle then tends to the intensive value

$$k_B T\left[\ln(\rho\lambda_D^3) - 1\right] \qquad (4.48)$$

depending only on density and temperature. The presence of indistinguishability factor $1/N!$ in counting the different microscopic states therefore enforces this fundamental property at the heart of macroscopic thermodynamics.

Finite size effects Beyond its interest for the concept of extensivity itself, asymptotic expansion (4.47) also allows to estimate the importance of finite size effects. The first correction to expression (4.48) is $\ln(2\pi N)/(2N)$. It is of the order of a percent for $N = 10^3$. This simple estimate suggests that it is possible to obtain accurate results on thermodynamic quantities of an infinite system by making numerical simulations with much smaller numbers of particles than Avogadro's number! Note that the intensive contribution (4.48) depends only on the maximum at the saddle point in integral expression (4.44) of $N!$, while the first finite size correction comes entirely from the contributions of fluctuations in the neighbourhood of the saddle point.

4.2.2   *Equivalence of canonical and microcanonical ensembles*

◇ **Presentation** ◇

In statistical mechanics, the description of systems at thermodynamic equilibrium is based in general on three different ensembles. The first of these ensembles is the microcanonical ensemble describing isolated systems, for which the number of particles N, volume V and energy E are fixed. The second is the canonical ensemble: N and V are still fixed but it is now temperature T which is fixed instead of the energy. This means that the system can exchange energy with a large thermostat that sets the temperature. Finally the last ensemble commonly used, especially for quantum statistics, is the grand canonical ensemble for which volume, temperature and chemical potential are set but particles can be exchanged between the system and the reservoir. The issue addressed here is that of the equivalence between the microcanonical and canonical descriptions[8]. Strictly speaking this question is extremely difficult, and requires the control of multiple integrals in the limit of an infinite number of integration variables! Here we take a very schematic view, based on a number of reasonable assumptions, but which remain to be proved.

Let us consider a system of N particles in a volume V. In the canonical ensemble, the system is in contact with a thermostat at temperature T_0. In the microcanonical ensemble, it is isolated and its energy is E_0. What should we do to show the equivalence of microcanonical and canonical descriptions? The answer is simple: we must show that the thermodynamic quantities derived independently in each ensemble actually become identical in the thermodynamic limit (TL). The latter is defined by $N, V \to \infty$ while density $\rho = N/V$ is fixed, temperature T_0 and energy per particle E_0/N being fixed in the canonical and microcanonical ensembles respectively. In the canonical ensemble the free energy reads

$$F_{\text{can}}(N, V, T_0) = -k_B T_0 \ln Z(N, V, T_0), \tag{4.49}$$

where k_B is Boltzmann's constant and Z is the canonical partition function. In the microcanonical ensemble entropy is given by

$$S_{\text{micro}}(N, V, E_0) = k_B \ln[\Omega(N, V, E_0)\Delta], \tag{4.50}$$

where Ω is micro-states density, while $\Omega\Delta$ is the number of micro-states with an energy between E_0 and $E_0 + \Delta$ where Δ is a constant. The microcanonical free

---

[8]Exercise 4.4, p. 257 is devoted to the equivalence of the microcanonical ensemble and a fourth one, the constant temperature and pressure ensemble.

energy then reads

$$F_{\text{micro}}(N, V, E_0) = E_0 - k_B T_{\text{micro}} S_{\text{micro}} \tag{4.51}$$

where the microcanonical temperature T_{micro} is defined by

$$T_{\text{micro}}(N, V, E_0) = \left[\frac{\partial S_{\text{micro}}}{\partial E_0} \right]^{-1}. \tag{4.52}$$

One has to show that

$$\lim_{\text{TL}} \frac{F_{\text{can}}(N, V, T_0)}{N} = \lim_{\text{TL}} \frac{F_{\text{micro}}(N, V, E_0)}{N} \tag{4.53}$$

by properly adjusting the free energy per particle E_0/N in the microcanonical ensemble for a given T_0 in the canonical ensemble.

As already announced, we adopt a minimalist point of view and start from the canonical partition function Z

$$Z(N, V, T_0) = \int_{-\infty}^{\infty} dE \, \Omega(N, V, E) e^{-\beta_0 E} \tag{4.54}$$

with $\beta_0 = 1/(k_B T_0)$. In this expression, all contributions of microscopic states with equal energy E are gathered via $\Omega(N, V, E)$, counting the number of states $\Omega(N, V, E) dE$ with an energy between E and $E + dE$. Using definition (4.50) for the microcanonical entropy, the partition function can be written as

$$\boxed{Z(N, V, T_0) = \int_{-\infty}^{\infty} \frac{dE}{\Delta} \, \exp\left(-\beta_0 F_0^*(N, V, E)\right),} \tag{4.55}$$

introducing the thermodynamic potential

$$F_0^*(N, V, E) = E - T_0 S_{\text{micro}}(N, V, E) \tag{4.56}$$

specific to the canonical ensemble. To show equivalence (4.53), we will apply the saddle-point method to find the asymptotic behaviour of expression (4.55) within TL.

$$\Diamond \ \textbf{Analysis and solution} \ \Diamond$$

The control parameter in the thermodynamic limit is the number of particles N. We must now show that integral (4.55) presents a very sharp peak when $N \to \infty$. To do so, we shall provide a physical argument at the origin of the equivalence between canonical and microcanonical descriptions. In the canonical ensemble, the system energy is not fixed as a result of exchanges with the thermostat. Experience shows however that it only fluctuates slightly around its mean value $\langle E \rangle$. Then integral (4.55) should present a peak near $\langle E \rangle$, so that the situation is similar to consider the microcanonical ensemble with parameters $(N, V, \langle E \rangle)$. Of course, and this is a crucial point, this assertion is correct only if fluctuations are negligible. We will see that up to some reasonable assumptions, this is indeed the case. Note, however that they may breakdown at critical points where fluctuations become very important.

Adjustment of the microcanonical energy   We then have to calculate integral (4.55) of generic form (4.1), p. 214, with $\lambda = N$, $x = E$, $\mathcal{D} = [-\infty, \infty]$ and

$$f(E, N) = \beta_0 F_0^*(N, V, E) \quad \text{with} \quad V = N\rho^{-1},$$

using the saddle-point method. The first step consists in looking for the energy corresponding to an extremum of $f(E, N)$. According to the previous argument, it is then natural to choose E_0 equal to this extremum. Then E_0 is such that

$$\boxed{\frac{1}{T_0} = \frac{\partial S_{\text{micro}}}{\partial E}(N, V, E_0).} \tag{4.57}$$

We admit this equation has one and only one solution E_0. Note that a necessary condition to ensure this uniqueness is that the microcanonical entropy is a monotonically increasing function of energy. We then characterise the nature of this extremum of $f(E, N)$ by computing the second derivative

$$\frac{\partial^2 f}{\partial E^2}(E_0, N) = -\frac{1}{k_B}\frac{\partial^2 S_{\text{micro}}}{\partial E^2}(N, V, E_0) = [k_B T_0^2 C_{\text{micro}}(N, V, E_0)]^{-1}, \tag{4.58}$$

where $C_{\text{micro}}(N, V, E_0)$ is the microcanonical specific heat at constant volume. When the entropy is a concave function of energy, the second derivative of S with respect to E is negative or equivalently, $C_{\text{micro}}(N, V, E_0) > 0$. This condition expresses the thermodynamic stability of the system and ensures that E_0 is indeed a minimum of $f = \beta_0 F^*$.

Asymptotic expression of Z in the thermodynamic limit   Let us now examine the conditions of validity (4.8), p. 218, ensuring that the saddle point formula gives the asymptotic behaviour of $Z(N, V, T_0)$. The width $\sigma(N)$ of the peak in E_0 is given by formula (4.58) and reads

$$\sigma(N) = \sqrt{k_B T_0^2 C_{\text{micro}}(N, V, E_0)} \propto \sqrt{N},$$

where we assumed that the microcanonical specific heat is indeed extensive. Furthermore, assuming that both entropy and energy are also extensive in the microcanonical ensemble, one finds

$$\frac{\partial^p f}{\partial E^p}(E_0, N) = -\frac{1}{k_B}\frac{\partial^p S_{\text{micro}}}{\partial E^p}(N, V, E_0) \propto N^{1-p}.$$

We then have, for $p > 2$,

$$\sigma^p(N)\frac{\partial^p f}{\partial E^p}(E_0, N) \propto \frac{1}{N^{p/2-1}} \to 0 \quad \text{when} \quad N \to \infty,$$

showing the asymptotic validity of the saddle point approximation. Formula (4.3) eventually leads to

$$\boxed{Z(N, V, T_0) \overset{\text{TL}}{\simeq} \sqrt{2\pi k_B T_0^2 C_{\text{micro}}(N, V, E_0)}\, e^{-\beta_0[E_0 - T_0 S_{\text{micro}}(N, V, E_0)]}.} \tag{4.59}$$

<h2 style="text-align:center">◇ Interpretation ◇</h2>

Let us comment on equation (4.57) allowing for the adjustment of energy E_0 to the parameters (N, V, T_0) defining the canonical ensemble. Definition (4.52) for the temperature in the microcanonical ensemble indicates that it must be identical to the given canonical temperature T_0. Note here that E_0 is the most probable energy. As probability distribution $\exp(-\beta_0 F^*(N, V, E))$ is peaked around E_0, the mean value $\langle E \rangle$ and the most probable value E_0 are very close, as expected. We shall now show that the equivalence of ensembles is achieved before checking explicitly that the underlying assumptions are met for an ideal gas, and conclude with brief remarks about the general case.

Equivalence of free energies per particle   Plugging asymptotic formula (4.59) into definition (4.49) we find:

$$\frac{F_{\text{can}}}{N} = \frac{E_0 - T_0 S_{\text{micro}}(N, V, E_0)}{N}$$
$$- \frac{k_B T_0}{2N} \ln(C_{\text{micro}}(N, V, E_0)) - \frac{k_B T_0}{2N} \ln(2\pi k_B T_0^2) + o(1/N). \quad (4.60)$$

The first term in the right-hand side is nothing but the microcanonical free energy per particle at the saddle energy value E_0. In proving formula (4.59), we have assumed that E_0, S_{micro} and C_{micro} are extensive. Therefore, in the thermodynamic limit, the microcanonical free energy per particle tends towards a finite value, while other terms in expression (4.60) go to zero. Equivalence (4.53) is then fulfilled.

Classical ideal gas   The reader is encouraged to explicitly check equivalence (4.53) in the case of a classical ideal gas. Calculations of the canonical partition function $Z(N, V, T)$ and the number of states $\Omega(N, V, E)$ are relatively elementary. An expression for Z has been obtained in the previous example, p. 233, while

$$\Omega(N, V, E) = \frac{V^N}{N!(2\pi\hbar)^{3N}} \frac{\pi^{3N/2}}{\Gamma(3N/2 + 1)} (2mE)^{3N/2},$$

where Γ is Euler's Gamma function (4.44), also introduced in the previous example. The reader will then eventually find[9]

$$\frac{F_{\text{micro}}(N, V, E_0)}{N} = \frac{F_{\text{can}}(N, V, T_0)}{N} + \frac{k_B T_0}{2N} \ln(3\pi N) + O(\frac{1}{N}),$$

showing that canonical and microcanonical free energies per particle are identical in the thermodynamic limit.

---

[9] One should use Stirling's formula (4.45), established in the previous example.

Gas with interactions Extensivity of thermodynamic quantities plays a crucial role in ensembles equivalence. We should emphasise that their proof starting from microscopic expressions for partition functions is a tour de force in the presence of interactions between particles. Corresponding proofs have been established in the literature for short-range interactions, and also for Coulomb interactions whose long range character is then attenuated by screening effects[10]. Note however that extensivity is lost for some long-range interactions such as gravitation for instance, and that there is no longer equivalence between microcanonical and canonical descriptions.

4.2.3 *Harmonic crystal at low temperature*

◊ **Presentation** ◊

This example deals with the low temperature stability of a harmonic crystal with respect to thermal fluctuations. In d dimensions, N identical particles moving in a domain $\mathcal{D}$ of $\mathbb{R}^d$ interact with each other via a potential V. At zero temperature, the spatial configuration of the particles minimises the potential and the system has a crystalline order. Does this crystalline order persist at very low temperature? How can one determine the specific heat at low temperature? We will see that the saddle-point method provides answers to these questions.

In a first step, the particles are treated classically and the domain is assumed to be one-dimensional. Results are easily generalised to an arbitrary dimension d. The system Hamiltonian is

$$H(\boldsymbol{x}, \mathbf{p}) = \sum_{n=1}^{Nd} \frac{p_n^2}{2m} + V(\boldsymbol{x}), \qquad (4.61)$$

where $\boldsymbol{x}$ and $\mathbf{p}$ are collective coordinates designating all positions x_n and momentums p_n of individual particles with $n = 1, \cdots, N$. The canonical partition function reads

$$Z = \frac{1}{N!} \int_{\mathcal{D}^N \otimes \mathbb{R}^{Nd}} \prod_{n=1}^{Nd} \left(\frac{\mathrm{d}x_n \, \mathrm{d}p_n}{(2\pi\hbar)} \right) e^{-\beta H(\boldsymbol{x}, \mathbf{p})}. \qquad (4.62)$$

Gibbs factor $e^{-\beta H(\boldsymbol{x}, \mathbf{p})}$ factorises as a part depending only on momenta, multiplied by Boltzmann factor $e^{-\beta V(\boldsymbol{x})}$. The integration over momenta is immediate and gives the same factor λ_D^{-Nd} as for an ideal gas (see p. 233). The partition function can then be written as

$$Z = \lambda_D^{-Nd} \, Z_{\text{conf}},$$

with a purely configurational partition function

$$\boxed{Z_{\text{conf}} = \frac{1}{N!} \int_{\mathcal{D}^N} \prod_{n=1}^{Nd} \mathrm{d}x_n \, e^{-\beta V(\boldsymbol{x})}.} \qquad (4.63)$$

---

[10]See example in Chapter 2 for another macroscopic example of these effects.

The crystal thermodynamic properties are entirely determined by the behaviour of Z_{conf} at low temperature $T = 1/(k_B\beta)$. We therefore propose to apply the saddle-point method to multiple integral (4.63) on Nd variables, taking generic form (4.29).

Here the control parameter β is multiplicative, i.e. $f(\boldsymbol{x};\lambda) \to \beta V(\boldsymbol{x})$. In this study, we deliberately omit any contribution from the domain boundaries⌖, and some steps will be carried out replacing $\mathcal{D}$ by $\mathbb{R}^d$. In addition, we assume that there are configurations corresponding to crystalline order minimising the potential $V(\boldsymbol{x})$.

⌖ **Comment:** In fact, in order to prevent the crystal from globally slipping as a result of translation invariance of the interaction potential between particles, it is essential to fix the boundaries, or to impose an external potential. The breaking of translation invariance in the crystalline phase is then preserved after first taking the thermodynamic limit and then removing the external potential.

◇ **Analysis and solution** ◇

We now consider the 1d case for simplicity reasons. We assume that the potential $V(\boldsymbol{x})$ has minima for configurations where the x_n form a regular chain of spacing a. Since the particles are identical, there are $N!$ minima giving equal contribution to the partition function. Neglecting overlap effects discussed in the general section, we multiply by $N!$ the contribution of one of these minima, denoted $\boldsymbol{x}_c$ where the particle positions take the form $x_n = na$. This contribution is estimated by using the saddle point formula and we then deduce the internal energy and the specific heat at constant volume.

Configurational partition function Consider then the minimum $\boldsymbol{x}_c = (a, 2a, ..., Na)$ such that

$$\frac{\partial V}{\partial x_n}(\boldsymbol{x}_c) = 0, \qquad n = 1, \cdots, N,$$

and the $N \times N$ matrix A of elements

$$A_{ln} = \frac{\partial^2 V}{\partial x_l \partial x_n}(\boldsymbol{x}_c),$$

defines a positive definite quadratic form. We then get, applying saddle point formula (4.34), p. 230 at this minimum,

$$Z_{\text{conf}} \sim e^{-\beta V(\boldsymbol{x}_c)}\frac{(2\pi)^{N/2}}{\sqrt{\text{Det}(\beta A)}} = e^{-\beta V(\boldsymbol{x}_c)}\beta^{-N/2}\frac{(2\pi)^{N/2}}{\sqrt{\text{Det}(A)}}. \qquad (4.64)$$

Note that since the control parameter is multiplicative, the saddle point formula provides the asymptotic behaviour of Z_{conf} at low temperature.

Internal energy and specific heat   Low temperature expressions of internal energy U and specific heat at constant volume C_V immediately derive from expression $Z = Z_{\text{conf}}/\lambda_D^N$ combined with saddle point formula (4.64)

$$U(N,V,T) = k_B T^2 (\partial \ln Z/\partial T)\,(N,V,T) = V(\boldsymbol{x}_c) + N k_B T + o(T),$$

$$C_V(N,V,T) = \qquad (\partial U/\partial T)\,(N,V,T) \qquad = \qquad N k_B + o(1).$$

Note that in partial derivatives with respect to temperature, quantities $V(\boldsymbol{x}_c)$ and A are constants because they only depend on N and V. The expression for U is easily interpreted. Term $V(\boldsymbol{x}_c)$ is the cohesive energy at zero temperature, also known as Madelung energy. The next contribution to U comes from equipartition of energy. The kinetic energy of each particle indeed accounts for $k_B T/2$. Moreover, in the Gaussian approximation, potential energy simply corresponds to N harmonic modes, each giving $k_B T/2$.

Introducing phonons   It is enlightening and useful to diagonalise the matrix A in order to interpret previous results within a simple picture, and also to access other physical quantities. As previously announced, we do not take into account boundary effects[11], and consider that A is translation invariant, $A_{nl} = A(n-l)$. Let us define deviation $\xi_n = x_n - na$ of particle n from its equilibrium position in the ground state (see Figure 4.8).

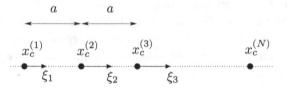

Fig. 4.8   The $x_c^{(n)} = na$ form a chain with regular spacing a. Quantities $\xi_n = x_n - na$ denote deviation of the particle n from its equilibrium position in the ground state.

The Gaussian approximation consists in writing

$$V(\boldsymbol{x}) \simeq V(\boldsymbol{x}_c) + \frac{1}{2}\boldsymbol{\xi}^T \cdot A \cdot \boldsymbol{\xi}$$

using matrix notation (4.31), p. 229. Owing to the translation invariance by any multiple of a, matrix A can be diagonalised by discrete Fourier transform, decomposing the ξ_n as

$$\xi_n = \frac{1}{\sqrt{N}} \sum_{k \in \text{BZ}} q_k \exp\left(inak\right).$$

Variables q_k in this decomposition are called phonons of wave number k in the first Brillouin zone $BZ = [-\pi/a, \pi/a]$. These wave numbers are regularly spaced by

---

[11]To overcome the boundary effects, one can also take periodic boundary conditions which consist in enforcing periodicity $x_{n+N} = x_n$. This procedure is similar to considering particles on a circle.

$2\pi/(Na)$. Since the ξ_n's are real, phonon variables satisfy identity $q_{-k} = q_k^*$ so that we eventually find

$$\boldsymbol{\xi}^T \cdot A \cdot \boldsymbol{\xi} = m \sum_{k\in} \omega_k^2 |q_k|^2,$$

where eigenfrequencies ω_k are simply given by the discrete Fourier transform of A, i.e.

$$m\omega_k^2 = \sum_{l\in\mathbb{Z}} A_{nl} e^{-ik(l-n)a}.$$

We have therefore diagonalised the original quadratic form. Note that writing the Fourier transform of A_{nl} as $m\omega_k^2$ is justified under the assumption that $\boldsymbol{x}_c$ is a minimum of potential $V(\boldsymbol{x})$. Note also that the frequency of mode $k = 0$, corresponding to an overall translation of the crystal, vanishes if one only takes into account interactions between particles. In the following, we omit the contributions from this mode, assuming implicitly for example that an external potential prevents the overall shift of the crystal, as already mentioned above.

The same transformation can be done for p_n and kinetic energy. Introducing the variable p_k canonically conjugate to q_k, Hamiltonian (4.61) can then be written within the harmonic approximation as

$$H(\boldsymbol{x}, \mathbf{p}) = \frac{1}{2} \sum_{k\in\mathrm{BZ}} \left(\frac{|p_k|^2}{m} + m\,\omega_k^2 |q_k|^2 \right), \qquad (4.65)$$

omitting the additive constant $V(\boldsymbol{x}_c)$. Phonons described by conjugate canonical variables (q_k, p_k) therefore correspond to N harmonic collective modes. In this approximation, the system then reduces to a set of N independent harmonic oscillators of mass m and frequencies ω_k. The generalisation to an arbitrary dimension $d \geq 1$ is immediate. The shape of the first Brillouin zone is determined by the symmetry of the crystal unit cell. The eigenfrequencies are determined by diagonalising the so-called $d \times d$ dynamical matrix formed with discrete Fourier transforms of partial second derivatives of V in the d possible directions. For each wave number $\mathbf{k}$, there are d eigenfrequencies denoted by $\omega_\alpha(\mathbf{k})$ where α is a polarisation index.

◇ Interpretation ◇

The saddle-point method and the introduction of phonon variables allow looking at the system stability around the zero temperature equilibrium configuration. This important question immediately raises another one regarding the validity of a classical treatment at very low temperatures.

Stability with respect to thermal fluctuations Let us determine the average deviation of a particle around its equilibrium position due to thermal agitation, defined by

$$\delta^2 = \frac{1}{N} \sum_{\mathbf{n}} \langle \xi_{\mathbf{n}}^2 \rangle = \frac{1}{N} \sum_{k\in\mathrm{BZ}} \langle q_{\mathbf{k}}^2 \rangle.$$

The calculation is elementary since phonon modes are independent and harmonic. In the infinite size limit, we can take variable $\mathbf{k}$ as continuous and obtain, up to proportionality constants,

$$\delta^2 \propto \int\limits_{BZ} \mathrm{d}^d\mathbf{k} \sum_{\alpha=1}^{d} \frac{k_B T}{m\,\omega_\alpha^2(\mathbf{k})}. \tag{4.66}$$

In order to compute this integral, one must know phonon dispersion relations $\omega_\alpha(\mathbf{k})$ which are specific to a particular crystal. In the case of short range interactions, there is however at least an acoustic mode for large wavelengths, with a linear dispersion relation

$$\omega_s(\mathbf{k}) \sim v_s\,k \qquad \text{for} \quad k \to 0,$$

where v_s is sound velocity. In this case integral (4.66) involves non-integrable divergent contributions for $k \to 0$ in one and two dimensions. This means that thermal fluctuations always break crystalline order at finite temperature. This result is actually a special case of Mermin-Wagner theorem on the breaking of continuous symmetries in low dimensions (see e.g. [6]), where the broken symmetry is nothing but translational invariance in the case of crystal formation.

Quantum effects To fix ideas, consider a three-dimensional crystal. At very low temperatures, we cannot neglect the quantum effects! We therefore have to replace the Hamiltonian (4.65) for classical harmonic oscillators by the Hamiltonian of a quantum system. In the framework of phonon variables however, we still have a set of independent harmonic oscillators which can be treated as bosons at the level of quantum statistics. So in addition to $V(\boldsymbol{x}_c)$, an additional constant term corresponding to zero-point energy appears in the internal energy. The temperature-dependent part is now controlled by the contribution of acoustic modes and is proportional to T^4. It results in a T^3 dependence of specific heat on temperature. The above argument regarding crystal stability is still valid at finite temperature because modes inducing instability are very low frequency modes: they can therefore be treated classically since $\hbar\omega_s(\mathbf{k}) \ll k_B T$.

4.2.4   *Ising model*

◇ **Presentation** ◇

The Ising model has played a fundamental role in understanding phase transitions in statistical mechanics, where the thermodynamic quantities are derived from a microscopic Hamiltonian. Beyond the original calculation by Ising and then the real tour de force by Onsager, who respectively solved exactly this model in one and two dimensions, many studies have been devoted to this model. We can distinguish on the one hand mean field approaches, and on the other hand implementations of methods from the renormalisation group. These efforts led to a fine and precise knowledge of this model critical properties. Such properties are also shared by a large class of systems of which it is the canonical representative.

As we shall see here, it turns out that equilibrium quantities of the Ising model can be represented by a functional integral via Hubbard-Stratanovitch transformation. The usual mean field approximation then corresponds to a saddle point estimate of the functional integral! This functional approach thus provides another perspective on this approximation presenting several interests: First it provides a deductive framework for the construction of Landau-Ginzburg free energy, usually introduced on the basis of symmetry arguments and *ad hoc* modelisation. It thus opens the way for a systematic study of contributions from fluctuations using renormalisation techniques. These contributions do play a fundamental role in the critical behaviour, which can be naively anticipated noting that saddle point approximation remains uncontrolled here.

The Ising model is defined by a set of N spins $\sigma_i = \pm 1$ on sites i of a d dimensional lattice, whose Hamiltonian reads

$$H = -\frac{1}{2} \sum_{i,j} J_{ij}\, \sigma_i \sigma_j, \tag{4.67}$$

where the sum runs over all lattice sites. Note that since $\sigma_i^2 = 1$, the terms $i = j$ give a constant contribution to H. Assume that J_{ij} defines a positive definite quadratic form, corresponding to an interaction which decays at large distances, namely $J_{ij} = f(|\mathbf{r}_i - \mathbf{r}_j|)$ where $f(r)$ is a positive and decreasing function. Note that other forms for J_{ij} are possible, for example $J_{ij} = J/N\ \forall\ i,j$ as discussed in exercise 4.7, p. 259, or $J_{ij} = J$ if i and j are nearest neighbours and $J_{ij} = 0$ otherwise. The latter is found more frequently in the literature and has been solved exactly in one and two dimensions. In practice the Ising model can describe a material whose magnetic properties are determined by an electron-induced spin $\pm 1/2$ for each atom. Coupling constants account for an effective interactions between spins. For example, they may be generated by exchange contributions to Coulomb interaction energy between electron clouds of nearest neighbour atoms.

In dimension greater than one, the previous model is expected to undergo a phase transition at a critical temperature T_c, between a ferromagnetic phase of finite spontaneous magnetisation for $T < T_c$, and an ordinary paramagnetic phase for $T > T_c$. We will study this transition in the context of a functional integral. We shall take implicitly the thermodynamic limit $N \to \infty$, assuming that boundary effects are irrelevant.

◇ Analysis and solution ◇

We consider the canonical partition function of the Ising model

$$Z = \sum_{\{\sigma_i\}} e^{\frac{\beta}{2} \sum_{i,j} J_{ij}\, \sigma_i \sigma_j}. \tag{4.68}$$

If the Hamiltonian looks very simple, the summation over discrete variables σ_i in formula (4.68) is by no means immediate. In fact it is useful to express Z in terms of continuous variables that are easier to handle, by performing a Hubbard-Stratonovitch transformation. We then explicit the transformation into a functional integral over a continuous field near the critical point. This naturally leads to the introduction of Ginzburg-Landau action. We then show that a saddle-point approximation gives back indeed the predictions of standard mean field approach.

Hubbard-Stratonovitch transformation   In practice the point is to use identity (E.12), p. 283, which reads here

$$(2\pi)^{N/2} \frac{1}{\sqrt{\mathrm{Det}\,A}} \, e^{\frac{1}{2}\mathbf{y}^T A^{-1}\mathbf{y}} = \int_{\mathbb{R}^N} \mathrm{d}\mathbf{x} \, e^{-\frac{1}{2}\mathbf{x}^T A\mathbf{x}+\mathbf{x}^T\mathbf{y}},$$

for any $N \times N$ real symmetric positive-definite matrix A. Choosing $A^{-1} = \beta J$ allows to write partition function (4.68) as

$$Z = \frac{1}{(2\pi\beta)^{N/2}\sqrt{\mathrm{Det}\,J}} \int_{\mathbb{R}^N} \mathrm{d}\boldsymbol{\phi} \, e^{-\frac{1}{2\beta}(\boldsymbol{\phi}^T J^{-1}\boldsymbol{\phi})} \sum_{\{\sigma\}} e^{\boldsymbol{\phi}^T\boldsymbol{\sigma}}$$

where we have introduced variables ϕ_i as well as collective notation $\boldsymbol{\sigma}$ and $\boldsymbol{\phi}$ for column vectors of components $\sigma_1, \cdots, \sigma_N$ and $\phi_1, \cdots, \phi_N$ respectively, the superscript T standing for transposition as defined in the Appendix. One can now carry out the sum over all spin configurations since it factorises as

$$\sum_{\{\sigma\}} e^{\boldsymbol{\phi}^T\boldsymbol{\sigma}} = \prod_i \left(e^{\phi_i} + e^{-\phi_i}\right),$$

so that

$$Z = \frac{1}{(2\pi\beta)^{N/2}\sqrt{\mathrm{Det}\,J}} \int \prod_i \mathrm{d}\phi_i \, e^{-\frac{1}{2\beta}\sum_{i,j}\left(\phi_i J_{ij}^{-1}\phi_j\right)-\sum_i U(\phi_i)}, \tag{4.69}$$

with

$$U(\phi_i) = -\ln \mathrm{ch}\,\phi_i - \ln 2.$$

Note that the transformation leading to expression (4.69) is exact. It shows that the original system is equivalent to a model from classical field theory on a lattice. Of course the problem remains extremely difficult, due to the simultaneous presence of a coupling between field values at different sites and an external potential $U(\phi_i)$. Field ϕ has a simple physical interpretation that one can highlight by expressing the spin mean value $\langle\sigma_l\rangle$ at site l as

$$\langle\sigma_l\rangle = \langle\mathrm{th}\,\phi_l\rangle.$$

Thus ϕ_l is directly related to the local magnetisation at site l, these two quantities having essentially the same average value in regimes where they are small compared to the saturation values corresponding to a total polarisation of the system.

Continuous limit and Ginzburg-Landau action In the vicinity of critical temperature T_c, we shall admit that only large scale behaviours are important, so that we can assume that the lattice structure is no longer relevant. The theory of second order phase transitions shows that this assumption is justified near the critical point, where the correlation length ξ becomes much larger than the lattice constant. It is then justified to take the continuous limit of the above expression for Z. Discrete lattice indices are then replaced by continuous variables, i.e.

$$J_{ij} \to J(\mathbf{r} - \mathbf{r}') \quad ; \quad \sum_i \to \frac{1}{a^d} \int d\mathbf{r} \quad ; \quad \delta_{ij} \to a^d \delta(\mathbf{r} - \mathbf{r}')$$

where a is the lattice constant, assuming a simple cubic structure. We now formally proceed to manipulations in the infinite system assuming translation invariance. Using previous notations and rules we find

$$\frac{1}{a^d} \int d\mathbf{r}' \, J(\mathbf{r} - \mathbf{r}') J^{-1}(\mathbf{r}' - \mathbf{r}'') = 2 a^d \delta(\mathbf{r} - \mathbf{r}''), \tag{4.70}$$

$$\sum_{i,j} \phi_i J_{ij}^{-1} \phi_j \to \frac{1}{a^{2d}} \int d\mathbf{r} \, d\mathbf{r}' \, \phi(\mathbf{r}) J^{-1}(\mathbf{r} - \mathbf{r}') \phi(\mathbf{r}') = \frac{1}{(2\pi)^d} \int d\mathbf{k} \, \frac{\left| \widehat{\phi}(\mathbf{k}) \right|^2}{\widehat{J}(\mathbf{k})}, \tag{4.71}$$

where the Fourier transform $\widehat{f}(\mathbf{k})$ of an arbitrary function $f(\mathbf{r})$ is defined by

$$\widehat{f}(\mathbf{k}) = \int d\mathbf{r} \, e^{i\mathbf{k}\cdot\mathbf{r}} \, f(\mathbf{r}).$$

To establish (4.71), we used Parseval-Plancherel formula, the identity $\widehat{\phi}(-\mathbf{k}) = \widehat{\phi}^*(\mathbf{k})$ resulting from the real nature of $\phi(\mathbf{r})$, as well as the identity

$$\frac{1}{a^d} \widehat{J}(\mathbf{k}) \widehat{J^{-1}}(\mathbf{k}) = a^d, \tag{4.72}$$

which is analogous to identity (4.70) expressing that function J^{-1} is the inverse of J.

In order to derive the large scale behaviours, and provided that $J(\mathbf{r})$ decays vast enough with distance, it is enough to assume that $J(\mathbf{r})$ is integrable and decays fast enough at infinity to admit a second moment. We then substitute $\widehat{J}(\mathbf{k})$ by its expansion at small wave number $\mathbf{k}$ in the vicinity of $\mathbf{k} = 0$,

$$\widehat{J}(\mathbf{k}) = \widehat{J}(0) \left(1 - \ell^2 k^2 + \cdots \right)$$

where we used the rotation invariance of $J(\mathbf{r}) = J(r)$, and with

$$\ell^2 = \frac{1}{2d} \frac{\int d\mathbf{r} \, r^2 J(r)}{\int d\mathbf{r} \, J(r)}.$$

The quadratic coupling term in field (4.71) then reads

$$\frac{1}{\widehat{J}(0)} \int d\mathbf{r} \left(\phi^2(\mathbf{r}) + \ell^2 (\boldsymbol{\nabla}\phi(\mathbf{r}))^2 + \cdots \right).$$

In this continuous limit, the external potential term can be written as

$$\sum_i \ln \operatorname{ch} \phi_i \rightarrow \frac{1}{a^d} \int d\mathbf{r} \ \ln \operatorname{ch} \phi(\mathbf{r}) = \frac{1}{a^d} \int d\mathbf{r} \ \left(\frac{1}{2}\phi(\mathbf{r})^2 - \frac{1}{12}\phi(\mathbf{r})^4 + \cdots\right).$$

The expansion in powers of ϕ is justified near the critical point because weak fields provide the dominant contributions.

Finally, introducing the functional measure $d[\phi(\cdot)]$ defined by the limit process

$$d[\phi(\cdot)] = \lim_{N\to\infty} \left\{ [(2\pi)^{N/2} \sqrt{\operatorname{Det}(\beta J)}]^{-1} \prod_{i=1}^{N} d\phi_i \right\},$$

we find for an infinite system

$$\boxed{Z = \int d[\phi(\cdot)] \ e^{-S[\phi(\cdot)]},} \tag{4.73}$$

with Ginzburg-Landau action

$$\boxed{S[\phi(\cdot)] = \int d\mathbf{r} \ \left(\frac{\alpha}{2} \left(\boldsymbol{\nabla}\phi(\mathbf{r})\right)^2 + \frac{b}{2}\phi^2(\mathbf{r}) + \frac{c}{4}\phi^4(\mathbf{r}) + \cdots \right)} \tag{4.74}$$

and coefficients

$$\alpha = \frac{\ell^2}{\beta \widehat{J}(0)} \quad ; \quad b = \frac{1}{\beta \widehat{J}(0)} - \frac{1}{a^d} \quad ; \quad c = \frac{1}{3a^d}. \tag{4.75}$$

Strictly speaking, expression (4.73) diverges, in agreement with the extensivity of Z of course, and other subtleties and mathematical difficulties appear in the continuum limit. We will go ahead nevertheless and proceed formally in the following, which is enough for our study. Let us also stress that if functional integral (4.73) takes general form (4.37), p. 231, it does not contain explicitly a well-identified control parameter λ.

Saddle point approximation   Despite the lack of control parameter, let us evaluate integral (4.73) by the saddle-point method. More precisely we simply look for the field $\phi_c(\mathbf{r})$ minimising Ginzburg-Landau action, and we do not study fluctuation contributions coming from neighbouring fields, a very difficult problem that goes well beyond our present goal. In agreement with the viewpoint which consists in ignoring any boundary effect, the functions to consider do not meet any particular constraint. Since term $(\boldsymbol{\nabla}\phi(\mathbf{r}))^2$ is positive definite, a first condition to minimise action $S[\phi(\cdot)]$ is that ϕ_c is homogeneous, i.e. $\phi_c(\mathbf{r}) = \phi_c$. One then obtains

$$b\phi_c + c\phi_c^3 = 0.$$

As shown by expressions (4.75), positive constant c is temperature independent, whereas coefficient b is negative for $T < \hat{J}(0)/(k_B a^d)$ and positive for $T > \hat{J}(0)/(k_B a^d)$. The number and the nature of saddle fields thus change at a temperature $\hat{J}(0)/(k_B a^d)$, that it is natural to identify the critical temperature T_c as justified in the following.

For $T > T_c$, the only uniform saddle field is given by $\phi_c = 0$. The corresponding value at the saddle $S[\phi_c(\cdot)]$ vanishes and constitutes a minimum of $S[\phi(\cdot)]$ which is always positive. For $T < T_c$, two opposite saddle fields appear corresponding to $\phi_c = \pm\sqrt{-b/c}$ and are minima of $S[\phi(\cdot)]$ while the identically zero field becomes a local maximum. This change is the signature of a phase transition in the system magnetic properties. This is confirmed by applying an external magnetic field proportional to h, giving an additional contribution $-\sum_i h\,\sigma_i$ to Hamiltonian (4.67). This magnetic field breaks the $\phi \to -\phi$ symmetry and a similar calculation to the derivation of formula (4.73) shows that, within the saddle point approximation,

$$\langle \phi(\mathbf{r}) \rangle_h \to \pm\sqrt{-b/c} \quad \text{when} \quad h \to 0^\pm \quad \text{for} \quad T < T_c.$$

As in the vicinity of the critical point, the mean value of $\phi(\mathbf{r})$ is proportional to the magnetisation, the low temperature phase $T < T_c$ is ferromagnetic with the appearance of a spontaneous magnetisation. Conversely we find that for $T > T_c$, $\langle \phi(\mathbf{r}) \rangle_h$ vanishes proportionally to h, showing that the high-temperature phase is paramagnetic.

$$\Diamond \ \textbf{Interpretation} \ \Diamond$$

We will show that the saddle point approximation on functional representation (4.73) happens to be equivalent to the usual mean-field approach. We then briefly discuss the role of fluctuations before discussing the universal nature of critical properties obtained for the Ising model.

Link with standard mean field approach In Landau theory it is assumed that a system state is completely determined by local magnetisation $M(\mathbf{r})$, which is the relevant order parameter for the transition under consideration. The thermodynamic potential $\Omega_T[M(\cdot)]$ for the system in contact with a thermostat setting the temperature T is then phenomenologically constructed for T near T_c using an expansion in powers of M and ∇M. Symmetry arguments then lead to the Ginzburg-Landau form for $\Omega_T[M(\cdot)]$.

Remarkably, action $S[\phi(\cdot)]$ has exactly the same structure as $\Omega_T[M(\cdot)]$, provided one identifies ϕ and M! Moreover the change of sign of coefficient b at $T = T_c$, which occurs naturally in the mathematical construction of $S[\phi(\cdot)]$, is here introduced heuristically in $\Omega_T[M(\cdot)]$, as a result of the competition between entropy on the one hand, promoting the state of maximum disorder with $M(\mathbf{r}) = 0$, and energy on the other hand, which is lower for ordered states with $M(\mathbf{r}) \neq 0$. Thus magnetisation equilibrium value M_{eq}, given by the minimisation of thermodynamic potential $\Omega_T[M(\cdot)]$ at fixed T, is simply proportional to saddle field ϕ_c minimising action $S[\phi(\cdot)]$. In other words, the saddle point approximation is equivalent to the Landau mean-field theory. The functional representation thus provides a constructive framework for this phenomenological approach, which is extremely useful, especially to take into account corrections due to fluctuations.

Fluctuations contributions Mean field theory does not reproduce exactly the transition critical properties. This is not surprising in the functional approach given the lack of control parameter: saddle point approximation can no longer be seen as asymptotically valid. In particular fluctuation contributions, Gaussian or not, in the neighbourhood of the saddle are not negligible compared with contributions of the saddle itself.

In fact the comparison with the exact solution in one dimension illustrates the limitations of a mean field approach. The predicted transition is then destroyed by the fluctuations, which are particularly important in low dimension. In three dimensions the value of T_c obtained here in terms of microscopic parameters is not exact, but the predicted ferromagnetic-paramagnetic transition remains qualitatively correct. The system thus presents a spontaneous magnetisation below a certain critical temperature. However the magnetisation does not vanish as $\sqrt{T_c - T}$ when $T \to T_c^-$ as in the saddle point approximation. The corresponding critical exponent depends crucially on fluctuations, and can be determined by advanced methods from the perturbative renormalisation group. Note however that mean field predictions become exact in dimension greater than 4.

Universality class By exploiting the fundamental idea that second-order phase transitions are governed by large scale behaviours, one can show that critical properties of the Ising model are common to a large class of systems with identical symmetries. This remarkable universality simply emerges from the construction of action $S[\phi(\cdot)]$, which is clearly invariant under transformation $\phi \to -\phi$. This obvious so-called $\mathbb{Z}_2$ symmetry is a consequence of the invariance of original Hamiltonian (4.67) under transformation $\sigma_i \to -\sigma_i$, $\forall i$. Instead of working with discrete variables, consider continuous variables weighted by a symmetric statistical weight $P(\sigma_i) = P(-\sigma_i)$ and replace $\sum_{\{\sigma_i\}}$ by $\int \prod_i P(\sigma_i) d\sigma_i$. Applying again Hubbard-Stratanovitch transformation and taking the limit $a \to 0$, we obtain a functional representation of the partition function, which involves an action presenting the same structure as expression (4.74). New coefficients a, b and c have the same temperature dependence as those related to discrete variables σ_i. Therefore all models corresponding to various possible choices of $P(\sigma)$ with the same $\mathbb{Z}_2$ symmetry should present identical critical properties.

4.2.5 *Semi-classical approximation*

◊ **Presentation** ◊

Taking up an idea from Dirac and developing an analogy with wave optics, Feynman was the first to rewrite the Green's function associated with Schrödinger equation in terms of path integral. This functional representation was soon extended to the thermal propagator associated with Bloch equation. Its countless applications show that it is a particularly effective tool to study quantum systems at equilibrium. Many of these applications are based on the famous semi-classical approximation, which amounts to a path integral calculation by the saddle-point method where the control parameter is $1/\hbar$. We consider here the tunnelling of a particle through a repulsive potential barrier as a simple illustration.

Consider a quantum particle of mass m and submitted to a potential $V(\mathbf{r})$ in a three dimensional space. Its Hamiltonian is

$$H = -\frac{\hbar^2}{2m}\Delta + V(\mathbf{r}),$$

and the corresponding thermal propagator reads

$$G(\mathbf{r}; \mathbf{r}'; \beta) = \langle \mathbf{r}|e^{-\beta H}|\mathbf{r}'\rangle$$

with the inverse temperature $\beta = 1/k_B T$. To fix ideas, assume that potential $V(\mathbf{r}) = Z^2 e^2/(4\pi\epsilon_0 r)$ describes a repulsive Coulomb barrier between two nuclei of charge Ze. The probability that the particle approaches at a distance r_0 from the origin is proportional to $G(\mathbf{r}_0; \mathbf{r}_0; \beta)$. At low temperatures such that $k_B T \ll V(\mathbf{r}_0)$, $G(\mathbf{r}_0; \mathbf{r}_0; \beta)$ is essentially controlled by tunnelling through the repulsive barrier, so that its value is largely increased compared to classical Boltzmann factor $e^{-\beta V(\mathbf{r}_0)}$. We will estimate the corresponding contribution by a semi-classical approximation.

We first construct a representation of $G(\mathbf{r}; \mathbf{r}'; \beta)$ in terms of path integrals, by exploiting perturbative expansions of Green's functions established in Chapter 3. Once a functional integral of generic form (4.37) is obtained, we give its saddle point estimation leading to the semi-classical expression of $G(\mathbf{r}_0; \mathbf{r}_0; \beta)$. The dominant term at low temperature can be recovered starting from a spectral representation of G, where eigenfunctions of H are evaluated using the Wentzel-Kramers-Brillouin (WKB) approximation.

◊ **Analysis and solution** ◊

First, it is essential to estimate the asymptotic form of $G(\mathbf{r}_a; \mathbf{r}_b; \tau)$ at high temperature, i.e. when $\tau \to 0^+$ with τ the inverse temperature. This form is then used to construct a path integral representation of $G(\mathbf{r}; \mathbf{r}'; \beta)$ at any finite temperature.

High temperature form of the propagator Let us fix $\mathbf{r}_a$ and $\mathbf{r}_b$ and study the behaviour of $G(\mathbf{r}_a; \mathbf{r}_b; \tau)$ when $\tau \to 0^+$. Here we can treat perturbatively the potential $V(\mathbf{r})$ itself, as ascertained later. A perturbative expansion of $G(\mathbf{r}_a; \mathbf{r}_b; \tau)$ in powers of V is derived from series (3.83), p. 157, with $H^{(0)} = -(\hbar^2/2m)\Delta$ and $W(\mathbf{r}) = V(\mathbf{r})$

$$G(\mathbf{r}_a; \mathbf{r}_b; \tau) = G^{(0)}(\mathbf{r}_a; \mathbf{r}_b; \tau)$$
$$- \int_0^\tau d\tau_1 \int d\mathbf{r}_1 \, G^{(0)}(\mathbf{r}_a; \mathbf{r}_1; \tau - \tau_1) \, V(\mathbf{r}_1) \, G^{(0)}(\mathbf{r}_1; \mathbf{r}_b; \tau_1) + \cdots . \quad (4.76)$$

Unperturbed propagator $G^{(0)}$ reduces here to the free propagator given by equation (3.84) in dimension $d = 3$. Therefore the first correction in expansion (4.76) becomes

$$-\left(\frac{m}{2\pi\hbar^2}\right)^3 \int_0^\tau d\tau_1 \, [(\tau - \tau_1)\tau_1]^{-3/2}$$
$$\int d\mathbf{r}_1 \, V(\mathbf{r}_1) \, \exp\left(-\frac{m(\mathbf{r}_1 - \mathbf{r}_a)^2}{2(\tau - \tau_1)\hbar^2} - \frac{m(\mathbf{r}_1 - \mathbf{r}_b)^2}{2\tau_1\hbar^2}\right). \quad (4.77)$$

In the limit $\tau \to 0^+$, $\tau - \tau_1$ and τ_1 become infinitely small. So in the spatial integral appearing in expression (4.77), Gaussian factors vary very quickly while $V(\mathbf{r}_1)$ varies over a much larger temperature-independent scale. We can therefore estimate this integral by the variant of the saddle-point method described on p. 227. The saddle point $\mathbf{r}_c$ is

$$\mathbf{r}_c = \frac{\tau_1 \mathbf{r}_a + (\tau - \tau_1)\mathbf{r}_b}{\tau},$$

and the Gaussian factor reads

$$\exp\left(-\frac{m(\mathbf{r}_b - \mathbf{r}_a)^2}{2\tau\hbar^2}\right) \exp\left(-\frac{m\tau(\mathbf{r}_1 - \mathbf{r}_c)^2}{2(\tau - \tau_1)\tau_1\hbar^2}\right).$$

Substituting $V(\mathbf{r}_1)$ by $V(\mathbf{r}_c)$ we obtain the high temperature asymptotic behaviour of expression (4.77),

$$-\left(\frac{m}{2\pi\tau\hbar^2}\right)^{3/2} \exp\left(-\frac{m(\mathbf{r}_b - \mathbf{r}_a)^2}{2\tau\hbar^2}\right) \int_0^\tau d\tau_1 \, V\big((\tau_1 \mathbf{r}_a + (\tau - \tau_1)\mathbf{r}_b)/\tau\big), \quad (4.78)$$

which therefore takes the form $G^{(0)}(\mathbf{r}_a; \mathbf{r}_b; \tau)$ multiplied by a factor of order τ. Corrections to the dominant term (4.78) when $\tau \to 0^+$ are computed by expanding $V(\mathbf{r}_1)$ in the vicinity of $V(\mathbf{r}_c)$ as a Taylor series of $(\mathbf{r}_1 - \mathbf{r}_c)$. The first non-zero correction comes from term $\frac{1}{2}[(\mathbf{r}_1 - \mathbf{r}_c) \cdot \nabla_{\mathbf{r}_c}]^2 V(\mathbf{r}_c)$, and is smaller than dominant term (4.78) by a factor of order τ, as shown by the simple change of variable $\tau_1 = u\tau$.

The above analysis can be extended to all the terms of perturbative series (4.76). One can easily find that the term of order V^n is asymptotically of the form $G^{(0)}(\mathbf{r}_a; \mathbf{r}_b; \tau)$ multiplied by a factor of order τ^n when $\tau \to 0^+$. So we can sum series (4.76) using identity

$$1 - \int_0^\tau d\tau_1 \, V((\tau_1 \mathbf{r}_a + (\tau - \tau_1)\mathbf{r}_b)/\tau) + O(\tau^2)$$

$$= \exp\left(-\int_0^\tau d\tau_1 \, V((\tau_1 \mathbf{r}_a + (\tau - \tau_1)\mathbf{r}_b)/\tau) + O(\tau^2)\right), \quad (4.79)$$

leading to the high temperature formula

$$G(\mathbf{r}_a; \mathbf{r}_b; \tau) = \left(\frac{m}{2\pi\tau\hbar^2}\right)^{3/2} \exp\left(-\frac{m(\mathbf{r}_b - \mathbf{r}_a)^2}{2\tau\hbar^2}\right)$$

$$\exp\left(-\int_0^\tau d\tau_1 \, V((\tau_1 \mathbf{r}_a + (\tau - \tau_1)\mathbf{r}_b)/\tau) + O(\tau^2)\right). \quad (4.80)$$

Path integral representation Consider now propagator $G(\mathbf{r}; \mathbf{r}'; \beta)$ at finite temperature. Identity

$$e^{-\beta H} = \left[e^{-\beta H/N}\right]^N$$

valid for any integer N leads to the following convolution formula

$$G(\mathbf{r}; \mathbf{r}'; \beta) = \int d\mathbf{r}_1 \, d\mathbf{r}_2 \cdots d\mathbf{r}_{N-1} \, G(\mathbf{r}; \mathbf{r}_1; \beta/N) \, G(\mathbf{r}_1; \mathbf{r}_2; \beta/N) \cdots$$

$$\times G(\mathbf{r}_{N-1}; \mathbf{r}'; \beta/N).$$

In the limit $N \to \infty$, it is legitimate to replace each propagator $G(\mathbf{r}_p; \mathbf{r}_{p+1}; \beta/N)$ by high-temperature formula (4.80). Here we have $\tau = \beta/N$ so that term $O(\tau^2)$ are of order $O(1/N^2)$. Carrying out the product of $N+1$ exponential factors, the sum of the $N+1$ terms is itself of order $O(1/N)$ and can therefore be neglected when $N \to \infty$. Thus we find

$$G(\mathbf{r}; \mathbf{r}'; \beta) = \lim_{N\to\infty} \left(\frac{mN}{2\pi\beta\hbar^2}\right)^{3N/2} \int d\mathbf{r}_1 \, d\mathbf{r}_2 \cdots d\mathbf{r}_{N-1}$$

$$\exp\left\{-\sum_{p=0}^{N-1}\left[\frac{mN(\mathbf{r}_{p+1} - \mathbf{r}_p)^2}{2\beta\hbar^2} + \frac{\beta}{N}\int_0^1 du \, V(u\mathbf{r}_p + (1-u)\mathbf{r}_{p+1})\right]\right\}, \quad (4.81)$$

with $\mathbf{r}_0 = \mathbf{r}$ and $\mathbf{r}_N = \mathbf{r}'$.

Expression (4.81) naturally leads to the path $\mathbf{r}(t)$ consisting of $N+1$ line segments successively connecting $\mathbf{r}_p$ to $\mathbf{r}_{p+1}$ with a fictitious time t ranging from 0 to $\beta\hbar$. Each point $\mathbf{r}_p$ is reached after a time $t_p = p\beta\hbar/N$. In addition velocity $\dot{\mathbf{r}}(t) = N(\mathbf{r}_{p+1} - \mathbf{r}_p)/(\beta\hbar)$ is uniform between t_p and t_{p+1}. Sum $\sum_{p=0}^{N-1}\cdots$ in expression (4.81) then identifies exactly to $S[\mathbf{r}(\cdot)]/\hbar$, the action associated with path $\mathbf{r}(t)$ in the potential of opposite sign $-V$,

$$\boxed{S[\mathbf{r}(\cdot)] = \int_0^{\beta\hbar} dt \left[\frac{m\dot{\mathbf{r}}^2(t)}{2} + V(\mathbf{r}(t))\right].}$$

By introducing the functional measure

$$d[\mathbf{r}(\cdot)] = \lim_{N \to \infty} \left(\frac{mN}{2\pi\beta\hbar^2} \right)^{3N/2} \int d\mathbf{r}_1 \, d\mathbf{r}_2 \cdots d\mathbf{r}_{N-1},$$

one finally gets

$$\boxed{G(\mathbf{r}; \mathbf{r}'; \beta) = \int d[\mathbf{r}(\cdot)] \, \exp\left(-\frac{S[\mathbf{r}(\cdot)]}{\hbar} \right).} \qquad (4.82)$$

In the presentation adopted to establish representation (4.82), we do not claim to absolute rigour provided that it would require, at a mathematical level, to control the convergence of perturbative series as well as the limit process defining functional measure $d[\mathbf{r}(\cdot)]$. The reader can find a proof of formula (4.82) for a large class of potentials in the book [85]. It shows in particular that measure $d[\mathbf{r}(\cdot)]$ is simply related to the Wiener measure governing the statistical properties of Brownian motion. Note finally that the path integral representation of dynamical Green's function, obtained by the replacement of β by $it/\hbar$, involves a pure phase factor with a phase proportional to the action in potential V. In this case, the presence of oscillations implies that the functional integral is not strictly speaking well defined.

Semi-classical approximation   Functional integral (4.82) takes the general form (4.37), p. 231, with substitutions $\phi \to \mathbf{r}$, $\zeta \to t$, $\Omega \to [0, \beta\hbar]$, and $S_\lambda \to S/\hbar$. Control parameter λ can be naturally identified with $1/\hbar$. The classical limit is reached by taking $\hbar \to 0$. As exponent $(S[\mathbf{r}(\cdot)]/\hbar)$ then varies very rapidly and can take very large values, it is *a priori* reasonable to use the saddle-point method to study $G(\mathbf{r}; \mathbf{r}'; \beta)$ in this limit.

A saddle path $\mathbf{r}_c(t)$ is therefore a minimum of action $S[\mathbf{r}(\cdot)]$ with initial and final constraints $\mathbf{r}_c(0) = \mathbf{r}$ and $\mathbf{r}_c(\beta\hbar) = \mathbf{r}'$ respectively, while the flight time is also imposed and equal to $\beta\hbar$. Functional equation (4.38) then reduces to Lagrange equations defining a classical trajectory according to the principle of least action. Note that there may be several classical trajectories satisfying the above constraints. We assume here that one gives the absolute minimum S_c of the action, and that contributions from other classical trajectories can be neglected[12].

The contribution from classical trajectory $\mathbf{r}_c(t)$ and its neighbourhood is given by saddle point formula (4.41), p. 232. As already commented generally, determining explicitly eigenvalues of the operator associated with covariance

$$C(t_1, t_2) = \hbar^{-1} \frac{\delta^2 S}{\delta \mathbf{r}(t_1) \delta \mathbf{r}(t_2)} [\mathbf{r}_c(\cdot)]$$

is not immediate, and we still have to perform a sum over the entire spectrum to get the total contribution of Gaussian fluctuations. Thanks to remarkable identities

---

[12]As outlined in this chapter general part in the case of a simple integral, it is difficult to take into account the contributions from several saddle paths.

from analytical mechanics, one can show that this sum is in fact expressed in a compact form in terms of classical action $S_c(\mathbf{r}, \mathbf{r}'; \beta\hbar)$ corresponding to classical trajectory $\mathbf{r}_c(t)$. The saddle point formula, also called semi-classical approximation or Van Vleck formula, eventually reads

$$
G_{\text{saddle}}(\mathbf{r}; \mathbf{r}'; \beta) = (2\pi\hbar)^{-3/2} \left[\text{Det} -\frac{\partial^2 S}{\partial \mathbf{r} \partial \mathbf{r}'}(\mathbf{r}, \mathbf{r}'; \beta\hbar) \right]^{1/2} \exp\left(-\frac{S_c(\mathbf{r}, \mathbf{r}'; \beta\hbar)}{\hbar} \right).
$$
(4.83)

Note that the square root of the determinant of partial derivatives of classical action with respect to initial and final points is in the numerator in this formula, not in the denominator as a tempting yet abusive identification with the determinant of the quadratic functional might suggest! In fact, it is this position at the numerator which guarantees that $G_{\text{saddle}}(\mathbf{r}; \mathbf{r}'; \beta)$ is indeed homogeneous to the inverse of a three-dimensional volume. Finally a still not completely convinced reader can apply formula (4.83) with $V(\mathbf{r}) = 0$: it will lead exactly to expression (3.84) for the free propagator. A determinant enthusiast could check that formula (4.83) also gives the exact propagator in the harmonic potential case.

◊ Interpretation ◊

In the formal limit $\hbar \to 0$, semi-classical approximation (4.83) becomes asymptotically exact. Note that control parameter $1/\hbar$ is not multiplicative, since $\hbar$ is also involved in the paths flight time. Nevertheless, it is possible to check it gives back Wigner-Kirkwood expansion of the diagonal part $G(\mathbf{r}; \mathbf{r}; \beta)$ near its classical value $(m/(2\pi\beta\hbar^2))^{3/2} \exp(-\beta V(\mathbf{r}))$ including terms of order $\hbar^2$. However, the semi-classical approximation has a much broader application range than Wigner-Kirkwood expansion, because it takes into account significant quantum effects non-perturbatively. This is well illustrated by the calculation of $G_{\text{saddle}}(\mathbf{r}_0; \mathbf{r}_0; \beta)$ at low temperature for repulsive Coulomb potential $V(\mathbf{r}) = Z^2 e^2/(4\pi\epsilon_0 r)$.

Application to tunnel effect Consider a temperature low enough so that $k_B T \ll V(r_0)$. The different saddle paths are classical trajectories in attractive Coulomb potential $-Z^2 e^2/(4\pi\epsilon_0 r)$. This set of trajectories therefore consists in ellipses of period $\beta\hbar$ passing through $\mathbf{r}_0$, and two straight trajectories starting from $\mathbf{r}_0$, and returning to $\mathbf{r}_0$ after a time $\beta\hbar$. The path giving the action absolute minimum is the straight one passing through the origin. The corresponding classical action behaves as

$$
S_c(\mathbf{r}_0, \mathbf{r}_0; \beta\hbar) \sim \left(\frac{27m Z^4 e^4 \hbar}{32\varepsilon_0^2 k_B T} \right)^{1/3} \quad \text{when} \quad T \to 0.
$$

Another much more involved calculation gives the determinant appearing in classical formula (4.83). The leading contribution to $G_{\text{saddle}}(\mathbf{r}_0; \mathbf{r}_0; \beta)$ at low temperature comes from factor $\exp(-S_c(\mathbf{r}_0, \mathbf{r}_0; \beta\hbar)/\hbar)$, and reduces to

$$
\exp\left[-\left(\frac{27m Z^4 e^4}{32\varepsilon_0^2 \hbar^2 k_B T} \right)^{1/3} \right].
$$
(4.84)

Penetration factor (4.84) takes into account a fundamental quantum mechanism, namely quantum tunnelling. Thus semi-classical approximation appears to be effective even away from the classical limit! Note in particular that Wigner-Kirkwood expansion is here completely useless: factor (4.84) becomes exponentially larger than classical Boltzmann factor $\exp(-\beta V(r_0))$ when $T \to 0$. It is however very difficult to check the validity of semi-classical formula (4.83) in situations where quantum effects are important. It turns out here that semi-classical factor (4.84) correctly describes the exact leading behaviour of $G(\mathbf{r}_0; \mathbf{r}_0; \beta)$ when $\beta \to \infty$.

Comparison with WKB approximation   Note finally that previous expression for the penetration factor can be recovered from WKB method. This method, also called semi-classical, consists in determining approximate eigenfunctions of Hamiltonian H through a perturbative resolution of Schrödinger equation. The corresponding thermal propagator is then inferred from spectral representation (3.81)

$$G(\mathbf{r}_0; \mathbf{r}_0; \beta) = \sum_n |\psi_n(\mathbf{r}_0)|^2 e^{-\beta E_n}.$$

The reader will not be surprised to learn that such verification requires the estimation of sums over energies by the saddle-point method!

4.3   Exercises

■ **Exercise 4.1. Asymptotic behaviour of Bessel function J_0**

Study the behaviour of Bessel function

$$J_0(\lambda) = \frac{1}{2\pi} \int_0^{2\pi} d\theta\, e^{i\lambda \cos\theta}$$

for λ real and positive, when $\lambda \gg 1$. To this end we start by writing J_0 as

$$J_0(\lambda) = \frac{1}{\pi} \mathrm{Re}\left[\int_{-\frac{\pi}{2}}^{\frac{\pi}{2}} d\theta\, e^{-i\varphi(\theta;\lambda)} \right]$$

with $\varphi(\theta; \lambda) = -\lambda \cos\theta$. Show then that

$$J_0(\lambda) \simeq \sqrt{\frac{2}{\pi\lambda}} \cos\left(\lambda - \frac{\pi}{4}\right) \qquad \text{when } \lambda \to +\infty \tag{4.85}$$

and that this approximation is justified.

Solution p. 325.

■ **Exercise 4.2. Binomial coefficients**

1. Show that

$$C_n^p = \frac{1}{2\pi i} \oint_C dz \, \frac{(1+z)^n}{z^{p+1}}$$

where C is an integration contour enclosing the origin of the complex plane.

2. Let us set $p = nx$. Show that in the limit $n \to \infty$,

$$C_n^{nx} \simeq \frac{1}{\sqrt{2\pi n x(1-x)}} \exp\Big\{-n\big[x \ln x + (1-x)\ln(1-x)\big]\Big\}. \qquad (4.86)$$

Solution p. 325.

■ **Exercise 4.3. Asymptotic behaviour of Helmholtz Green's function**

Let $G_\infty(\mathbf{r})$ be the Green's function of Helmholtz operator in d dimensions with BC at infinity, solution of

$$-\Delta G_\infty(\mathbf{r}) + m^2 G_\infty(\mathbf{r}) = \delta(\mathbf{r})$$

with $G_\infty(\mathbf{r}) \to 0$ when $r \to \infty$. The discussion presented in Chapter 3, p. 141, leads to

$$G_\infty(\mathbf{r}) = \tilde{G}_\infty^+(\mathbf{r}; s = m^2)$$

where $\tilde{G}_\infty^+(\mathbf{r}; s = m^2)$ is the Laplace transform (evaluated in $s = m^2$) of the causal Green's function for the diffusion equation with a diffusion coefficient $D = 1$:

$$G_\infty(\mathbf{r}) = \frac{1}{(4\pi)^{d/2}} \int_0^{+\infty} \frac{dt}{t^{d/2}} \exp(-m^2 t - r^2/4t).$$

Show that when $r \to \infty$:

$$G_\infty(\mathbf{r}) \simeq \sqrt{\frac{\pi}{2}} \frac{1}{(2\pi)^{d/2}} \frac{m^{\frac{d-3}{2}}}{r^{\frac{d-1}{2}}} e^{-mr} \left(1 + O\big(1/r\big)\right). \qquad (4.87)$$

Solution p. 325.

■ **Exercise 4.4. Isothermal-isobaric ensemble**

There are many experimental situations where the temperature and the pressure of a thermodynamic system are imposed through the contact with a temperature and pressure reservoir. This reservoir thus sets temperature and pressure of the system at equilibrium. As for the canonical ensemble, the isothermal-isobaric ensemble corresponds to a situation where temperature, pressure and number of atoms in the system are fixed. In particular, the system volume and energy are free to vary. The isothermal-isobaric partition function for this ensemble is defined as

$$Q_N(P_0, T_0) = \int dV e^{-\beta_0 P_0 V} Z_N(V, T_0),$$

with $\beta_0 = 1/k_B T_0$ and where $Z_N(V, T_0)$ is the canonical partition function for a system of N atoms in a volume V at temperature T_0. Using expression for the canonical partition function, introduced for instance p. 237, $Q_N(P_0, T_0)$ can be written as

$$Q_N(P_0, T_0) = \int dV dE \, e^{-\beta_0(P_0 V + E - T_0 S)},$$

where $S = S(N, V, E)$ is the micro-canonical entropy. Finally Gibbs free energy is defined in this ensemble as:

$$G_N(P_0, T_0) = -k_B T_0 \ln Q_N(P_0, T_0). \tag{4.88}$$

Evaluate $G_N(P_0, T_0)$ and the difference between micro-canonical and isothermal-isobaric expressions of this thermodynamic function in the large N limit.

Solution p. 326.

■ Exercise 4.5. Evolution of a wave packet and group velocity

Consider a one-dimensional wave packet defined at $t = 0$ as:

$$P(x, t = 0) = \int_{-\infty}^{+\infty} dk \, e^{ikx} F(k). \tag{4.89}$$

Furthermore, the dispersion relation is given by a function $\omega(k)$ such that the evolution of the wave packet is

$$P(x, t) = \int_{-\infty}^{+\infty} dk \, e^{i(kx - \omega(k)t)} F(k). \tag{4.90}$$

We stay in $x(t) = vt$. Evaluate $P(x(t), t)$ for t large.

Solution p. 327.

■ Exercise 4.6. From Cattaneo to diffusion Green's function

We recall that the Green's function $G_\infty^+(x; t)$ for Cattaneo equation in one dimension introduced on p. 194 reads $G_\infty^+(x; t) = e^{-\frac{c^2 a^2 t}{2}} g_\infty^+(x; t)$ with (see p. 197)

$$g_\infty^+(x; t) = \theta(ct - |x|) \frac{c}{4\pi} \int_{-\pi+i\alpha}^{\pi+i\alpha} d\beta \, e^{mc\sqrt{t^2 - x^2/c^2} \, \sin\beta}$$

with $m = a^2 c/2$. Recover the Green's function for diffusion operator in one dimension from this result.

Solution p. 327.

■ **Exercise 4.7. Ising model with long-range interactions**

We consider an Ising model consisting of N spins $S_i = \pm 1$ located on N sites, whose Hamiltonian is:

$$H = \frac{-1}{2N} \sum_{i,j} S_i S_j - h \sum_i S_i$$

where the sum runs on all sites $i, j = 1, \cdots, N$.

1. Show that it is possible to write $\exp(-\beta H)$ as

$$\exp(-\beta H) = \left(\frac{N\beta}{2\pi}\right)^{1/2} \int_{-\infty}^{\infty} d\lambda \, \exp[-N\beta\lambda^2/2 + \sum_i (\beta\lambda + \beta h)S_i].$$

Conclude that the model partition function Z is

$$Z = \left(\frac{N\beta}{2\pi}\right)^{1/2} \int_{-\infty}^{\infty} d\lambda \, \exp(-N\beta A(\lambda)),$$

where $A(\lambda)$ is a function to identify.

2. Show that the free energy per site then reads

$$f = \frac{-1}{N\beta} \ln Z = A(\lambda_0) + O(1/N),$$

where λ_0 is the value of λ corresponding to a minimum of $A(\lambda)$.

3. Show that magnetisation

$$m = -\frac{\partial f}{\partial h}$$

for fixed β and N is given by $m = \lambda_0 + O(1/N)$. Deduce that the equation setting λ_0 gives for m:

$$\text{th}(\beta(m + h)) = m,$$

which is the mean field equation for this model. Give β_c, the critical point of transition from a paramagnetic phase to a ferromagnetic phase. Show that a mean field approach gives the exact result for this model in the limit $N \to \infty$.

Solution p. 327.

■ **Exercise 4.8. Bernoulli random walk**

Consider a particle moving in a discrete space in one dimension. We can therefore identify its position by an integer m while time is also discretised. This particle changes randomly position at each time. It can only jump of one unit by time interval, to the right or left with the same probability. So if $P_N(m)$ is the probability to find the particle in m at time N, it satisfies the recurrence relation:

$$P_N(m) = \frac{1}{2}\left[P_{N-1}(m-1) + P_{N-1}(m+1)\right].$$

1. Defining respectively the Fourier transform and its inverse by

$$\widehat{P}_N(k) = \sum_{m=-\infty}^{\infty} e^{ikm} P_N(m),$$

and

$$P_N(m) = \int_{-\pi}^{\pi} \frac{dk}{2\pi} e^{-ikm} \widehat{P}_N(k),$$

find a relation between $\widehat{P}_N(k)$ and $\widehat{P}_{N-1}(k)$. Then give an expression for $\widehat{P}_N(k)$ assuming that initially $P_0(m) = \delta_{m,0}$.

2. Show that one can write $P_N(m)$ as

$$P_N(m) = \left[\frac{1 + (-1)^{N+m}}{2\pi}\right] I_N(m)$$

where $I_N(m)$ is an integral over $-\pi/2$ and $\pi/2$ to be defined.

3. Give then the expression of $P_N(m)$ for large N.

Solution p. 329.

■ **Exercise 4.9. Harmonic oscillator and number theory**

Consider a system consisting of an infinite number of independent harmonic oscillators of eigenfrequencies ω_i such that

$$\hbar\omega_1 = 1 \; ; \; \hbar\omega_2 = 2 \; ; \; \hbar\omega_k = k.$$

The energy reference is chosen so that the ground state energy of each oscillator is 0 and not $\hbar\omega_i/2$.

1. Show that the system partition function $Z(\beta)$ can be written in two ways, first as an infinite product to be determined and second as

$$Z(\beta) = \sum_{n=1}^{\infty} e^{-\beta n}\,\Omega(n) \tag{4.91}$$

where $\Omega(n)$ is the number of partitions of integer n. We have for instance $\Omega(4) = 5$ since

$$4 = 4$$
$$= 3 + 1$$
$$= 2 + 2$$
$$= 2 + 1 + 1$$
$$= 1 + 1 + 1 + 1.$$

2. Write $\Omega(n)$ as an integral in the complex plane using the residue theorem.

3. Using Euler-MacLaurin formula

$$\sum_{n=1}^{\infty} \ln\left[1 - e^{-\beta n}\right] = \frac{-\pi^2}{6\beta} - \frac{1}{2}\ln(\beta) + \frac{1}{2}\ln(2\pi) + O(\beta),$$

give the asymptotic expression for $\Omega(n)$ for large n. This result is known as Hardy-Ramanudjan formula.

Solution p. 329.

Appendix A

Functions of a complex variable

This appendix recalls some definitions and properties related to functions of a complex variable.

◇ Analyticity ◇

Holomorphic functions   A function f is analytic (or holomorphic) in $z_0 \in \mathbb{C}$ if it is differentiable in a neighbourhood of z_0. Then writing $z = x + iy$, $P = \mathrm{Re}\, f$, and $Q = \mathrm{Im}\, f$, one has (Cauchy's condition):

$$\frac{\partial P}{\partial x}(x_0, y_0) = \frac{\partial Q}{\partial y}(x_0, y_0),$$

$$\frac{\partial Q}{\partial x}(x_0, y_0) = -\frac{\partial P}{\partial y}(x_0, y_0).$$

A consequence of Cauchy's conditions is that functions $P(x, y)$ and $Q(x, y)$ are harmonic, that is to say their Laplacians are zero:

$$\Delta P = 0 \qquad \text{and} \qquad \Delta Q = 0.$$

An entire function is a function analytic on all $\mathbb{C}$.

Singularities   Points at which a function is not holomorphic are called singularities. We recall below various types of singularities, simply by means of a corresponding typical example:

- Artificial or apparent singularity: The standard example is to start with a function $f(z)$ holomorphic in z_0 and consider

$$g(z) = \frac{f(z) - f(z_0)}{z - z_0}.$$

  z_0 is an apparent singularity for $g(z)$ and it is possible to define g in z_0 by $g(z_0) = f'(z_0)$.
- Pole: The standard example is

$$f(z) = \frac{1}{(z - z_0)^n}.$$

  More generally $f(z)$ has a pole in z_0 if the function $(z - z_0)^n f(z)$ has an apparent singularity in z_0. Integer n is then the order of the pole.

- Essential singularity: The standard example is $f(z) = \exp\frac{1}{z}$ which has an essential singularity in $z = 0$.
- Branch points: They are associated with cuts of multivalued functions such as the logarithm or a non-integer power z^α. In the latter case, the origin is a branch point from which start necessarily a branch cut. Indeed, when one makes a complete turn around the origin, the argument of z varies by 2π, so that z^α is multiplied by $e^{2i\pi\alpha}$. For non-integer values of α, $e^{2i\pi\alpha} \neq 1$, and it is mandatory to introduce a cut to prevent function z^α to be multivalued: function values for points infinitely close on each side of the branch cut are different. For a function of form $[P(z)]^\alpha$, where $P(z)$ is a polynomial in z, we must first identify the zeros of the polynomial, which will be branch points. For the previous reason, each of them is the starting point of a branch cut. Its position is completely arbitrary. It is often convenient in practice to choose a branch cut joining two branch points.

Meromorphic functions   A function f is meromorphic on a domain D if all its singularities in D are isolated poles.

Laurent series   Let f be a holomorphic function on $0 \leq r_1 < |z-z_0| < r_2$. Then f admits a unique expansion in this domain, of the form $f(z) = \sum_{n=-\infty}^{+\infty} c_n(z-z_0)^n$. If f has a pole (respectively an essential singularity) in z_0, then there is a finite (respectively infinite) number of finite coefficients with negative index.

$\lozenge$ Residue theorem and Jordan's lemma $\lozenge$

Residue theorem   Let $D \subset \mathbb{C}$ be a simply connected open set, f a meromorphic function on D and $\mathcal{C}$ a simple closed path contained in D and avoiding the singularities of f. Then

$$\oint_{\mathcal{C}} dz\, f(z) = \pm 2i\pi \sum_k \operatorname{Res}(f, P_k).$$

Signs $+$ and $-$ in this expression correspond to path oriented respectively in the counterclockwise and clockwise direction. The sum runs over the poles of f inside $\mathcal{C}$ and $\operatorname{Res}(f, P_k)$ denotes the residue of f in p_k. This residue is the coefficient of $1/(z - z_k)$ in the Laurent series of function $f(z)$ in the neighbourhood of z_k.

Jordan's lemma   Let D be the domain $\{z = re^{i\theta}; r > 0\,,\ 0 \leq \theta_1 \leq \theta \leq \theta_2 \leq \pi\}$ and $\mathcal{C}_r$ the arc $\{re^{i\theta}; \theta_1 \leq \theta \leq \theta_2\}$.

- Let $f : \mathbb{C} \to \mathbb{C}$ be a continuous function of D such that $zf(z) \to 0$ when $|z| \to \infty$. Then

$$\int_{\mathcal{C}_r} dz\, f(z) \to 0 \quad \text{when} \quad r \to +\infty.$$

- Let $f : \mathbb{C} \to \mathbb{C}$ be a continuous function of D such that $f(z) \to 0$ when $z \to \infty$ with $z \in D$. Then

$$\int_{C_r} dz f(z) e^{iz} \to 0 \quad \text{when} \quad r \to +\infty.$$

$\Diamond$ **Cauchy principal part and Dirac distribution** $\Diamond$

Principal part   Let f be a function with a real simple pole z_0 with $z_0 \in [a, b]$. Cauchy principal part is defined as:

$$\text{PP} \int_a^b dz f(z) = \lim_{\epsilon \to 0} \left\{ \int_a^{z_0 - \epsilon} dz f(z) + \int_{z_0 + \epsilon}^b dz f(z) \right\}.$$

For instance

$$\text{PP} \int_{-1}^1 \frac{dz}{z} = 0.$$

We finally have, in the sense of distributions,

$$\lim_{\epsilon \to 0^+} \frac{1}{x \pm i\epsilon} = \text{PP} \frac{1}{x} \mp i\pi\delta(x). \tag{A.1}$$

Dirac distribution   We take advantage of this appendix to recall the following property of Dirac distribution

$$\int_a^b dx g(x) \delta[f(x)] = \sum_{x_0} \frac{g(x_0)}{|f'(x_0)|}, \tag{A.2}$$

where the sum runs over all points $x_0 \in [a, b]$ such that $f(x_0) = 0$. This identity is simply obtained by a change of variable.

Appendix B

Laplace transform

This appendix provides brief reminders about the Laplace transform.

Definition   Let $F(t)$ be a function defined for $t \in [0, +\infty[$. The Laplace transform of F, denoted $\mathcal{L}[F]$, is a function of a complex variable s defined by:

$$\mathcal{L}[F](s) = \int_0^\infty \mathrm{d}t \, e^{-st} F(t).$$

Domaine of definition   If there are finite $\alpha_0 > 0$, $M > 0$ and t_0 such that $\forall t > t_0$, $|e^{-\alpha_0 t} F(t)| \leq M$, then $\mathcal{L}[F]$ is analytic for $\mathrm{Re}\, s > \alpha_0$.

Examples

- Function $F(t) = e^{kt}$ admits a Laplace transform

$$\mathcal{L}[F](s) = \frac{1}{s - k},$$

  for $\mathrm{Re}\, s > \mathrm{Re}\, k$. This function can be extended in the whole complex plane where it is analytic, except at the point $s = k$ which is a simple pole.
- Function $F(t) = \exp(t^2)$ does not admit a Laplace transform.
- Note finally that if $F(t) = O(1/t^n)$ in $t = 0$ with $n \geq 1$, then F has no Laplace transform.
- We indicate in the table below some Laplace transforms of usual functions. We write this table for the argument $s = -iz$.

Table B.1   Laplace transform of some usual functions.

$F(t)$	1	t^n	$\cos \omega t$	$\sin \omega t$	$\mathrm{ch}\, \omega t$	$\mathrm{sh}\, \omega t$
$\mathcal{L}[F](-iz)$	$\frac{i}{z}$	$\frac{n!}{(-iz)^{n+1}}$	$\frac{iz}{z^2 - \omega^2}$	$\frac{-\omega}{z^2 - \omega^2}$	$\frac{iz}{z^2 + \omega^2}$	$\frac{-\omega}{z^2 + \omega^2}$

Properties Provided that the corresponding Laplace transforms are well defined, one has the following basic properties:

(1) Linearity : $\mathcal{L}[aF + bG] = a\mathcal{L}[F] + b\mathcal{L}[G]$.

(2) The Laplace transform of the derivative of a function is:

$$\boxed{\mathcal{L}[F'](s) = s\mathcal{L}[F](s) - F(0^+).}$$ (B.1)

(3) Generalisation to derivative of order n:

$$\mathcal{L}[F^{(n)}](s) = s^n\mathcal{L}[F] - s^{n-1}F(0^+) - s^{n-2}F'(0^+) - \cdots - F^{(n-1)}(0^+).$$

(4) $\mathcal{L}[e^{at}F(t)](s) = \mathcal{L}[F](s - a)$.

(5) $\mathcal{L}[\delta(t - t_0)] = e^{-st_0}$ for $t \geq t_0$.

Convolution Let F_1 and F_2 be two functions defined on $[0, \infty[$. Define

$$C_{F_1 F_2}(t) = \int_0^t dt' \, F_1(t - t')F_2(t').$$

The Laplace transform of $C_{F_1 F_2}$ is the product of Laplace transforms of F_1 and F_2:

$$\boxed{\mathcal{L}[C_{F_1 F_2}] = \mathcal{L}[F_1]\mathcal{L}[F_2].}$$

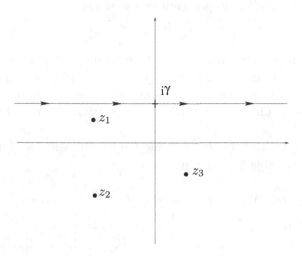

Fig. B.1   Integration domain used for the calculation of the inverse Laplace transform. Points z_1, z_2, z_3 represent the poles of $f(z)$ and the contour is selected above all these points.

Inverse transform Let $f(z) = \mathcal{L}[F](s = -iz)$ be the Laplace transform of $F(t)$, for $s = -iz$. Then

$$F(t) = \frac{1}{2\pi} \int_{-\infty+i\gamma}^{+\infty+i\gamma} dz\, e^{-izt} f(z) \qquad \text{(B.2)}$$

where γ satisfies $\gamma > \text{Sup}[\text{Im}(\text{Poles of } f)]$ and $t > 0$. The domain of integration is represented on Figure B.1.

Proof The idea is to draw upon the inverse Fourier transform. For this define

$$G(t) = \begin{cases} e^{-\gamma t} F(t) & \text{for } t > 0, \\ 0 & \text{for } t < 0, \end{cases}$$

where γ is chosen such that G has a Fourier transform. One then has

$$G(t) = \frac{1}{2\pi} \int_{-\infty}^{+\infty} d\omega\, e^{-i\omega t} \int_{-\infty}^{+\infty} dx\, e^{i\omega x} G(x).$$

It follows that

$$\begin{aligned} F(t) &= e^{\gamma t} G(t), \\ &= \frac{e^{\gamma t}}{2\pi} \int_{-\infty}^{+\infty} d\omega\, e^{-i\omega t} \int_{0}^{\infty} dx\, e^{i\omega x - \gamma x} F(x), \\ &= \frac{1}{2\pi} \int_{-\infty+i\gamma}^{+\infty+i\gamma} dz\, e^{-izt} \mathcal{L}[F](-iz) \end{aligned}$$

where we set $z = \omega + i\gamma$ and used the definition for the Laplace transform of F.

Appendix C

One-variable differential operators

In this appendix, we provide a method for computing the causal Green's function associated with a one-variable differential operator. We then recall the method of variation of the constant.

◇ Causal Green's function ◇

Let $G^+(t;t')$ be the causal Green's function of a linear differential operator

$$\mathcal{O}_t = a_n(t)\frac{\mathrm{d}^n}{\mathrm{d}t^n} + a_{n-1}(t)\frac{\mathrm{d}^{n-1}}{\mathrm{d}t^{n-1}} + \cdots + a_0(t).$$

By definition it satisfies

$$\mathcal{O}_t G^+(t;t') = \delta(t-t') \qquad (C.1)$$

with

$$G^+(t;t') = 0 \quad \text{for} \quad t < t'.$$

This Green's function reads

$$\boxed{G^+(t;t') = \theta(t-t')Z(t;t')} \qquad (C.2)$$

where Z is the solution of homogeneous equation $\mathcal{O}_t Z(t;t') = 0$ with initial conditions

$$\boxed{\begin{array}{l} Z(t=t';t') = Z'(t=t';t') = \cdots = Z^{(n-2)}(t=t';t') = 0, n \geq 2 \\[3mm] Z^{(n-1)}(t=t';t') = \dfrac{1}{a_n} \end{array}}$$

using notation $Z^{(j)} = (\partial^j Z/\partial t^j)$. In order to prove this result note first that $G^+(t;t')$ defined in this way is causal. We then get:

$$\begin{aligned} \frac{\partial}{\partial t}G^+(t;t') &= \frac{\partial \theta(t-t')}{\partial t}Z(t;t') + \theta(t-t')\frac{\partial Z(t;t')}{\partial t}, \\ &= \delta(t-t')Z(t;t') + \theta(t-t')Z'(t;t'), \\ &= \theta(t-t')Z'(t;t') \end{aligned}$$

given the initial conditions verified by Z. Mathematical induction then allows to show that for $k \leq n - 1$

$$\frac{\partial^{(k)}}{\partial t^{(k)}} G^+(t; t') = \theta(t - t') Z^{(k)}(t; t').$$

Differentiating this result with respect to t for $k = n - 1$ and taking into account the initial condition we eventually get

$$\frac{\partial^{(n)}}{\partial t^{(n)}} G^+(t; t') = \theta(t - t') Z^{(n)}(t; t') + \frac{1}{a_n} \delta(t - t').$$

It is then clear that $G^+(t; t')$ is a solution of differential equation (C.1).

◇ Variation of the constant method ◇

Let us consider the second order linear differential equation:

$$\ddot{\Phi} + \alpha(t) \, \dot{\Phi} + \beta(t) \, \Phi = F(t) \tag{C.3}$$

with $\dot{\Phi} = (d\Phi/dt)$. Assume that Φ_1 and Φ_2 are two known linearly independent solutions of the corresponding homogeneous equation. Their Wronskian is defined as

$$W = \Phi_1 \dot{\Phi}_2 - \Phi_2 \dot{\Phi}_1.$$

Then a solution of equation (C.3) reads

$$\boxed{\Phi(t) = -\Phi_1(t) \int_{t_0}^t dx \frac{\Phi_2(x) F(x)}{W(x)} + \Phi_2(t) \int_{t_1}^t dx \frac{\Phi_1(x) F(x)}{W(x)}} \tag{C.4}$$

where t_0 and t_1 are constants. These constants are adjusted according to the initial conditions imposed on Φ. To prove this result we look for Φ of the form

$$\Phi(t) = C_1(t) \Phi_1(t) + C_2(t) \Phi_2(t)$$

with

$$\dot{C}_1 \Phi_1 + \dot{C}_2 \Phi_2 = 0. \tag{C.5}$$

Differential equation (C.3) is then equivalent to the linear system

$$\begin{pmatrix} \Phi_1 & \Phi_2 \\ \dot{\Phi}_1 & \dot{\Phi}_2 \end{pmatrix} \begin{pmatrix} \dot{C}_1 \\ \dot{C}_2 \end{pmatrix} = \begin{pmatrix} 0 \\ F \end{pmatrix}$$

which can easily be inverted as $\dot{C}_1 = -(\Phi_2 F/W)$ and $\dot{C}_2 = (\Phi_1 F/W)$. It is then sufficient to integrate these expressions.

Remember that this method can be generalised to a differential equation of order n whose n independent homogeneous solutions ϕ_i are known. To do so, look for Φ as $\Phi(t) = \sum_{i=1}^N C_i(t) \Phi_i(t)$ with

$$\sum_i \dot{C}_i \Phi_i = 0 \quad ; \quad \sum_i \dot{C}_i \dot{\Phi}_i = 0 \quad ; \quad \cdots \quad ; \quad \sum_i \dot{C}_i \Phi_i^{(n-1)} = F.$$

Application Suppose we look for the Green's function $G(t; t')$ associated with equation (C.3) on interval $[a, b]$ with boundary conditions $G(a; t') = 0$ and $G(b, t') = 0$. We can then choose $t_0 = b$ and $t_1 = a$ in result (C.4) provided to also impose that $\Phi_1(a) = 0$ and $\Phi_2(b) = 0$. Since $F(x) = \delta(x - t')$ in the case of Green's function, expression (C.4) becomes

$$G(t; t') = \Phi_1(t) \int_t^b dx \frac{\Phi_2(x)\delta(x - t')}{W(x)} + \Phi_2(t) \int_a^t dx \frac{\Phi_1(x)\delta(x - t')}{W(x)}.$$

One then has to distinguish between the cases $t < t'$ and $t > t'$. The results read:

$$G(t; t') = \theta(t' - t) \frac{\Phi_1(t)\Phi_2(t')}{W(t')} + \theta(t - t') \frac{\Phi_2(t)\Phi_1(t')}{W(t')}. \tag{C.6}$$

Appendix D

Hilbert spaces and Dirac notation

This appendix is intended to recall some properties related to Hilbert spaces, particularly Dirac notation and closure relation.

◇ Kets in a Hilbert space ◇

Very schematically, a Hilbert space is a vector space of possibly infinite dimension equipped with a scalar product. Thus a vector in this space is nothing but a function $\psi(\mathbf{r})$ of a domain $\mathcal{D}$ of $\mathbb{C}$. It is written as $|\psi\rangle$, pronounced « ket ψ », according to the terminology introduced by Dirac. Originally introduced in the context of quantum mechanics, this notation may be used for any other problem where the physical quantity of interest belongs to a Hilbert space. Note that $|\psi\rangle$ is the strict analogue of a vector $\mathbf{u}$ in a finite dimension vector space.

◇ Scalar product ◇

The scalar product of ket $|\psi_b\rangle$ with ket $|\psi_a\rangle$, denoted as $\langle\psi_b|\psi_a\rangle$, where $\langle\psi_b|$ is called « bra ψ_b », is defined by the bracket

$$\boxed{\langle\psi_b|\psi_a\rangle = \int_{\mathcal{D}} d\mathbf{r}\ \psi_b^*(\mathbf{r})\psi_a(\mathbf{r})}$$

where $\psi_b^*(\mathbf{r})$ is the complex conjugate of $\psi_b(\mathbf{r})$. Note that here the equivalent of the usual scalar product $\mathbf{u}_b \cdot \mathbf{u}_a$ is not invariant in the exchange of both vectors since $\langle\psi_a|\psi_b\rangle = (\langle\psi_b|\psi_a\rangle)^*$. Other usual properties of the scalar product are preserved.

◇ Basis $|\mathbf{r}\rangle$ ◇

The Dirac distribution of $\delta(\mathbf{r} - \mathbf{r}')$, seen as a function of $\mathbf{r}$ for a given $\mathbf{r}'$, is a particular vector denoted $|\mathbf{r}\rangle$. The set of vectors $|\mathbf{r}\rangle$ where $\mathbf{r}$ runs through $\mathcal{D}$ forms a complete orthonormal basis of the space under consideration. In particular one easily gets, with the definition of the scalar product,

$$\boxed{\langle\mathbf{r}''|\mathbf{r}'\rangle = \delta(\mathbf{r}'' - \mathbf{r}').}$$

Moreover, the scalar product of vector $|\mathbf{r}\rangle$ by vector $|\psi\rangle$ reduces to the value of function ψ at point $\mathbf{r}$, i.e.

$$\boxed{\langle \mathbf{r}|\psi\rangle = \psi(\mathbf{r}).}$$

Any vector $|\psi\rangle$ can be written as a unique linear combination of $|\mathbf{r}\rangle$, i.e.

$$|\psi\rangle = \int_{\mathcal{D}} d\mathbf{r}\, c(\mathbf{r})\, |\mathbf{r}\rangle,$$

where coefficients $c(\mathbf{r})$ are analogous to coefficients c_i for the decomposition of any vector $\mathbf{v}$ on a basis $\{\mathbf{u}_i, i = 1, ..., d\}$ of a vector space of finite dimension d. Exploiting the orthonormality of basis $\{|\mathbf{r}\rangle\ , \ \mathbf{r} \in \mathcal{D}\}$, one can easily show that coefficients $c(\mathbf{r})$ of this linear combination identify with $\psi(\mathbf{r})$, i.e.

$$\boxed{|\psi\rangle = \int_{\mathcal{D}} d\mathbf{r}\, \psi(\mathbf{r})\, |\mathbf{r}\rangle.}$$

◊ **Orthogonal projectors** ◊

The image of a vector $|\psi\rangle$ by the linear operator $\mathcal{P}_a$ corresponding to an orthogonal projection along vector $|\psi_a\rangle$ is a vector $\mathcal{P}_a|\psi\rangle$ collinear to $|\psi_a\rangle$ with a proportionality factor which is nothing but scalar product $\langle\psi_a|\psi\rangle$:

$$\mathcal{P}_a|\psi\rangle = \langle\psi_a|\psi\rangle\, |\psi_a\rangle.$$

It is useful to introduce notation

$$\mathcal{P}_a = |\psi_a\rangle\langle\psi_a|,$$

the image $\mathcal{P}_a|\psi\rangle$ then being obtained simply by combining bra ψ_a with ket ψ to form scalar product $\langle\psi_a|\psi\rangle$ multiplying the vector $|\psi_a\rangle$. Note that this bra-ket notation $|\psi_a\rangle\langle\psi_a|$ defines an operator and not a vector.

◊ **Closure relation** ◊

Given an orthonormal basis composed of vectors $|\psi_a\rangle$, where index a describes an even set A of (discrete and/or continuous) values, the component along $|\psi_a\rangle$ of a vector $|\psi\rangle$ is given by scalar product $\langle\psi_a|\psi\rangle$. Thus the corresponding decomposition of $|\psi\rangle$ can be seen as the action of operator $\sum_{a \in A} \mathcal{P}_a$ on $|\psi\rangle$. As it applies to any vector, the sum over a of orthogonal projection operators along $|\psi_a\rangle$ is the identity operator $\mathcal{I}$:

$$\boxed{\sum_{a \in A} |\psi_a\rangle\langle\psi_a| = \mathcal{I}\,.} \tag{D.1}$$

This closure relation is valid for any arbitrary orthonormal basis.

◇ **Operators** ◇

Let finally consider a linear operator $\mathcal{O}$, originally defined by its action on any function $\psi(\mathbf{r})$. Then the image of corresponding vector $|\psi\rangle$ by $\mathcal{O}$, denoted as $\mathcal{O}|\psi\rangle$, is defined by the set of its components $\langle\mathbf{r}'|\mathcal{O}|\psi\rangle$ on basis $\{|\mathbf{r}'\rangle, \mathbf{r}' \in \mathcal{D}\}$, each component being identified with $\mathcal{O}\psi(\mathbf{r}')$. More generally, any linear operator $\mathcal{O}$ is defined from an orthonormal basis $\{|\psi_a\rangle, a \in A\}$ by its matrix elements $\langle\psi_b|\mathcal{O}|\psi_a\rangle$. Operator $\mathcal{O}$ can therefore be written as:

$$\mathcal{O} = \sum_{a,b} \langle\psi_b|\mathcal{O}|\psi_a\rangle \, |\psi_b\rangle\langle\psi_a|.$$

Appendix E

Gaussian integrals

In this appendix we show how to compute Gaussian integrals using examples frequently appearing in physics. Techniques used in each of the examples can be combined to address the most general possible case of Gaussian integral. In all three cases, the idea is to get back to integral

$$\int_{-\infty}^{\infty} \mathrm{d}x \, e^{-ax^2} = \sqrt{\frac{\pi}{a}}. \tag{E.1}$$

◇ Fourier transform of a real Gaussian ◇

The first example corresponds to the Fourier transform of a real Gaussian function. This type of integral appears for example in the calculation of Green's functions for diffusion equation. To calculate integrals of the type:

$$I_\sigma(x) = \int_{-\infty}^{+\infty} \mathrm{d}k \, \exp(ikx - \sigma k^2),$$

with $\sigma > 0$, we start by writing it as

$$I_\sigma(x) = \exp[-x^2/(4\sigma)] \int_{-\infty}^{+\infty} \mathrm{d}k \, \exp\{-\sigma[k - (ix/2\sigma)]^2\}.$$

Note then that integral

$$\oint_{\mathcal{C}_R} \mathrm{d}z \, \exp\{-\sigma[z - (ix/2\sigma)]^2\}, \tag{E.2}$$

defined on the contour $\mathcal{C}_R$ in the complex plane shown in Figure E.1, vanishes because the integrand is analytic inside the contour.

We then take the limit $R \to \infty$:

- It is easy to show that the contribution from the contour vertical segments vanishes.
- The contribution from the real axis is the integral $I_\sigma(x)$ we wish to evaluate.
- Finally the contribution from the horizontal segment with $\mathrm{Im}\, z = x/(2\sigma)$ is simply a Gaussian integral leading to $-\sqrt{\pi/\sigma}$.

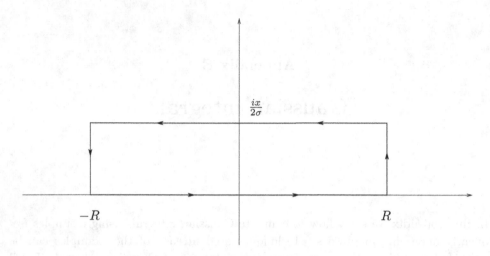

$$\frac{ix}{2\sigma}$$

$-R$ $\qquad$ R

Fig. E.1 Contour $\mathcal{C}_R$ used to compute integral (E.2).

The final result then reads:

$$\int_{-\infty}^{+\infty} dk \ \exp(ikx - \sigma k^2) = \sqrt{\frac{\pi}{\sigma}} \ \exp[-x^2/(4\sigma)]. \tag{E.3}$$

◊ **Purely imaginary quadratic term** ◊

Consider as a second example Gaussian integral with a purely imaginary coefficient of the quadratic part. This type of integral is found in quantum mechanics, for example in the calculation of the free particle propagator. Consider then an integral of the form

$$J_\sigma = \int_{-\infty}^{\infty} dx \ \exp(i\sigma x^2), \tag{E.4}$$

with $\sigma \neq 0$. To evaluate J_σ we shall compute the integral

$$\oint_{\mathcal{D}_R} dz \ \exp(i\sigma z^2) \tag{E.5}$$

on the closed contour $\mathcal{D}_R$ consisting of the portion $L_1 = [-R, R]$ of the real axis, circular arcs C_1 and C_2, respectively defined by $z = Re^{i\theta}$, $0 \leq \theta \leq \pi/4$ and $\pi \leq \theta \leq 5\pi/4$, and finally the element L_2 defined by $z = \rho e^{i\pi/4}$, $-R \leq \rho \leq R$ (see Figure E.2).

- Since function $e^{i\sigma z^2}$ does not have any singularity within the integration contour, it is clear that its integral on this closed contour vanishes.
- Moreover the integral on segment L_1 identifies with J_σ in the limit $R \to \infty$.

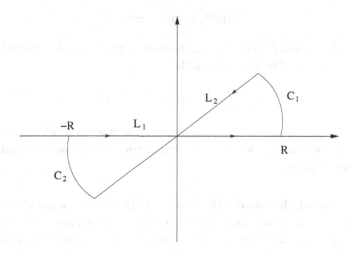

Fig. E.2   Contour $\mathcal{D}_R$ used to compute integral (E.5).

- In addition, we know how to calculate the contribution of L_2 segment in the limit $R \to \infty$, which is:

$$-i^{1/2} \int_{-\infty}^{\infty} \mathrm{d}x e^{\sigma x^2} = -(i\pi/\sigma)^{1/2}.$$

- It remains to show that the integrals over arcs C_1 and C_2 tend to 0 for $R \to \infty$. We shall show that the modulus of integral on C_1 is bounded by a function going to 0 when $R \to \infty$ (the extension to the case of C_2 is immediate). Indeed,

$$\left| \int_{C_1} \mathrm{d}z \, e^{i\sigma z} \right| = R \left| \int_0^{\pi/4} \mathrm{d}\theta \, e^{i\theta + i\sigma R^2 e^{2i\theta}} \right| \leq 4R \int_0^{\pi/4} \mathrm{d}\theta \, \cos\theta \, e^{-\sigma R^2 \sin(2\theta)},$$

where we used inequality $\cos\theta \geq \sin\theta$ for $0 \leq \theta \leq \pi/4$. On the same domain we also have inequality $\sin(2\theta) \geq \sin\theta$, leading to

$$4R \int_0^{\pi/4} \mathrm{d}\theta \, \cos\theta \, e^{-\sigma R^2 \sin(2\theta)} \leq 4R \int_0^{\pi/4} \mathrm{d}\theta \, \cos\theta \, e^{-\sigma R^2 \sin\theta}$$

$$\leq \frac{4}{R} \int_0^{\frac{R^2}{\sqrt{2}}} \mathrm{d}x \, e^{-\sigma x}$$

which goes to zero for $R \to \infty$.

This completes the proof and thus

$$\boxed{\int_{-\infty}^{\infty} \mathrm{d}x e^{i\sigma x^2} = \left(\frac{i\pi}{\sigma} \right)^{1/2}.} \tag{E.6}$$

◊ **Multiple integrals** ◊

We conclude this appendix with the calculation of two multiple integrals on d real variables $(x_1, \cdots, x_d)$. The first integral is

$$I_C = \int_{\mathbb{R}^d} \mathrm{d}\mathbf{x} \ \exp(-\mathbf{x}^T \cdot C \cdot \mathbf{x}), \tag{E.7}$$

where we wrote variables $(x_1, \cdots, x_d)$ as a column vector $\mathbf{x}$ and $\mathbf{x}^T$ is the transposed of $\mathbf{x}$. In this expression C is a $d \times d$ real symmetric positive definite matrix (all its eigenvalues are positive).

Diagonalisation of the quadratic form   A direct integration of the Gaussian factor over variables x_i is not immediate since all these variables are coupled in the quadratic form defined by matrix C. It is useful to diagonalise C to decouple these variables. This matrix is diagonalisable since it is real and symmetric. So there is a real orthogonal matrix R, $R^T R = \mathcal{I}$, diagonalising C:

$$R^T C R = \begin{pmatrix} \mu_1 & 0 & 0 & . \\ 0 & \mu_2 & 0 & . \\ . & . & . & . \\ . & . & . & \mu_d \end{pmatrix}. \tag{E.8}$$

We perform the change of variables $(x_1, \cdots, x_d) \to (\xi_1, \cdots, \xi_d)$ defined by $\boldsymbol{\xi} = R^T \mathbf{x}$. Its Jacobian is $|\operatorname{Det} R|$, and it reduces to 1 since R is orthogonal. Indeed, identity $R^T R = \mathcal{I}$ implies $[\operatorname{Det} R]^2 = 1$. The new integration domain is still $\mathbb{R}^d$. The original quadratic form eventually becomes, in terms of the new variables,

$$\mathbf{x}^T \cdot C \cdot \mathbf{x} = \sum_i \mu_i \, \xi_i^2 \ . \tag{E.9}$$

Integrations on the new variables are now completely decoupled, so that Gaussian integral (E.7) becomes a product of simple integrals

$$I_C = \prod_{i=1}^d \left[\int_{\mathbb{R}} \mathrm{d}\xi_i \ \exp\left(-\mu_i \xi_i^2\right) \right].$$

Note here that the d eigenvalues μ_i are real and strictly positive by assumption. We are then brought back to the simple evaluation of Gaussian integrals over a single variable, leading to

$$I_C = \sqrt{\frac{\pi^d}{\mu_1 \mu_2 \cdots \mu_d}}. \tag{E.10}$$

The latter can be written as:

$$\boxed{\int_{\mathbb{R}^d} \mathrm{d}\mathbf{x} \ e^{-\mathbf{x}^T C \mathbf{x}} = \sqrt{\frac{\pi^d}{\operatorname{Det} C}},}$$

where $\mathrm{Det}\, C$ is the determinant of matrix C. Note that one can write I_C, given by equation (E.10), as

$$I_C = \sqrt{\frac{\pi^d}{\mu_1 \mu_2 \cdots \mu_d}} = \exp\left[\frac{1}{2}\sum_{i=1}^{d} \ln(\pi/\mu_i)\right]. \tag{E.11}$$

Let us now generalise this calculation to integral

$$\int_{\mathbb{R}^d} d\mathbf{x}\ \exp(-\mathbf{x}^T \cdot C \cdot \mathbf{x} + \mathbf{x}^T \cdot \mathbf{y})$$

where $\mathbf{y}$ is a given vector and C is as before a real symmetric positive definite matrix. As for the corresponding integral over a single variable, this integral is calculated noticing that

$$-\mathbf{x}^T \cdot C \cdot \mathbf{x} + \mathbf{x}^T \cdot \mathbf{y} = -(\mathbf{x} - \frac{1}{2}C^{-1}\mathbf{y})^T \cdot C \cdot (\mathbf{x} - \frac{1}{2}C^{-1}\mathbf{y}) + \frac{1}{4}\mathbf{y}^T \cdot C^{-1} \cdot \mathbf{y}.$$

The result is then

$$\boxed{\int_{\mathbb{R}^d} d\mathbf{x}\ \exp\left[-\mathbf{x}^T C \mathbf{x} + \mathbf{x}^T \mathbf{y}\right] = \sqrt{\frac{\pi^d}{\mathrm{Det}\, C}}\ \exp\left[\frac{1}{4}\mathbf{y}^T C^{-1} \mathbf{y}\right].} \tag{E.12}$$

Appendix F

Overview of coordinate transformations

This appendix provides brief reminders on changes of coordinates.

Metric   Denote by $\{\xi^i\}$ with $i = 1, \cdots, d$ the coordinates of a domain (manifold) in d dimensions, and $\{\mathbf{e}_i\}$ the basis of vectors (defined in the tangent space) associated with these coordinates. The infinitesimal line element d$\mathbf{s}$ is given by:

$$\mathrm{d}\mathbf{s} = \mathrm{d}\xi^i \mathbf{e}_i = \sum_{i=1}^d \mathrm{d}\xi^i \mathbf{e}_i,$$

where we used Einstein's convention implying summation on repeated indices. Defining the interior product for elements in this basis as

$$\mathbf{e}_i \cdot \mathbf{e}_j = g_{ij},$$

then symmetric matrix g_{ij} is the metric associated with coordinates $\{\xi^i\}$. We have in particular

$$\boxed{\mathrm{d}s^2 = g_{ij}\,\mathrm{d}\xi^i\mathrm{d}\xi^j.}$$

Change of coordinates   Imagine now that one performs a change of coordinates $\xi^i \mapsto \widetilde{\xi}^j$ with a transformation matrix

$$J_{ij} = \frac{\partial \xi^i}{\partial \widetilde{\xi}^j}$$

and its inverse $(\partial \widetilde{\xi}^j / \partial \xi^i)$. The corresponding transformation rules are then

$$\mathrm{d}\widetilde{\xi}^i = \frac{\partial \widetilde{\xi}^i}{\partial \xi^j}\mathrm{d}\xi^j.$$

We have, for the new basis elements $\{\widetilde{\mathbf{e}}_i\}$, associated with coordinates $\widetilde{\xi}^i$, and defined by d$\mathbf{s} = \mathrm{d}\widetilde{\xi}^i\widetilde{\mathbf{e}}_i$:

$$\widetilde{\mathbf{e}}_i = \frac{\partial \xi^j}{\partial \widetilde{\xi}^i}\,\mathbf{e}_j.$$

One then gets in the new coordinates:

$$\mathrm{d}s^2 = \widetilde{g}_{ij}\,\mathrm{d}\widetilde{\xi}^i\mathrm{d}\widetilde{\xi}^j$$

with $\widetilde{g}_{ij}$ elements of the metric matrix in the new coordinates $\{\widetilde{\xi}^i\}$:

$$\boxed{\widetilde{g}_{ij} = g_{lm}\,\frac{\partial \xi^l}{\partial \widetilde{\xi}^i}\,\frac{\partial \xi^m}{\partial \widetilde{\xi}^j}.}$$

Infinitesimal volume element   If transformation $\xi^i \mapsto \widetilde{\xi}^j$ is oriented, that is to say if $\mathrm{Det}\, J > 0$ where $\mathrm{Det}\, J$ denotes the determinant of matrix J, then

$$\mathrm{d}\xi^1 \mathrm{d}\xi^2 \cdots \mathrm{d}\xi^n = \mathrm{Det}\, J \, \mathrm{d}\widetilde{\xi}^1 \mathrm{d}\widetilde{\xi}^2 \cdots \mathrm{d}\widetilde{\xi}^n.$$

From this transformation rule and previous results, it is possible to define a volume element invariant under coordinate transformations, i.e.

$$\sqrt{\mathrm{Det}\, g} \; \mathrm{d}\xi^1 \mathrm{d}\xi^2 \cdots \mathrm{d}\xi^n = \sqrt{\mathrm{Det}\, \widetilde{g}} \; \mathrm{d}\widetilde{\xi}^1 \mathrm{d}\widetilde{\xi}^2 \cdots \mathrm{d}\widetilde{\xi}^n. \tag{F.1}$$

Dirac distribution   Result (F.1) allows to establish the transformation rule

$$\frac{1}{\sqrt{\mathrm{Det}\, g}} \prod_{i=1}^{d} \delta(\xi^i) = \frac{1}{\sqrt{\mathrm{Det}\, \widetilde{g}}} \prod_{i=1}^{d} \delta(\widetilde{\xi}^i) \tag{F.2}$$

for Dirac distribution.

Gradient   Expression of the gradient of a scalar function ϕ in any coordinate system is:

$$\boldsymbol{\nabla}\phi = g^{ij} \frac{\partial \phi}{\partial \xi^j} \, \mathbf{e}_i = \widetilde{g}^{ij} \frac{\partial \phi}{\partial \widetilde{\xi}^j} \, \widetilde{\mathbf{e}}_i \tag{F.3}$$

where g^{ij} are elements of the inverse matrix of g, $g^{ij} g_{jk} = \delta_i^k$ where δ_i^k is Kronecker symbol.

Divergence   The divergence of a vectorial function $\mathbf{A} = A^i \mathbf{e}_i = \widetilde{A}^i \widetilde{\mathbf{e}}_i$ reads:

$$\boldsymbol{\nabla} \cdot \mathbf{A} = \frac{1}{\sqrt{\mathrm{Det}\, g}} \frac{\partial}{\partial \xi^i} \left(\sqrt{\mathrm{Det}\, g}\, A^i \right) = \frac{1}{\sqrt{\mathrm{Det}\, \widetilde{g}}} \frac{\partial}{\partial \widetilde{\xi}^i} \left(\sqrt{\mathrm{Det}\, \widetilde{g}}\, \widetilde{A}^i \right). \tag{F.4}$$

Laplacian   To get the Laplacian of a scalar function ϕ, one just takes the divergence of its gradient, leading to

$$\Delta \phi = \frac{1}{\sqrt{\mathrm{Det}\, g}} \frac{\partial}{\partial \xi^i} \left(\sqrt{\mathrm{Det}\, g}\, g^{ij} \frac{\partial \phi}{\partial \xi^j} \right). \tag{F.5}$$

Appendix G

Spherical harmonics

In this appendix we recall some properties related to spherical harmonics.

Let Ψ be a solution of Laplace equation, $\Delta\Psi = 0$. In spherical coordinates (r, θ, φ), this equation becomes

$$\frac{1}{r}\frac{\partial^2}{\partial r^2}(r\Psi) + \frac{1}{r^2 \sin\theta}\frac{\partial}{\partial\theta}\left(\sin\theta\frac{\partial\Psi}{\partial\theta}\right) + \frac{1}{r^2 \sin^2\theta}\frac{\partial^2\Psi}{\partial\varphi^2} = 0. \qquad \text{(G.1)}$$

Separation of variables This equation is separable by taking Ψ as a product of functions of each variables

$$\Psi(r, \theta, \varphi) = U(r)P(\theta)Q(\varphi).$$

In fact, using this form for Ψ and multiplying equation (G.1) by $(r^2 \sin^2\theta/\Psi)$, the last member of this equation becomes a function of φ only, while this variable does not appear in other terms[1]. This equation only has a solution when the last term, a function of φ only, is equal to a constant written as $-m^2$:

$$\frac{1}{Q}\frac{\mathrm{d}^2 Q}{\mathrm{d}\varphi^2} = -m^2.$$

One then gets $Q(\varphi) = e^{\pm im\varphi}$. Since $\Psi(r, \theta, \varphi + 2\pi) = \Psi(r, \theta, \varphi)$, m must be an integer.

We can continue this reasoning and also separate the part depending on r and θ introducing another separation constant c leading to:

$$\frac{1}{\sin\theta}\frac{\mathrm{d}}{\mathrm{d}\theta}\left(\sin\theta\frac{\mathrm{d}P}{\mathrm{d}\theta}\right) + \left(c - \frac{m^2}{\sin^2\theta}\right)P = 0$$

and

$$\frac{\mathrm{d}^2 V}{\mathrm{d}r^2} - \frac{c\,V}{r^2} = 0 \qquad \text{(G.2)}$$

for $V(r) = rU(r)$.

---

[1]This situation is similar to the one encountered in analytical mechanics by the method of separation variables for the Hamilton-Jacobi equation (see e.g. [25] or [39]).

Associated Legendre functions   The equation for $P(\theta)$ admits finite solutions if and only if $c = l(l + 1)$ with l integer and $l \geq |m|$. The corresponding solutions are the associated Legendre functions $P_l^m(x)$ with $x = \cos\theta$. These functions then satisfy the differential equation

$$\frac{\mathrm{d}}{\mathrm{d}x}\left[(1 - x^2)\frac{\mathrm{d}P_l^m}{\mathrm{d}x}\right] + \left[l(l + 1) - \frac{m^2}{1 - x^2}\right]P_l^m(x) = 0.$$

In the case $m = 0$, functions $P_l(x) = P_l^0(x)$ correspond to Legendre polynomials

$$P_l(-x) = (-1)^l P_l(x),$$

$$P_0(x) = 1, \qquad P_1(x) = x, \qquad P_2(x) = \frac{1}{2}(3x^2 - 1),$$

$$\int_{-1}^{1} \mathrm{d}x\ P_l(x)P_{l'}(x) = \frac{2\delta_{ll'}}{2l + 1},$$

$$\sum_{l=0}^{+\infty} \frac{2l + 1}{2} P_l(x)P_l(x') = \delta(x - x').$$

The last two properties respectively show that these polynomials are orthogonal and form a complete basis of functions defined on $[-1, 1]$.

Associated Legendre functions satisfy the following properties:

$$P_l^m(x) = (1 - x^2)^{\frac{m}{2}} \frac{\mathrm{d}^m}{\mathrm{d}x^m} P_l(x) \qquad \text{for} \quad m > 0,$$

$$P_l^{-m}(x) = (-1)^m \frac{(l - m)!}{(l + m)!} P_l^m(x),$$

$$\int_{-1}^{1} \mathrm{d}x\ P_l^m(x)P_{l'}^m(x) = \frac{2}{2l + 1}\frac{(l + m)!}{(l - m)!}\delta_{ll'}.$$

Spherical harmonics   Spherical harmonics are defined by

$$Y_{lm}(\theta, \varphi) = (-1)^m \sqrt{\frac{2l + 1}{4\pi}\frac{(l - m)!}{(l + m)!}} P_l^m(\cos\theta)\ e^{im\varphi}.$$

They constitute complete basis of the set of functions defined on the sphere:

$$\sum_{l=0}^{\infty} \sum_{m=-l}^{l} Y_{lm}^*(\theta', \varphi')\ Y_{lm}(\theta, \varphi) = \delta(\varphi - \varphi')\ \delta(\cos\theta - \cos\theta'). \qquad (G.3)$$

They are orthonormal:

$$\int_0^{2\pi} \mathrm{d}\varphi \int_0^{\pi} \sin\theta\mathrm{d}\theta\ Y_{l'm'}^*(\theta, \varphi)\ Y_{lm}(\theta, \varphi) = \delta_{mm'}\ \delta_{ll'}.$$

Harmonic functions Since $c = l(l+1)$, equation (G.2) leads to

$$U(r) = A\,r^l + B\,r^{-l-1}.$$

Any harmonic function can be written as a superposition of the solutions obtained, or

$$\Psi(r, \theta, \varphi) = \sum_{l=0}^{\infty} \sum_{m=-l}^{l} \left[A_{l,m}\,r^l + B_{l,m}\,r^{-l-1} \right] Y_{lm}(\theta, \varphi)$$

where $A_{l,m}$ and $B_{l,m}$ are constants.

Appendix H

Functional derivative

Consider a space of functions $f(x_i)$ of N variables x_i defined on a domain which is a subset of $\mathbb{R}^N$ (usually the domain is $\mathbb{R}^N$ itself). We will assume that all functions $f(x_i)$ vanish at the domain boundary or decay fast enough to zero if the domain is $\mathbb{R}^N$. Let $F[f]$ be a functional of f. Its functional derivative denoted

$$\frac{\delta}{\delta f(x_i)} F[f]$$

is defined from the following fundamental properties:

- Linearity:

$$\frac{\delta}{\delta f(x_i)} \left(F_1[f] + F_2[f]\right) = \frac{\delta}{\delta f(x_i)} F_1[f] + \frac{\delta}{\delta f(x_i)} F_2[f].$$

- Distributive property:

$$\frac{\delta}{\delta f(x_i)} \left(F_1[f] F_2[f]\right) = F_2[f] \frac{\delta}{\delta f(x_i)} F_1[f] + F_1[f] \frac{\delta}{\delta f(x_i)} F_2[f].$$

- Derivation rule:

$$\frac{\delta f(y_i)}{\delta f(x_i)} = \prod_{i=1}^{N} \delta(y_i - x_i).$$

The latter property is a generalisation to the case of continuous functions of the derivation rule over discrete variables: $\frac{\partial r_i}{\partial r_j} = \delta_{ij}$.

The function $f(x_i)$ evaluated at each point of the domain parametrised by variables x_i plays the role of one of the components of a vector with an infinite and continuous number of components. The functional derivative can be considered as the generalisation to continuous variables of the derivative of a function $h(r_i)$ over its variables r_i: $\frac{\partial h}{\partial r_i}$. With the rules we have just stated, it is possible for example to calculate:

$$\frac{\delta}{\delta f(x_i)} \int \prod_{i=1}^{N} \mathrm{d}y_i \, \Phi(y_i) f(y_i) = \Phi(x_i).$$

Using that functions f vanish at the domain boundaries and a simple integration by parts, it is also possible to show that:

$$\frac{\delta}{\delta f(x_i)} \int \prod_{i=1}^{N} dy_i \, g(y_i) \frac{\partial}{\partial x_k} [f(y_i)] = -\frac{\partial}{\partial x_k} [g(x_i)] \,.$$

Example   To conclude this appendix, we give as an example a way to get the equations of motion for the harmonic oscillator from its action $S[q(t)]$:

$$S[q(t)] = \int_{t_i}^{t_f} dt \, \left(\frac{m\dot{q}^2}{2} - \frac{m\omega^2 q^2}{2} \right) \,.$$

Here $q(t)$ is the particle position at time t, m its mass and ω the oscillator pulsation. In this example q plays the role of the function f and time t, that of variables x_i. $q(t_i)$ and $q(t_f)$ are not necessarily zero, but are assumed to remain constant. The boundary terms do not give a contribution to the functional derivative of the action with respect to function $q(t)$. Euler-Lagrange equations, equivalent to the equations of motion, are obtained by imposing that the action is stationary relative to functional variable $q(t)$:

$$\frac{\delta}{\delta q(t)} S[q(t)] = 0,$$

leading to the harmonic oscillator equations of motion:

$$\ddot{q}(t) + \omega^2 q(t) = 0.$$

Appendix I

Usual Green's functions

We give in this appendix Green's functions in infinite volume G_∞ computed in Chapters 2 and 3.

$$\Diamond \textbf{ Laplacian operator } -\Delta \Diamond$$

Dimension	Green's function	Page
$d \geq 3$	$G_\infty(\mathbf{r}) = \frac{\Gamma(d/2)}{(d-2)2\pi^{d/2}|\mathbf{r}|^{d-2}}$	303, 321
3	$G_\infty(\mathbf{r}) = \frac{1}{4\pi|\mathbf{r}|}$	68
2	$G_\infty(\mathbf{r}) = -\frac{1}{2\pi}\ln\big(|\mathbf{r}|/\ell\big)$	84
1	$G_\infty(x) = -\frac{1}{2}|x|$	84

◇ **Helmholtz operator** $-\Delta + m^2$ ◇

Dimension	Green's function	Page
3	$G_\infty(\mathbf{r}) = \frac{e^{-m\lvert\mathbf{r}\rvert}}{4\pi\lvert\mathbf{r}\rvert}$	68
2	$G_\infty(\mathbf{r}) = \frac{1}{2\pi}K_0(m\lvert\mathbf{r}\rvert)$	83
1	$G_\infty(x) = \frac{1}{2m}e^{-m\lvert x\rvert}$	83

◇ **Diffusion: operator** $\partial/\partial t - D\Delta$ ◇

Dimension	Green's function	Page
d	$G_\infty^+(\mathbf{r};t) = \frac{1}{(4\pi Dt)^{d/2}}\,\exp\left[-\frac{\mathbf{r}^2}{4Dt}\right]$	141

◇ **Free quantum particle: operator** $\partial/\partial t - (i\hbar/2m)\Delta$ ◇

Dimension	Green's function	Page
3	$G_\infty^+(\mathbf{r};t) = \left(\frac{m}{2\pi i\hbar t}\right)^{3/2}\exp\left[\frac{im\mathbf{r}^2}{2\hbar t}\right]$	149

◊ D'Alembert operator $(1/c^2)(\partial^2/\partial t^2) - \Delta$ ◊

Dimension	Green's function	Page
3+1	$G_\infty^+(\mathbf{r};t) = \frac{c}{4\pi|\mathbf{r}|}\,\delta(ct - |\mathbf{r}|)$	166
2+1	$G_\infty^+(\mathbf{r},t) = \frac{c\,\theta(ct-|\mathbf{r}|)}{2\pi\sqrt{c^2t^2-|\mathbf{r}|^2}}$	319
1+1	$G_\infty^+(x,t) = \frac{c}{2}\theta(ct - |x|)$	321

●

Appendix J

Solutions of exercises

◊ Chapter 1 ◊

□ **Solution of exercise 1.1. Response functions associated with linear operators**

To determine the response functions, it is enough to use result (C.2), p. 271. For the first differential operator, the result is:

$$K_1(t - t') = \theta(t - t')e^{-a(t-t')} \qquad \text{and} \qquad \chi_1(z) = \frac{1}{a - iz}.$$

For $a < 0$, the Laplace transform is only defined for $\text{Im } z > -a$ but the result can be analytically continued on $\mathbb{C} \setminus \{-ia\}$.

The result for the second operator reads:

$$K_2(t - t') = \frac{1}{\omega}\theta(t - t')\sin(\omega(t - t')) \qquad \text{and} \qquad \chi_2(z) = \frac{-1}{z^2 - \omega^2},$$

with $b = \omega^2$. For $b = -\omega^2$, one just replaces the sine function by a hyperbolic sine. For χ_2 the result is then $\chi_2(z) = -1/(z^2 + \omega^2)$.

□ **Solution of exercise 1.2. Response function for a RLC circuit**

The first two calculations do not present any particular difficulty. Note that one must first calculate the response function related to the capacitor charge Q and then derive this function to get the response function of the current. The inverse Laplace transform of $\chi(z)$ can be computed by decomposing $\chi(z)$ into simple elements and then, using table B.1 on p. 267 of Laplace transforms of the usual functions. We thus find the results shown on p. 21. Rather than detailing the calculation, let us indicate how inversion formula (B.2), p. 269 operates for the simple element

$f(z) = 1/(z - z_1)$ with $\operatorname{Im} z_1 < 0$. Let $F(t)$ be the inverse Laplace transform of $f(z)$. The application of formula (B.2) leads to:

$$F(t) = \frac{1}{2\pi} \int_{-\infty+i\gamma}^{+\infty+i\gamma} dz \, \frac{e^{-izt}}{z - z_1}$$

where $\gamma > \operatorname{Im} z_1$. Let us fix for example $\gamma = 0$. One just has to deform the integration contour as presented in Figure J.1 and apply both Jordan's lemma and residue theorem to get

$$F(t) = \frac{1}{2\pi}(-2i\pi)e^{-iz_1 t} = -ie^{-iz_1 t}.$$

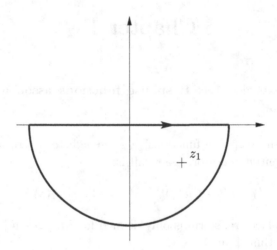

Fig. J.1 Integration contour for the calculation of $F(t)$ by application of Jordan lemma and residue theorem.

□ **Solution of exercise 1.3. Charged Brownian particle**

1. One immediately gets

$$\chi(z) = \frac{1}{\gamma - imz} \qquad \text{and} \qquad -iz\mu(z) = \chi(z).$$

Susceptibility χ satisfies K.K. relations under their usual form, since $\chi(z)$ is analytic on the real axis. On the other hand, susceptibility μ satisfies generalised K.K., as $\mu(z)$ has a simple pole in $z = 0$.

2. Response functions are easily obtained using the methods from exercise 1.2:

$$R(\tau) = \frac{1}{\gamma}(1 - e^{-\frac{\gamma}{m}\tau}) \qquad \text{and} \qquad V(\tau) = \frac{dR}{d\tau}.$$

3. The constant term in $R(\tau)$ when $\tau \to \infty$ leads to a pole in $z = 0$ for $\mu(z)$. This term disappears if one adds a restoring force in the equation of motion, for example by placing the particle in a harmonic potential. The situation is then analogous to that of the RLC circuit studied in section 1.2.1, p. 18 or to Thomson model seen p. 26.

□ **Solution of exercise 1.4. Absorption line**

1. One has to take this combination of two Dirac distributions centred in ω_0 and $-\omega_0$ because $\chi'(\omega)$ is an odd function of ω.

2. Kramers-Kronig relations lead to:

$$\chi'(\omega) = \frac{-2\sigma\omega_0}{\pi(\omega^2 - \omega_0^2)}. \tag{J.1}$$

3. Susceptibility χ_ϵ then reads:

$$\chi_\epsilon(z) = \frac{-\sigma}{\pi} \left(\frac{1}{z - \omega_0 + i\epsilon} - \frac{1}{z + \omega_0 + i\epsilon} \right).$$

Its real and imaginary parts are then:

$$\chi_\epsilon'(\omega) = -\frac{\sigma}{\pi} \left(\frac{\omega - \omega_0}{(\omega - \omega_0)^2 + \epsilon^2} - \frac{\omega + \omega_0}{(\omega + \omega_0)^2 + \epsilon^2} \right),$$

$$\chi_\epsilon''(\omega) = \frac{\sigma}{\pi} \left(\frac{\epsilon}{(\omega - \omega_0)^2 + \epsilon^2} - \frac{\epsilon}{(\omega + \omega_0)^2 + \epsilon^2} \right).$$

Note that $\chi_\epsilon(z)$ presents singularities in $\mathbb{C}^-$ for $z = \pm\omega_0 - i\epsilon$. Result (J.1) of the previous question is found in the limit $\epsilon \to 0^+$.

4. To perform this calculation, simply decompose expressions appearing in the Kramers-Kronig relation into simple elements,

$$\chi_\epsilon''(\omega) = \frac{\sigma}{\pi^2} \text{PP} \int_{-\infty}^{+\infty} d\omega' \frac{1}{\omega' - \omega} \left(\frac{\omega' - \omega_0}{(\omega' - \omega_0)^2 + \epsilon^2} - \frac{\omega' + \omega_0}{(\omega' + \omega_0)^2 + \epsilon^2} \right).$$

□ **Solution of exercise 1.5. Application of Kramers-Kronig relations in astrophysics**

1. This alternative way of writing Kramers-Kronig relations corresponds in fact to their original formulation by Kramers. It simply comes from relations (1.23), p. 15, and the even (resp. odd) nature of $\chi'(\omega)$ (resp. $\chi''(\omega)$).

2. We first write relation (1.96) for $\omega = 0$:

$$\chi'(0) = \frac{2}{\pi} \, \text{PP} \int_0^\infty d\omega' \, \frac{\chi''(\omega')}{\omega'}. \tag{J.2}$$

Since in the integrand $\chi''(\omega') > 0$, the volume fraction of grains pr^3 satisfies

$$pr^3 \geq \frac{1}{2\pi^2} \frac{\epsilon_g + 2}{\epsilon_g - 1} \int_{\omega_1}^{\omega_2} d\omega \, \frac{\chi''(\omega)}{\omega}. \tag{J.3}$$

This kind of argument has been developed and used by E.M. Purcell in 1969 in his study of interstellar medium.

□ Solution of exercise 1.6. Sum rules

To get $\tilde{\chi}(z)$ and $\tilde{\tilde{\chi}}(z)$, we first use equation (1.13), p. 11, directly leading to $\tilde{\chi}(z)$:

$$\tilde{\chi}(z) = -K_0(0) - iz\chi(z).$$

Repeating this result one gets:

$$\tilde{\tilde{\chi}}(z) = -K_0'(0) + izK_0(0) - z^2\chi(z).$$

By assumption, $\tilde{\chi}(z)$ and $\tilde{\tilde{\chi}}(z)$ both satisfy K.K. relations as well as sum rule (1.27), p. 16, with their respective response function. The requested sum rules are then obtained by expressing the real parts of $\tilde{\chi}$ and $\tilde{\tilde{\chi}}$ in terms of χ' and χ'', and then writing sum rule (1.27) for $\tilde{\chi}$ and $\tilde{\tilde{\chi}}$.

□ Solution of exercise 1.7. Response to noise

1. One easily shows that $\langle x^2(t) \rangle = \alpha \int_{-\infty}^t dt_1 K^2(t; t_1)$.

2. The response function of this operator is easy to compute and reads

$$K(\tau) = \frac{1}{m(\alpha_- - \alpha_+)} \left(e^{-\alpha_+ \tau} - e^{-\alpha_- \tau} \right),$$

with $\alpha_\pm = \frac{\gamma}{2m} \left(1 \pm \sqrt{1 - 4\frac{m^2\omega^2}{\gamma^2}} \right)$. A straightforward integration of $K(\tau)^2$ then leads to $\langle x^2(t) \rangle$.

□ Solution of exercise 1.8. Kramers-Kronig relations for a metal

1. To consider the field as homogeneous in the sample, it is necessary that the size of the latter is much smaller than the electric field wavelength, given by the ratio of the speed of light on frequency.

2. The current density then reads:

$$\mathbf{j} = \frac{\partial \mathbf{P}}{\partial t} = -i\omega\varepsilon_0(\varepsilon(\omega) - 1)\mathbf{E}.$$

On the other hand, in the static limit, the current density is related to the electric field by the conductivity σ:

$$\mathbf{j} = \sigma\mathbf{E}.$$

Comparing these two results, we obtain the expression for the low-frequency dielectric constant.

3. This is a special case of susceptibility with simple poles on the real axis. We can therefore use generalised KK relations (1.24), p. 16, leading to:

$$\varepsilon''(\omega) = -\frac{1}{\pi} \, \mathrm{PP} \int_{-\infty}^{\infty} d\omega' \, \frac{\varepsilon'(\omega')}{(\omega' - \omega)} + \frac{\sigma}{\varepsilon_0 \omega}.$$

The other relation remaining unchanged.

4. The expression for the dielectric constant at high frequency is obtained by assuming the electrons are free. They then satisfy the differential equation

$$m\frac{d^2\mathbf{r}}{dt^2} = q\mathbf{E},$$

leading to:

$$\varepsilon(\omega) - 1 \to -\frac{nq^2}{\varepsilon_0 m\omega^2} \qquad \text{when} \qquad \omega \to \infty.$$

As this term $1/\omega^2$ is real, it is clear that $\varepsilon''(\omega)$ decreases faster than $1/\omega^2$. By comparison with expansions (1.15) and (1.16), p. 12, it follows that $K_0(0) = 0$ and $K_0'(0) = nq^2/(\varepsilon_0 m)$ where $K_0(\tau)$ is the response function associated with $\varepsilon(z) - 1$. Sum rule (1.28), p. 17 then leads to:

$$\int_{-\infty}^{\infty} d\omega \, \omega\varepsilon(\omega) = \frac{\pi nq^2}{\varepsilon_0 m}.$$

5. $\varepsilon''(\omega)$ satisfies the constraints we have mentioned for small and large ω as well as the positivity condition for $\omega > 0$. The sum rule then sets $\tau = (m\sigma/nq^2)$.

6. This is the so-called Drude model wherein each charge is submitted to a simple viscous friction force. Note that this model can be derived from the microscopic Thomson model studied in section 1.2.2 (see p. 24) in the limit $\omega_0 \to 0$. We can superimpose over this susceptibility the one for bound electrons equal to $\omega_0 \neq 0$ and simulate the contribution of different electronic bands (conduction, valence, etc.).

□ Solution of exercise 1.9. Signal propagation in dielectric media

1. It is necessary to impose that $G(\tau)$ decreases sufficiently fast at infinity.

2. Amplitude g_z is the Laplace transform of $F(0,t)$. As $F(0,t)$ is bounded and causal, g_z is analytic in the upper half complex plane, which justifies that $\gamma > 0$ in the formula for inverse Laplace transform (see equation (B.2), p. 269).

3. General form of $F(x,t)$ follows from $F(0,t)$ and the wave equation:

$$F(x,t) = \int_{-\infty+i\gamma}^{\infty+i\gamma} dz\, g_z\, e^{-iz\left(t-\frac{\sqrt{\epsilon(z)}}{c}x\right)}.$$

Let us calculate this integral for $x > ct$: as $\epsilon(z) \to 1$ when $z \to \infty$, we can close the contour from the top, with a semi-circle whose radius tends towards infinity, and choose γ large enough so that $\sqrt{\epsilon(z)}$ is analytic inside the contour. Applications of Jordan's lemma's and Cauchy's theorem then ensure that this integral vanishes. Of course the result is different for $x < ct$ because in this case the domain of integration is closed from the bottom: the integration contour then contains singularities. Note that we use similar arguments in the example in section 3.2.5 p. 196.

◇ **Chapter 2** ◇

□ Solution of exercise 2.1. Green's function G_∞ for the Laplacian in 3d

1. Since $G_\infty(\mathbf{r})$ does not depend on r, equation (2.120) reads for $r \neq 0$:

$$\frac{1}{r^2}\frac{d}{dr}\left(r^2\frac{dG_\infty}{dr}\right) = \frac{d^2G_\infty}{dr^2} + \frac{2}{r}\frac{dG_\infty}{dr} = 0.$$

2. Two independent solutions of this equation are A/r and B where A and B are constants. The second solution is however incompatible with the boundary conditions at infinity, so that $G_\infty(r) = A/r$.

3. Constant A can be determined using Gauss's flux theorem, given by integration of equation (2.120) on a volume V including the origin:

$$-\int_V d\mathbf{r}\,\Delta G_\infty = -\int_{\partial V} dS\,\mathbf{n}\cdot\boldsymbol{\nabla}G_\infty = 1.$$

Let us consider as volume V the sphere centred at the origine: since $\mathbf{n}\cdot\boldsymbol{\nabla}G_\infty = (\partial G_\infty/\partial r) = -(A/r^2)$, we get the expected result $4\pi A = 1$.

□ **Solution of exercise 2.2. Green's function G_∞ for the Laplacian in dimension $d \geq 3$**

For $d > 2$, two independent solutions of equation $\Delta f(r) = 0$ are $f(r) = (A/r^{d-2})$ and $f(r) = B$. Only the former of these solutions meets the boundary condition. Finally, the constant A is fixed, as in exercise 2.1 by Gauss's flux theorem so that

$$G_\infty(r) = \frac{1}{(d-2)\Omega_d r^{d-2}}.$$

Note that exercise 3.7, p. 208, allows to compute this Green's function too, starting from the Green's function for diffusion. One then finds that $\Omega_d = (2\pi^{d/2}/\Gamma(d/2))$ where Euler's function Γ satisfies $\Gamma(x+1) = x\Gamma(x)$, $\Gamma(1) = 1$ and $\Gamma(1/2) = \sqrt{\pi}$.

□ **Solution of exercise 2.3. Green's functions for the Laplacian in 1d and 2d**

1. It is clear that the origin can be chosen in x' without loss of generality. An integration by parts then leads to

$$-\frac{1}{2}\int_{-\infty}^{\infty} dx f(x)\frac{d^2}{dx^2}|x| = -\frac{1}{2}\int_{-\infty}^{0} dx \frac{df(x)}{dx}\frac{d}{dx}x + \frac{1}{2}\int_{0}^{\infty} dx \frac{df(x)}{dx}\frac{d}{dx}x = -f(0),$$

where we assumed that $\lim_{x\to\pm\infty} f(x) = 0$. This result is a confirmation that

$$\frac{1}{2}\Delta(|x|) = \delta(x).$$

2. In the bidimensionnal case we start by writing the Laplacian of $G(r)$ in polar coordinates:

$$\Delta G(r) = -\frac{1}{2\pi}\left[\frac{d^2}{dr^2}\ln r + \frac{1}{r}\frac{d}{dr}\ln r\right].$$

We then apply a test-function $f(r,\theta)$ and evaluate:

$$-\frac{1}{2\pi}\int_{0}^{2\pi} d\theta \int_{0}^{\infty} dr\, rf(r,\theta)\left[\frac{d^2}{dr^2}\ln r + \frac{1}{r}\frac{d}{dr}\ln r\right].$$

We can then integrate by parts the term in $(d^2/dr^2)\ln r$, leading after simplifications to:

$$\frac{1}{2\pi}\int_{0}^{2\pi} d\theta \int_{0}^{\infty} dr\, r\frac{\partial}{\partial r}(f(r,\theta))\frac{d}{dr}\ln r = -f(\mathbf{r} = 0).$$

Here we also imposed that f vanishes at infinity. This result shows that in 2d

$$\frac{1}{2\pi}\Delta(\ln r) = \delta(\mathbf{r}).$$

□ **Solution of exercise 2.4. Symmetry of Laplacian Green's functions with homogeneous Dirichlet BC**

We apply the second Green's formula (2.20), p. 62, to $u(\mathbf{r}) = G(\mathbf{r}; \mathbf{r}_1)$ and $v(\mathbf{r}) = G(\mathbf{r}; \mathbf{r}_2)$ where G is an arbitrary Green's function for the Laplacian. One gets:

$$G(\mathbf{r}_1; \mathbf{r}_2) - G(\mathbf{r}_2; \mathbf{r}_1) = \oint_{\partial \mathcal{D}} d\Sigma\, \mathbf{n} \cdot [G(\mathbf{r}; \mathbf{r}_1) \nabla G(\mathbf{r}; \mathbf{r}_2)$$
$$- G(\mathbf{r}; \mathbf{r}_2) \nabla G(\mathbf{r}; \mathbf{r}_1)]. \quad (J.4)$$

If G satisfies homogeneous Dirichlet boundary conditions, the right-hand side in the above equation vanishes and G is then symmetric.

□ **Solution of exercise 2.5. Special Neumann Green's functions of the Laplacian**

1. Function F is easily obtained from result (J.4) of exercise 2.4.

$$F(\mathbf{r}) = -\frac{1}{s} \oint_{\partial \mathcal{D}} d\Sigma'' G_{\bar{N}}(\mathbf{r}''; \mathbf{r}).$$

Since the difference between $\tilde{G}_{\bar{N}}$ and $G_{\bar{N}}$ is a function depending only on $\mathbf{r}'$, $\tilde{G}_{\bar{N}}(\mathbf{r}; \mathbf{r}')$ is also a Green's function for the Laplacian. It furthermore satisfies BC (2.122) and BC (2.123) with $\tilde{c}(\mathbf{r}') = c(\mathbf{r}') + F(\mathbf{r}')$. It is then a special Neumann Green's functions, which is moreover symmetric.

2. The difference between the two expressions (2.39) obtained for $\tilde{G}_{\bar{N}}$ and $G_{\bar{N}}$ is:

$$\delta\phi(\mathbf{r}) = F(\mathbf{r}) \left(\int_{\mathcal{D}} d\mathbf{r}'\, \rho(\mathbf{r}') + \oint_{\partial \mathcal{D}} d\Sigma'\mathbf{n}' \cdot \nabla_{\mathbf{r}'} \phi(\mathbf{r}') \right).$$

It vanishes by application of Green-Ostrogradki theorem.

□ **Solution of exercise 2.6. Sum rules and resolvent**

1. One gets

$$\int_0^1 \prod_{i=1}^m dy_i\ G_\lambda(y_1, y_2) \cdots G_\lambda(y_{m-1}, y_m) G_\lambda(y_m, y_1) = \sum_n \frac{1}{(\lambda_n + \lambda)^m} \quad (J.5)$$

simply from spectral representation (2.77), p. 88, and orthonormality of the eigenfunctions ψ_n. Note that the resulting expression is nothing but the trace of operator $1/(\mathcal{O} + \lambda)^m$. We thus find a set of sum rules for Green's function G_λ.

2. First, the Green's function thus obtained meets homogeneous Dirichlet boundary conditions. It is easy to check that it is a solution of PDE (2.124) by rewriting this function in terms of $\theta(\pm(x - x'))$. A possibility is to make a suitable change of variables in expression (2.52), p. 77. Indeed, for $\lambda = \omega^2\pi^2$, G_λ is the homogeneous Dirichlet Green's function for Helmholtz operator on the segment $[0, 1]$. Another method consists in using result (C.6), p. 273.

3. To answer this question, just write sum rule (J.5) for $\lambda = \omega^2\pi^2$ and $m = 1$, given that $\lambda_n = n^2\pi^2$.

4. The Green's function in $\lambda = 0$ is easily obtained and reads

$$G_0(x; x') = -x_<(x_> - 1).$$

The expected result is obtained by applying again sum rule (J.5) with $m = 1$.

□ Solution of exercise 2.7. Conductive plane

1. $\widehat{G}_{\text{HD}}(x; x'; \mathbf{k})$ is a solution of differential equation

$$\left(-\frac{\partial^2}{\partial x^2} + k^2\right) \widehat{G}_{\text{HD}}(x; x'; \mathbf{k}) = \delta(x - x').$$

Moreover $\widehat{G}_{\text{HD}}(x; x'; \mathbf{k})$ vanishes at $x = 0$ and $x \to +\infty$. $\widehat{G}_{\text{HD}}$ is then the homogeneous Dirichlet Green's function of Helmholtz operator in the one-dimensional domain $x \geq 0$.

2. Take as independent solutions of homogeneous equation

$$-\frac{d^2\phi}{dx^2} + k^2\phi = 0$$

the functions $\phi_\pm(x) = e^{\pm kx}$ of Wronskian $-2k$. The method of variation of constants (see p. 272) then gives:

$$\widehat{G}_{\text{HD}}(x, x', k) = \frac{1}{2k}\left(e^{-k|x-x'|} - e^{-k|x+x'|}\right). \tag{J.6}$$

This intermediate result illustrates of course the method of images as the first and second terms in equation (J.6) are the Green's functions (2.66) of the one dimensional Helmholtz operator with BC at infinity, for a source in x' and its image with respect to the origin in $(-x')$! Note that this result can also be obtained by taking an appropriate limit of the homogeneous Dirichlet Green's function of Helmholtz operator on a segment of length L, given by equation (2.52).

3. Let us give some steps of the calculation of the inverse Fourier transform of $\widehat{G}_{\text{HD}}$. After switching to polar coordinates (k, θ) and integration on k, we have to evaluate the integral

$$\int_0^{2\pi} d\theta \frac{1}{i\rho\cos\theta - |x \pm x'|},$$

where $\rho = \sqrt{(y - y')^2 + (z - z')^2}$. This integral is computed in the complex plane by setting $z = e^{i\theta}$. It leads to the integral

$$\int_0^{2\pi} d\theta \frac{1}{i\rho \cos\theta - |x \pm x'|} = \oint_C dz \frac{2}{z^2 + \frac{2i|x \pm x'|}{\rho} z + 1}, \tag{J.7}$$

where the integration contour C is the unit circle. One can compute this integral using the residue theorem. The integrand has two poles

$$z_\pm = -\frac{i}{\rho}\left(|x \pm x'| \pm \sqrt{|x \pm x'|^2 + \rho^2}\right),$$

but only z_- lies inside the integration contour (see Figure J.2). The final result is

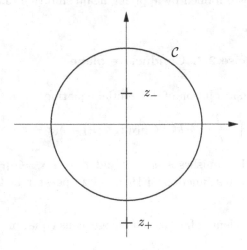

Fig. J.2 Integration contour C used to compute integral (J.7): only the pole z_- is inside C.

of course

$$G_{HD}(\mathbf{r}; \mathbf{r}') = G_\infty(\mathbf{r} - \mathbf{r}') - G_\infty(\mathbf{r} - \mathbf{r}'_{im}) \tag{J.8}$$

where $\mathbf{r}'_{im}$ is the image of $\mathbf{r}$. This exercise illustrates the efficiency of the method of images. Result (J.8) is indeed obtained far less easily by Fourier transform!

□ **Solution of exercise 2.8. Green's functions for Laplace operator in spherical coordinates**

1. To answer this question we must write the Laplacian of G in spherical coordinates and identify term by term with the expansion of $\delta(\mathbf{r} - \mathbf{r}')$ in spherical harmonics. It then follows that $g_{lm}(r; r')$ does not depend on m and is solution of the differential equation

$$\frac{1}{r^2}\frac{\partial}{\partial r}\left(r^2 \frac{\partial g_l}{\partial r}\right) - \frac{l(l + 1)}{r^2} g_l = -\frac{1}{r^2}\delta(r - r'). \tag{J.9}$$

Solutions of the homogeneous PDE associated with PDE (J.9) are of course the radial part of harmonic functions. They are obtained for instance by posing $h_l = r g_l$ and take the form $A_l r^l + B_l r^{-(l+1)}$.

2. Result (2.125) is obtained by applying formula[1] (C.6) to

$$\Phi_1(r) = r^l - \frac{a^{2l+1}}{r^{l+1}} \qquad \text{and} \qquad \Phi_2(r) = \frac{1}{r^{l+1}} - \frac{r^l}{b^{2l+1}},$$

which are solutions of the homogeneous equation, vanishing respectively in $r = a$ and $r = b$, and remaining finite respectively when $a \to 0$ and $b \to \infty$. Their Wronskian is $-\frac{2l+1}{r^2}\left[1 - \left(\frac{a}{b}\right)^{2l+1}\right]$.

3. Taking the limit $a \to 0$ in expression (2.125) for Green's function $G_{a,b}$ leads to the result (2.126) for G_b.

4. Similarly, the limit $b \to \infty$ in expression (2.125) of Green's function $G_{a,b}$ gives the result (2.127) for G_a.

5. To conclude this exercise, one simply has to take the limit $a \to 0$ in the expression for G_a, or the limit $b \to \infty$ in G_b.

☐ **Solution of exercise 2.9. Point charge in a conducting sphere**

1. The desired boundary conditions are guaranteed if the fictitious charge, of charge $-a/r$, is located in $r_{\text{im}} = a^2/r'$ and with the same orbital and azimuthal angles as $\mathbf{r}'$ (see Figure J.3). We then find result[2] (2.126).

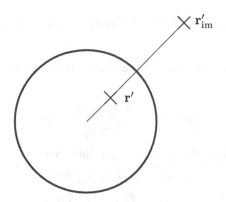

Fig. J.3   Charge in $\mathbf{r}'$ and image charge in $\mathbf{r}'_{\text{im}}$.

2. Simply place additional fictitious charges symmetric relative to the plane $x = 0$, i.e. with $\tilde{\varphi} = -\varphi' + \pi$ (see Figure J.4).

---

[1]One must be careful in applying this formula to the factor $-1/r^2$ multiplying the Dirac function in the right-hand side of equation (J.9).

[2]Of course b should be replaced with a in expression (2.126).

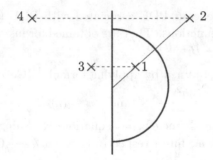

Fig. J.4 In this figure, the charge in **r** is called 1. The image charge 2 has the value $-a/r$, the one in 3 the value -1 and that in 4, a/r'.

3. The limit to take is simply $a \to \infty$, which is easily obtained. One then recognises the expression corresponding to two identical charges, symmetric with respect to the plane $x = 0$.

□ Solution of exercise 2.10. Point charge and dielectric sphere

1. Function $\psi(\mathbf{r})$ is harmonic inside and outside of the sphere and tends to zero at infinity. Its form is:

$$\psi(\mathbf{r}) = \sum_{l=0}^{\infty} \frac{B_l}{r^{l+1}} P_l(\cos\theta) \qquad \text{for} \quad r > R, \tag{J.10}$$

$$\psi(\mathbf{r}) = \sum_{l=0}^{\infty} A_l r^l P_l(\cos\theta) \qquad \text{for} \quad r < R. \tag{J.11}$$

2. Function ϕ_0 and its derivative are continuous on the sphere. In addition, as $R < r_0$,

$$\phi_0(R,\theta) = \frac{q}{4\pi} \sum_{l=0}^{\infty} \frac{R^l}{r_0^{l+1}} P_l(\cos\theta).$$

Continuity conditions at the sphere surface then read:

$$B_l = A_l R^{2l+1}, \tag{J.12}$$

$$\frac{q}{4\pi} \frac{l R^{l-1}}{r_0^{l+1}} - (l+1)\frac{B_l}{R^{l+2}} = \frac{q}{4\pi} \epsilon \frac{l R^{l-1}}{r_0^{l+1}} + \epsilon l R^{l-1} A_l. \tag{J.13}$$

They are easily solved to give

$$A_l = \frac{q}{4\pi} \frac{1}{r_0^{l+1}} \frac{(1-\epsilon)l}{[(1+\epsilon)l+1]}.$$

The reader may notice that $A_0 = 0$, indicating that the monopole term in the expression for the electric potential is not affected by the presence of the sphere.

This is actually the manifestation of its electric neutrality. Let us conclude by giving the expression of the electric potential for $r > r_0$:

$$\phi(\mathbf{r}) = \frac{q}{4\pi} \sum_{l=0}^{\infty} \frac{1}{r^{l+1}} \left(r_0^l - \frac{R^{2l+1}}{r_0^{l+1}} \frac{(\epsilon-1)l}{[(1+\epsilon)l+1]} \right) P_l(\cos\theta).$$

□ **Solution of exercise 2.11. Green's function G_∞ for Laplace operator in cylindrical coordinates**

1. To establish the requested result, we must first apply formula (F.2), p. 286, leading to the expression of Dirac distribution after a change of coordinates:

$$\delta(\mathbf{r} - \mathbf{r}') = \frac{1}{\rho} \delta(\rho - \rho')\delta(\varphi - \varphi')\delta(z - z').$$

Expression (2.130) then comes from the rewriting of Dirac distributions on φ and z.

2. The expected differential equation is:

$$\frac{\partial}{\partial\rho} \left(\rho \frac{\partial g_{mk}}{\partial\rho} \right) - \left(k^2 \rho + \frac{m^2}{\rho} \right) g_{mk}(\rho; \rho') = -\delta(\rho - \rho').$$

It is obtained by writing the Laplacian operator in cylindrical coordinates.

3. Let us assume that $\rho < \rho'$ and use result (C.4), p. 272, with

$$F(x) = -\frac{1}{x}\delta(x - \rho'), \qquad \Phi_1(\rho) = I_m(k\rho), \qquad \Phi_2(\rho) = K_m(k\rho),$$

and then $t_0 \to \infty$, $t_1 = 0$. One then gets $g_{mk}(\rho; \rho') = I_m(k\rho)K_m(k\rho')$, leading to the final result by symmetry between ρ and ρ'.

□ **Solution of exercise 2.12. Oseen's tensor**

1. Pressure p satisfies equation

$$-\Delta p = \boldsymbol{\nabla} \cdot \mathbf{f}.$$

It is uniquely determined if Dirichlet or Neumann BC are imposed upon $\partial \mathcal{D}$. The velocity is then solution of equation

$$\eta \Delta \mathbf{v} = \mathbf{f} + \boldsymbol{\nabla} p$$

and here again, Dirichlet or Neumann BC ensure unicity of the solution. Of course, these BC must be compatible with the requirement $\boldsymbol{\nabla} \cdot \mathbf{v} = 0$!

2. Pressure can be written in terms of f:

$$p(\mathbf{r}) = \int d\mathbf{r}' G_\infty(\mathbf{r} - \mathbf{r}')\boldsymbol{\nabla}_{\mathbf{r}'} \cdot \mathbf{f}(\mathbf{r}'). \tag{J.14}$$

The pressure is invariant under the proposed transformation.

3. The velocity is simply given by:

$$\eta v_i(\mathbf{r}) = -\int d\mathbf{r}' G_\infty(\mathbf{r} - \mathbf{r}')\big[f_i(\mathbf{r}') + \partial_i' p(\mathbf{r}')\big]$$

with $\partial_i' = (\partial/\partial x_i')$. One must then use result (J.14). The expression thus obtained may be written as

$$\eta v_i(\mathbf{r}) = -\int d\mathbf{r}' d\mathbf{r}'' G_\infty(\mathbf{r} - \mathbf{r}') \left[\delta_{ij}\delta(\mathbf{r}' - \mathbf{r}'') + \partial_i'\partial_j' G_\infty(\mathbf{r}' - \mathbf{r}'')\right] f_j(\mathbf{r}''),$$

which brings up the so-called Oseen's tensor. Again, $\mathbf{v}$ is invariant under the proposed transformation.

□ Solution of exercise 2.13. Green's function in elasticity theory

1. Function $\mathbf{g}_0^i(\mathbf{r})$ is easily derived from the Green's function G_∞ for the Laplacian:

$$\mathbf{g}_0^i(\mathbf{r}) = \frac{1}{4\pi r}\mathbf{e}_i.$$

The PDE satisfied by $\mathbf{g}_1^i(\mathbf{r})$ is:

$$(1 - 2\sigma)\Delta\mathbf{g}_1^i(\mathbf{r}) + \boldsymbol{\nabla}\left(\boldsymbol{\nabla} \cdot \mathbf{g}_1^i(\mathbf{r})\right) = -\boldsymbol{\nabla}\left(\boldsymbol{\nabla} \cdot \mathbf{g}_0^i(\mathbf{r})\right). \tag{J.15}$$

2. One then gets, taking the curl of the above equation:

$$\Delta\left(\boldsymbol{\nabla} \wedge \mathbf{g}_1^i(\mathbf{r})\right) = 0.$$

Combined with BC, it imposes $\boldsymbol{\nabla} \wedge \mathbf{g}_1^i(\mathbf{r}) = 0$. The curl of function $\mathbf{g}_1^i$ is zero, so that this function can be written as the gradient of a scalar function.

3. ϕ^i satisfies the equation

$$\Delta\phi^i(\mathbf{r}) = -\frac{1}{8\pi(1 - \sigma)}\frac{\partial}{\partial x_i}\frac{1}{r}. \tag{J.16}$$

The reader can easily verify that the expression for ϕ^i given in the text is a solution that produces adequate BC for $\mathbf{g}_1^i$.

4. With the expression for $\mathbf{g}_0^i(\mathbf{r})$ and $\mathbf{g}_1^i(\mathbf{r})$, we can compute the components of Green's tensor:

$$G^{ij}(\mathbf{r}) = \frac{1}{4\pi}\left[\frac{\delta^{ij}}{r} - \frac{1}{4(1 - \sigma)}\frac{\partial^2}{\partial x_i\partial x_j}r\right]. \tag{J.17}$$

$$\mathbf{r} - \mathbf{e}_1 \qquad \mathbf{r} \qquad \mathbf{r} + \mathbf{e}_1$$

Fig. J.5   Balance of currents arriving at the point **r** in the one-dimensional case.

□ Solution of exercise 2.14. Discrete Laplacian and resistors network

1. The current flowing from $\mathbf{r} + \mathbf{e}_i$ to $\mathbf{r}$, two neighbouring sites in the network is given by $[V(\mathbf{r}+\mathbf{e}_i) - V(\mathbf{r})]/R$ (see Figure J.5). In the absence of ohmmetre the total current entering each vertex must be zero, automatically leading to $\Delta V(\mathbf{r}) = 0$.

2. As currents I and $-I$ are respectively injected at vertices $\mathbf{r}_0$ and $\mathbf{0}$, Kirchhoff's law gives (see Figure J.6)

$$\Delta V(\mathbf{r}) = RI \left(\delta_{\mathbf{r},\mathbf{r}_0} - \delta_{\mathbf{r},0} \right). \tag{J.18}$$

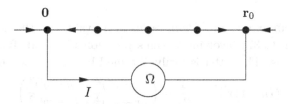

Fig. J.6   Representation of the one-dimensional case. An ohmmetre measures the resistance between the two points **0** and $\mathbf{r}_0$.

3. The resistance between $\mathbf{r}_0$ and 0 is given by $R(\mathbf{r}_0) = (V(0) - V(\mathbf{r}_0))/I$. However, the solution of equation (J.18) is

$$V(\mathbf{r}) = RI \left[G(\mathbf{r}) - G(\mathbf{r} - \mathbf{r}_0) \right].$$

Therefore we have

$$R(\mathbf{r}_0) = 2R \left[G(0) - G(\mathbf{r}_0) \right]$$

where we used the result $G(\mathbf{r}) = G(-\mathbf{r})$.

4. Writing

$$G(\mathbf{r}) = \int_{[-\pi,\pi]^d} \frac{\mathrm{d}^d k}{(\sqrt{2\pi})^d} \widehat{G}(\mathbf{k}) e^{i\mathbf{k}\cdot\mathbf{r}},$$

one gets

$$\widehat{G}(\mathbf{k}) = \frac{1}{(\sqrt{2\pi})^d} \left(\frac{1}{2(d - \sum_i \cos k_i)} \right).$$

The final result for the resistance reads

$$R(\mathbf{r}_0) = R \int_{[-\pi,\pi]^d} \frac{\mathrm{d}^d k}{(2\pi)^d} \frac{1 - e^{i\mathbf{k}\cdot\mathbf{r}_0}}{d - \sum_i \cos k_i}.$$

In one dimension, this integral is computed thanks to the residue theorem, leading to the expected result: $R(r_0) = Rr_0$.

5. For $|\mathbf{r}_0| \gg 1$, dominant terms in the integral defining $R(\mathbf{r}_0)$ are given by small $|\mathbf{k}|$. We can therefore write $2(d - \sum_i \cos k_i) \simeq d\mathbf{k}^2$ and find the expression for the Green's function of the Laplacian in two dimensions. So the dominant term in $R(\mathbf{r}_0)$ is $(R/\pi) \ln r_0$. Furthermore, the expression for the resistance between two neighbouring sites can easily be obtained by observing that $R(\mathbf{e}_1) = R(\mathbf{e}_2)$ and

$$R(\mathbf{e}_1) + R(\mathbf{e}_2) = R \int_{[-\pi,\pi]^2} \frac{d^2 k}{(2\pi)^2} = R,$$

so that $R(\mathbf{e}_1) = (R/2)$.

☐ **Solution of exercise 2.15. Method of image charges for a bidimensional problem**

1. To obtain an expression of Φ in terms of its values on the domain boundary, we use expression (2.37) involving Green's function G_{HD} satisfying homogeneous Dirichlet conditions. The latter is easily obtained by the method of images:

$$G_{HD}(\mathbf{r}; \mathbf{r}') = -\frac{1}{2\pi} \ln \left(\frac{(x - x')^2 + (y - y')^2}{(x + x')^2 + (y - y')^2} \right).$$

We first compute the normal derivative of the Green's function and the final result reads:

$$\Phi(x, y) = \frac{x}{\pi} \int_{-\infty}^{\infty} dy' \left[\frac{f(y')}{x^2 + (y - y')^2} \right].$$

2. Verifying that the resulting expression satisfies imposed boundary conditions amounts to notice that function

$$d(x, y) = \frac{x}{\pi(x^2 + y^2)}$$

tends in the sense of distributions to $\delta(y)$ for $x \to 0$. To this end one can for example apply test functions, or less formally note that

$$d(x, y) = \frac{x}{\pi(x^2 + y^2)},$$

for all $y \neq 0$ and

$$\int_{-\infty}^{\infty} dy \, d(x, y) = 1 \qquad \forall x.$$

3. To show that $\Phi(x, y)$ is a harmonic function for $x > 0$, we can notice that the function $d(x, y)$ is the real part of analytic function $\mathcal{F}(z) = 1/(\pi z)$.

□ Solution of exercise 2.16. Semi-cylindrical warehouse exposed to wind

1. The BC met by **u** are: $\mathbf{u}_\perp = 0$ on the ground and on the warehouse (Neumann BC), and $\mathbf{u} = \mathbf{u}_0$ at infinity.

2. In order to find the requested conformal transformation, we can note that
$$\cos\theta = \frac{1}{2}(e^{i\theta} + \frac{1}{e^{i\theta}}),$$
so that when $\exp(i\theta)$ describes the upper semicircle, $x = \cos\theta$ describes segment $[-1, 1]$. It guides us to transformation
$$Z = z + \frac{R^2}{z}.$$

3. Proceeding as in section 2.2.5, we find $\phi(r, \theta) = u_0 r \cos\theta(1 + \frac{R^2}{r^2})$. Furthermore,
$$u_x = u_0 \left(1 - \frac{R^2}{r^2}\cos(2\theta)\right) \quad ; \quad u_y = -u_0\frac{R^2}{r^2}\sin 2\theta,$$
$$u_r = u_0 \cos\theta(1 - \frac{R^2}{r^2}) \quad ; \quad u_\theta = -u_0 \sin\theta(1 + \frac{R^2}{r^2}). \tag{J.19}$$

4. Starting from the decomposition of $\phi(r, \theta)$ in Fourier series,
$$\phi(r, \theta) = \sum_{l=0}^{+\infty} c_l(r) \cos(l\theta),$$
obtained by exploiting that ϕ is real, we find that BC impose that only $c_0(r)$ and $c_1(r)$ are non-zero. Moreover $c_0(r)$ is a pure constant not affecting the velocity field, while $c_1(r) = u_0 r[1 + (R^2/r^2)]$, giving back the previous velocity field.

5. To compute this force we use Bernoulli relation and find $F_y = \frac{8}{3}\rho u_0^2 LR$. There is a depression and the roof is ripped off!

□ Solution of exercise 2.17. Dirac operator

1. This question does not pose any difficulty. It is enough to apply the operator under consideration on the proposed expression for $f_i(\mathbf{r})$.

2. In order to determine Green's matrix G we note that
$$\begin{pmatrix} \partial_x & \partial_y \\ \partial_y & -\partial_x \end{pmatrix}\begin{pmatrix} \partial_x & \partial_y \\ \partial_y & -\partial_x \end{pmatrix} = \Delta \mathcal{I}$$
where Δ is the two-dimensional Laplace operator. Thus matrix G satisfies the matrix equation
$$\Delta_{\mathbf{r}} G(\mathbf{r} - \mathbf{r}') = \begin{pmatrix} \partial_x & \partial_y \\ \partial_y & -\partial_x \end{pmatrix}\delta(\mathbf{r} - \mathbf{r}').$$

This equation is solved using Green's function (2.67), p. 84, of two-dimensional Laplace operator, i.e.

$$G(\mathbf{r} - \mathbf{r}') = \int_{\mathbb{R}^2} d\mathbf{r}'' \frac{1}{2\pi} \ln |\mathbf{r} - \mathbf{r}''| \begin{pmatrix} \partial_x'' & \partial_y'' \\ \partial_y'' & -\partial_x'' \end{pmatrix} \delta(\mathbf{r}'' - \mathbf{r}'),$$

leading to

$$G(\mathbf{r} - \mathbf{r}') = \frac{1}{2\pi} \begin{pmatrix} \partial_x & \partial_y \\ \partial_y & -\partial_x \end{pmatrix} \ln |\mathbf{r} - \mathbf{r}'|.$$

3. Integration by parts leads to the result:

$$f_1 \to f_1 + h \qquad \text{and} \qquad f_2 \to f_2.$$

□ **Solution of exercise 2.18. Mercury perihelion precession**

1. The Green's function can easily be calculated:

$$G(\phi - \phi') = \theta(\phi - \phi') \sin(\phi - \phi').$$

2. The derivation of integral equation (2.137) does not present any difficulty. Function $h(\phi)$ just imposes the initial conditions. Note that we thus obtain an integral equation for the solution of a non-linear differential equation.

3. The desired function is $h(\phi) = \alpha(1 + e) \cos \phi$.

4. To answer this question, we start from integral equation (2.137), p. 122, and apply the method seen on p. 89 in another context. Note that $u_0(\phi) = \alpha(1 + e \cos \phi)$. One then gets:

$$u(\phi) \simeq u_0(\phi) + \beta \int_0^\phi d\phi' \, G(\phi - \phi') u_0^2(\phi').$$

The calculation is then carried out without major difficulty and leads to:

$$A = \tfrac{1+e^2}{2}, \qquad B = -\frac{1 + e^2}{2} - \frac{e^2}{6},$$
$$C = -\tfrac{e^2}{6}, \qquad D = e.$$

The solution shows a correction term with respect to u_0. This term exhibits a behaviour in $\phi \sin \phi$ whose amplitude increases with time, which is the main limitation of this perturbative expansion.

5. To answer this question it is enough to study equation $\dot{u}(\phi) = 0$.

6. Since $D = e$, we find

$$\epsilon = 2\pi\beta\alpha = \frac{6\pi G^2 M^2}{L^2 c^2},$$

also written as

$$\epsilon = \frac{6\pi MG}{ac^2(1 - e^2)}.$$

The computation gives 43.05 seconds of arc per century, while the observation gives 43.11 ± 0.45 seconds of arc per century. Note that several effects explain why planet Mercury has historically played an important role: first, its semi-major axis a is small and therefore the effect is larger compared to other planets; then it makes 415 revolutions per century.

□ **Solution of exercise 2.19. Harmonic oscillator in the presence of an impurity**

1. By inserting two closure relations and then taking a Laplace transform, we find

$$\hat{G}_0(x_a, x_b; z) = \sum_{n=0}^{+\infty} \frac{\psi_n(x_a)\psi_n^*(x_b)}{z + E_n}.$$

We then get, using the value of Hermite polynomials in zero

$$\hat{G}_0(0, 0; z) = \frac{1}{\sqrt{\pi}l} \sum_{p=0}^{+\infty} \frac{(2p - 1)!!}{2^p p!} \frac{1}{z + (2p + \frac{1}{2})\hbar\omega}.$$

For $G_0(0, 0, \beta)$, one performs the inverse Laplace transform of each term in the series using table B.1 in Appendix B and sum the resulting series to get

$$G_0(0, 0, \beta) = \frac{1}{\sqrt{2\pi \, \text{sh}(\beta\hbar\omega)l}}.$$

2. For this question we can apply the methods of paragraph 2.2.3, p. 99, leading to:

$$\rho_0(E) = \sum_{n=0}^{+\infty} \delta\left(E - (n + \frac{1}{2})\hbar\omega\right).$$

3. Equation (2.138) is obtained in a similar way to the example in paragraph 2.2.3. It leads to

$$\hat{G}(x_a, x_b; z) = \hat{G}_0(x_a, x_b; z) - Vl\frac{\hat{G}_0(x_a, 0; z)\hat{G}_0(0, x_b; z)}{1 + Vl\hat{G}_0(0, 0; z)}. \tag{J.20}$$

4. The fact that $H_{2p+1}(0) = 0$ shows immediately that the second term of $\hat{G}$ in equation (J.20) is regular in $z = -(2p + 1 + \frac{1}{2})\hbar\omega$, ensuring that the total expression always has poles is $z = -(2p+1+\frac{1}{2})\hbar\omega$. It is however enough to take the dominant term in equation (J.20) for z near $-(2p+\frac{1}{2})\hbar\omega$ to see that this singularity is compensated. These singularities are replaced by new ones, originating from the cancellation of the denominator of the last part of equation (J.20). We thus find a new set of eigenenergies whose values $E(V)$ must satisfy equation

$$\sum_{p=0}^{+\infty} \frac{(2p-1)!!}{2^p p!} \frac{1}{E(V) - (2p + \frac{1}{2})\hbar\omega} = \frac{\sqrt{\pi}}{V}. \qquad (J.21)$$

5. When $V \to 0$, we write $E_0(V)$ as $\frac{1}{2}\hbar\omega + \delta E$ and expand in equation (J.21) to first order in δE. We then get $\delta E = \frac{V}{\sqrt{\pi}} + O(V^2)$.

When $V \to -\infty$, we have $\hbar\omega \ll |V|$ and thus end up in the case of a particle in a δ potential, corresponding to the example in paragraph 2.2.3. The ground state energy has been obtained in equation (2.107), p. 102, which gives the desired result using the correspondence $Vl = -V_0$. When $V \to +\infty$, we set $E(V) = E_0^\infty + \delta E$. E_0^∞ clearly meets the relation

$$\sum_{p=0}^{+\infty} \frac{(2p-1)!!}{2^p p!} \frac{1}{E_0^\infty - (2p + \frac{1}{2})\hbar\omega} = 0.$$

It is clear that in order to get zero, we must have positive and negative terms in this sum, which implies that $E_0^\infty > \frac{1}{2}\hbar\omega$. Then the first order term in the Taylor expansion in δE gives

$$-S(E_0^\infty)\delta E + O(\delta E^2) = \frac{\sqrt{\pi}}{V}$$

with

$$S(E_0^\infty) = \sum_{p=0}^{+\infty} \frac{(2p-1)!!}{2^p p!} \frac{1}{\left[E_0^\infty - (2p + \frac{1}{2})\hbar\omega\right]^2}.$$

◊ **Chapter 3** ◊

□ **Solution of exercise 3.1. Uniqueness of solutions of diffusion and wave equations**

1. To answer this question, we must begin exactly as in the static case, p. 63 and consider two solutions ϕ_1 and ϕ_2 of the diffusion equation. The difference $\alpha = \phi_1 - \phi_2$ is then solution of the homogeneous diffusion equation

$$\frac{\partial}{\partial t'}\alpha(\mathbf{r}', t') - D\Delta_{\mathbf{r}'}\alpha(\mathbf{r}', t') = 0.$$

We then multiply this equation by $\alpha^*(\mathbf{r}', t')$, and integrate over domain $\mathcal{D}$ and between the initial time t_0 and $t > t_0$. Repeating this procedure for the homogeneous diffusion equation satisfied by $\alpha^*(\mathbf{r}', t')$ multiplied by $\alpha(\mathbf{r}', t')$ yields

$$\int_{t_0}^{t} dt' \int_{\mathcal{D}} d\mathbf{r}' \left[\frac{\partial}{\partial t'}|\alpha(\mathbf{r}', t')|^2 - D\alpha^*(\mathbf{r}', t')\Delta_{\mathbf{r}'}\alpha(\mathbf{r}', t') - D\alpha(\mathbf{r}', t')\Delta_{\mathbf{r}'}\alpha^*(\mathbf{r}', t') \right] = 0.$$

For the first term of this equation, the integral over t' can be computed. For the second and third terms, we use the first Green's formula (2.10), p. 62 respectively with $(u = \alpha^*, v = \alpha)$ and $(u = \alpha, v = \alpha^*)$. One thus obtains equation

$$\int_{\mathcal{D}} d\mathbf{r}' \, |\alpha(\mathbf{r}', t)|^2 + 2D \int_{t_0}^t dt' \int_{\mathcal{D}} d\mathbf{r}' \, |\boldsymbol{\nabla}_{\mathbf{r}'}\alpha(\mathbf{r}', t')|^2 = \int_{\mathcal{D}} d\mathbf{r}' \, \alpha^2(\mathbf{r}', t_0)$$

$$+ D \int_{t_0}^t dt' \int_{\partial\mathcal{D}} d\Sigma' \, [\alpha^*(\mathbf{r}', t') \, \mathbf{n}' \cdot \boldsymbol{\nabla}_{\mathbf{r}'}\alpha(\mathbf{r}', t') + \alpha(\mathbf{r}', t') \, \mathbf{n}' \cdot \boldsymbol{\nabla}_{\mathbf{r}'}\alpha^*(\mathbf{r}', t')].$$

Boundary conditions are such that the right-hand side of this equality is zero. It then implies $\alpha(\mathbf{r}, t) = 0$ and therefore shows the uniqueness of the solution.

2. For d'Alembert operator, multiply homogeneous equation

$$\frac{1}{c^2}\frac{\partial^2}{\partial t'^2}\alpha(\mathbf{r}', t') - \Delta_{\mathbf{r}'}\alpha(\mathbf{r}', t') = 0,$$

by $(\partial \alpha^*/\partial t')$ instead of α^*. The argument is then similar to the one presented for diffusion.

□ **Solution of exercise 3.2. Reciprocal relations**

This exercise is a generalisation of exercise 2.4, p. 112. Note $G_1 = G_H^+(\mathbf{r}; \mathbf{r}_1; t_1 - t)$ and $G_2 = G_H^+(\mathbf{r}; \mathbf{r}_2; t - t_2)$. They are respectively solutions of differential equations:

$$-\Box_{\mathbf{r}, t} G_1 = \delta(\mathbf{r} - \mathbf{r}_1)\delta(t - t_1),$$
$$-\Box_{\mathbf{r}, t} G_2 = \delta(\mathbf{r} - \mathbf{r}_2)\delta(t - t_2). \tag{J.22}$$

Then multiply the first of these equations by G_2 and the second by G_1, take the difference and finally integrate over $\mathbf{r} \in \mathcal{D}$ and over t between t_i and t_f with $t_i < t_2 < t_1 < t_f$. This procedure leads to

$$G_H^+(\mathbf{r}_1; \mathbf{r}_2; t_1 - t_2) - G_H^+(\mathbf{r}_2; \mathbf{r}_1; t_1 - t_2) = \int_{\mathcal{D}} d\mathbf{r} \int_{t_i}^{t_f} dt \left\{ \frac{1}{c^2}\frac{\partial}{\partial t}\left(G_2\dot{G}_1 - G_1\dot{G}_2\right) \right\}$$

$$+ \oint_{\partial\mathcal{D}} d\Sigma \int_{t_i}^{t_f} dt \, \mathbf{n} \cdot \{G_1\boldsymbol{\nabla}_{\mathbf{r}}G_2 - G_2\boldsymbol{\nabla}_{\mathbf{r}}G_1\}.$$

The second term on the right-hand side of this equation vanishes because G_H^+ satisfies homogeneous Dirichlet or Neumann boundary conditions. Integration of the first term gives four terms. It is easy to show that each of these terms is zero since Green's function G_H^+ is causal and $t_i < t_2 < t_1 < t_f$.

□ **Solution of exercise 3.3. Equation for long cables**

To express $V(x,t)$ in terms of $\rho(x,t)$, we start by writing:

$$V(x,t) = v(x,t) + V_0$$

so that $v(x,t)$ satisfies equation

$$\lambda^2 \frac{\partial^2 v(x,t)}{\partial x^2} - \tau_0 \frac{\partial v(x,t)}{\partial t} - v(x,t) = \rho(x,t).$$

This equation is solved using causal Green's function $G_\infty^+(x - x'; t - t')$ associated with this PDE. This Green's function is itself derived from spectral representation (3.15), p. 131. Here operator $\mathcal{O}_x$ is $[\lambda^2(\partial^2/\partial x^2) - 1]$ with eigenfunctions[3] e^{ikx} so that:

$$G_\infty^+(x - x'; t - t') = \theta(t - t')\frac{1}{2\pi}\int_{-\infty}^{+\infty} dk\, Z_k(t - t')\, e^{ik(x-x')}.$$

Functions $Z_k(t - t')$ are easily obtained as solutions of equation

$$\left[-\tau_0(\partial/\partial t) - (1 + \lambda^2 k^2)\right] Z_k(t - t') = 0,$$

with the initial condition $Z_k(0) = -(1/\tau_0)$. This eventually leads to:

$$G_\infty^+(x - x'; t - t') = -\frac{\theta(t - t')}{(2\tau_0\pi)}\int_{-\infty}^{+\infty} dk\, \exp[ik(x - x') - \frac{1 + \lambda^2 k^2}{\tau_0}(t - t')].$$

One then gets, using (E.3), p. 280,

$$G_\infty^+(x - x'; t - t') = -\frac{\theta(t - t')e^{-(t-t')/\tau_0}}{(4\tau_0\pi\lambda^2(t - t'))^{1/2}}\exp\left[-\frac{\tau_0(x - x')^2}{4\lambda^2(t - t')}\right].$$

The damping factor appearing in Green's function G_∞^+ of course comes from the first order term proportional to τ_0 in PDE (3.165). Note that this result can also be obtained by setting

$$G_\infty^+(x - x'; t - t') = \frac{1}{\tau_0}e^{-(t-t')/\tau_0}g_\infty^+(x - x'; t - t').$$

Indeed, $g_\infty^+(x - x'; t - t')$ is then the causal Green's function for diffusion equation in one dimension with $D = \lambda^2$ and $t \to t/\tau_0$.

□ Solution of exercise 3.4. Neumann conditions in diffraction theory

It is even simpler to work with Neumann BC than with Dirichlet BC! Equation (3.118), p. 179, indeed gives

$$A_N(\mathbf{r}) = -\int_0^\infty d\tau \int_{S_h} d\Sigma'\, e^{i\omega\tau} G_{HN}^+(\mathbf{r}; \mathbf{r}'; \tau)\frac{\partial A_N}{\partial z'}(\mathbf{r}').$$

The integral over τ is immediate from expression (3.120) for G_{HN}^+ and G_∞^+, p. 166. One thus finds the result (3.123).

---

[3]Strictly speaking one should consider a domain $\mathcal{D} = [-L, L]$ and then take the limit $L \to \infty$.

□ **Solution of exercise 3.5. Green's function of d'Alembert operator in $2+1$ dimensions**

1. The result (3.166) may be found in two ways. First one can apply directly d'Alembert operator in $2+1$ dimensions to the proposed expression for G_2^+, leading to

$$\int_{-\infty}^{+\infty} dz \, \left(\frac{1}{c^2} \partial_t^2 - \partial_x^2 - \partial_y^2 - \partial_z^2 \right) G_3^+ (x - x_0, y - y_0, z, t - t_0)$$

$$+ \int_{-\infty}^{+\infty} dz \, \partial_z^2 G_3^+ (x - x_0, y - y_0, z, t - t_0).$$

One must then use that G_3^+ is a Green's function of d'Alembert operator in $3+1$ dimensions and that its derivative $\partial_z G_3^+ (x - x_0, y - y_0, z, t - t_0)$ goes to zero at infinity. The second method appeals to spatio-temporal Fourier transforms G_3^+ and G_2^+, linked by the relation

$$\widehat{G}_2^+ (k_x, k_y, \omega) = \widehat{G}_3^+ (k_x, k_y, k_z = 0, \omega).$$

Identity (3.166) is then obtained by inverse Fourier transform.

2. The calculation of G_2^+ does not pose any major difficulty since G_3^+ is proportional to a Dirac distribution. The reader should set

$$\rho = \sqrt{(x - x_0)^2 + (y - y_0)^2}$$

and first show that $G_2^+ (x - x_0, y - y_0, t - t_0) = 0$ for $\rho > c(t - t_0)$. For $\rho < c(t - t_0)$, one should use rule (A.2), p. 265 with

$$f(z) = \sqrt{\rho^2 + z^2} - c(t - t_0).$$

The final result is

$$\boxed{G_2^+ (|\mathbf{r} - \mathbf{r}_0|, t - t_0) = \frac{c \, \theta(c(t - t_0) - |\mathbf{r} - \mathbf{r}_0|)}{2\pi \sqrt{c^2(t - t_0)^2 - |\mathbf{r} - \mathbf{r}_0|^2}}.} \qquad (J.23)$$

3. Unlike in the case of the space-time in $3+1$ dimensions, the Green's function of d'Alembert operator is not proportional to $\delta(c(t - t_0) - |\mathbf{r} - \mathbf{r}_0|)$. Thus it is not located on the light cylinder in $2+1$ dimensions. It has a forward wavefront advancing at the velocity c but no rear front (see Figure J.7). This property has been announced at section 3.1.6, p. 168.

4. The discussion proposed in this question is similar to the one carried out in Chapter 2, p. 84. As shown in Figure J.8 we can work for example in the plane $z = 0$ for such a source ρ and the problem becomes two-dimensional. Indeed, the solution of the wave equation associated with this source ρ is given by (3.103), p. 169, and reduces to the source term

$$\phi(\mathbf{r}, t) = \int d\mathbf{r}' \int_{-\infty}^{t} dt' \lambda \delta(x' - x_0) \delta(y' - y_0) \delta(t' - t_0) G_3^+ (\mathbf{r} - \mathbf{r}', t - t')$$

$$= \lambda G_2^+ (x - x_0, y - y_0, t - t_0).$$

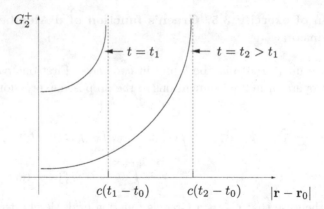

Fig. J.7   Representation of Green's function $G_2^+(|\mathbf{r} - \mathbf{r}_0|, t - t_0)$ for d'Alembert operator in $2 + 1$ dimensions, given by equation (J.23), as a function of $|\mathbf{r} - \mathbf{r}_0|$, for two times t_1 and $t_2 > t_1$. There is a forward wavefront but no rear one.

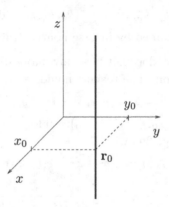

Fig. J.8   For a line source, the situation is identical for any plane orthogonal to the source and one can therefore work in the plane $z = 0$ to analyse the situation.

This solution thus identifies with the Green's function of d'Alembert operator in $2 + 1$ dimensions. The interpretation of the property from the previous question in dimensions $3 + 1$ goes as follows: at a given point in space, an observer begins to receive a signal from the nearest point on the line where the sources are located, and from that time on, the observer then continuously receives signals from more distant emission points, with a decreasing amplitude.

□ **Solution of exercise 3.6. Green's function of d'Alembert operator in $1 + 1$ dimensions**

Spectral representation (3.93) of $G_1^+(x - x_0, t - t_0)$ is:

$$G_1^+(x - x_0, t - t_0) = \frac{c}{2\pi} \int_{-\infty}^{+\infty} \frac{dk}{k} e^{ik(x-x_0)} \sin(kc(t - t_0)).$$

It remains to compute this integral, which poses no difficulty. One gets:

$$\boxed{G_1^+(|x - x_0|, t - t_0) = \frac{c}{2} \theta \left[c(t - t_0) - |x - x_0| \right].}$$
(J.24)

The discussion carried out in the previous exercise for d'Alembert operator in $2 + 1$ dimensions is also valid in the $1 + 1$ dimensions case. The analogue of Figure J.7 is reproduced on Figure J.9.

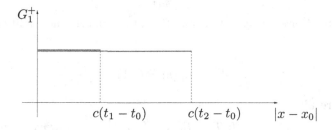

Fig. J.9   Representation of Green's function $G_1^+(|x - x_0|, t - t_0)$ for d'Alembert operator in $1 + 1$ dimensions given by equation (J.24) as a function of $|x - x_0|$, for two times t_1 and $t_2 > t_1$. There is a forward wavefront but no rear one.

□ **Solution of exercise 3.7. Laplacian Green's function G_∞ in $d \geq 3$**

1. For this question, one only has to follow exactly the method of exercise 3.5. The easiest path is here to note, following the discussion on p. 137 that

$$G_\infty(\mathbf{r} - \mathbf{r}') = D\widetilde{G}_\infty^+(\mathbf{r} - \mathbf{r}'; s = 0).$$

2. The change of variable is $u = (|\mathbf{r} - \mathbf{r}'|^2/4Dt)$. We then find:

$$\boxed{G_\infty(\mathbf{r} - \mathbf{r}') = \frac{\Gamma\left(\frac{d}{2} - 1\right)}{4\pi^{\frac{d}{2}} |\mathbf{r} - \mathbf{r}'|^{d-2}}.}$$

We recover the result obtained p. 303 with the expression of the solid angle Ω_d given on the same page.

□ **Solution of exercise 3.8. Heat diffusion in a ball**

Given the problem symmetry, the easiest solution here is a direct calculation of $T(r,t)$. Thus function $[T(r,t) - T_b]$ is a solution of homogeneous diffusion equation. In addition, it must vanish in $r = R$ and reach the value $T_0 - T_b$ at $t = 0$ for any $r \neq R$. One then follows the argument led on p. 130 for the spectral representation of causal homogeneous Green's functions. Since

$$\Delta\psi_n(r) = \frac{1}{r^2}\frac{\mathrm{d}}{\mathrm{d}r}\left(r^2\frac{\mathrm{d}\psi_n}{\mathrm{d}r}(r)\right) = -\frac{n^2\pi^2}{R^2}\psi_n(r),$$

functions $\psi_n(r)$ are eigenfunctions of the Laplacian, with eigenvalues $-(n^2\pi^2/R^2)$. The analogue of equation (3.34), p. 139, is then:

$$T(r,t) - T_b = \sum_{n=1}^{\infty} C_n e^{-\frac{Dn^2\pi^2}{R^2}t}\psi_n(r).$$

Coefficients C_n are set by initial condition at $t = 0$, orthonormality of $\{\psi_n\}$ and the property

$$\int_0^R \mathrm{d}r\, r^2\,\psi_n(r) = (-1)^{n+1}\sqrt{\frac{2}{R}}\frac{R^2}{n\pi}.$$

The final result is:

$$T(r,t) = T_b - \frac{2R(T_b - T_0)}{\pi r}\sum_{n=1}^{\infty}\frac{(-1)^{n+1}}{n}\sin\left(\frac{n\pi r}{R}\right)e^{-\frac{Dn^2\pi^2}{R^2}t}.$$

Note that the IC $T(r,t) = T_0$ for $t \to 0^+$ and $r \neq R$ leads to a sum rule.

□ **Solution of exercise 3.9. From Dirichlet to Robin boundary conditions**

1. One finds the intermediate result:

$$\phi(x) = c(x)e^{hx} \qquad \text{with} \quad c(x) = -\int_{+\infty}^{x} \mathrm{d}u\, e^{-hu}b(u).$$

2. In the case of the diffusion equation, the procedure from exercise 3.1 leads to

$$\int_{\mathcal{D}} \mathrm{d}\mathbf{r}'\,|\alpha(\mathbf{r}',t)|^2 + 2D\int_{t_0}^{t}\mathrm{d}t'\int_{\mathcal{D}}\mathrm{d}\mathbf{r}'|\boldsymbol{\nabla}_{\mathbf{r}'}\alpha(\mathbf{r}',t')|^2$$

$$+ 2D\int_{t_0}^{t}\mathrm{d}t'\int_{\partial\mathcal{D}}\mathrm{d}\Sigma'\,h(\mathbf{r}',t')|\alpha(\mathbf{r}',t')|^2 = 0$$

where α is the difference between two solutions. As h is a positive function, we have indeed $\alpha = 0$, proving the uniqueness of the solution. The proof is similar for the wave equation.

Note that this proof also applies to Poisson equation and thus shows that Dirichlet and Neumann BC are not the only BC ensuring the uniqueness for this PDE.

3. Function b_D is solution of PDE

$$(\mathcal{O}_{\mathbf{r}} + \mathcal{O}_t)\, b_D(x,t) = -\frac{\partial}{\partial x}\rho(x,t) + h\rho(x,t).$$

It satisfies by definition homogeneous Dirichlet BC on $\partial\mathcal{D}$ and vanishing IC at t_0.

4. One simply has:

$$b_D(x,t) = \int_0^\infty dx'' \int_{t_0}^t dt''\, G_{\mathrm{HD}}^+(x;x'';t-t'')\left(-\frac{\partial}{\partial x''}\rho(x'',t'') + h\rho(x'',t'')\right),$$

$$\phi_R(x,t) = \int_0^\infty du\, e^{-hu}\, b_D(x+u).$$

□ **Solution of exercise 3.10. Robin conditions for heat equation**

1. On one hand Fourier law implies that the heat crossing surface element $d\Sigma$ per unit time is

$$\frac{dQ}{dt} = -\mathbf{n}\cdot\mathbf{j}_Q\, d\Sigma = \lambda\, \mathbf{n}\cdot\boldsymbol{\nabla}\phi\, d\Sigma.$$

On the other hand, considering $|T-T_0| \ll T_0$, the linearisation of Stefan-Boltzmann law leads to

$$\frac{dQ}{dt} \simeq -4aT_0^3\phi\, d\Sigma.$$

Comparing these two expressions gives the boundary condition with $h = (4aT_0^3/\lambda)$.

2. The expression for $G_R^+(x;x';\tau)$ is obtained by applying the results of question **4.** of the previous exercise to

$$\rho(x'',t'') = \delta(x'' - x')\delta(t'' - t').$$

3. To get the final expression of G_R^+, we use the result

$$G_{\mathrm{HD}}^+(x+u;x';\tau) = G_{\mathrm{HN}}^+(x+u;x';\tau) - 2G_\infty^+(x+u;-x';\tau),$$

which stems from the method of images. Another useful relation to answer this question is

$$\frac{\partial}{\partial x'}G_{\mathrm{HD}}^+(x+u;x';\tau) = -\frac{\partial}{\partial u}G_{\mathrm{HN}}^+(x+u;x';\tau).$$

Note that as expected, Green's functions G_R^+ and G_{HN}^+ coincide when h vanishes.

□ **Solution of exercise 3.11. Cattaneo equation in 3D**

1. To answer the first question, it is enough to perform a Fourier transform exactly as on p. 195.

2. The integration on angles is immediate. Note that a similar calculation was performed on p. 68.

3. It is easy to connect the result to the calculation in 1d. We find:

$$g_\infty^+(\mathbf{r};t) = \frac{c}{4\pi r}\delta(ct - r) + \frac{a^2 c}{8\pi\sqrt{t^2 - r^2/c^2}}\theta(ct - r)I_1\Big(\frac{a^2 c^2}{2}\sqrt{t^2 - r^2/c^2}\Big) \quad \text{(J.25)}$$

where I_1 is the derivative of I_0.

☐ Solution of exercise 3.12. Klein-Gordon equation

1. The calculation is similar to the one performed in section 3.2.5 for Cattaneo equation. We successively get:

$$G_1^+(x;t) = \frac{ic}{4\pi}\int_{-\infty+i\gamma}^{+\infty+i\gamma} dz \frac{e^{-izt+i(-m^2c^2+z^2)^{1/2}\frac{|x|}{c}}}{(-m^2c^2 + z^2)^{1/2}}$$

$$= \theta(ct - |x|)\frac{c}{4\pi}\int_{-1}^{1} dw \left[\frac{e^{mc(iwt-\sqrt{1-w^2}\frac{|x|}{c})}}{\sqrt{1-w^2}} + \frac{e^{mc(-iwt-\sqrt{1-w^2}\frac{|x|}{c})}}{\sqrt{1-w^2}}\right].$$

A change of variable similar to the one of section 3.2.5 eventually leads to

$$\boxed{G_1^+(x;t) = \theta(ct - |x|)\frac{c}{2}J_0\big[mc\sqrt{t^2 - (x^2/c^2)}\big].} \quad \text{(J.26)}$$

2. The Green's function in three dimensions is obtained from the result

$$G_3^+(\mathbf{r};t) = -\frac{1}{2\pi r}\frac{\partial}{\partial r}G_1^+(r;t)$$

esablished in the previous exercise. One then finds

$$\boxed{G_3^+(\mathbf{r};t) = \frac{c}{4\pi r}\delta(ct - r) - \frac{m}{4\pi}\frac{\theta(ct - r)}{\sqrt{t^2 - r^2/c^2}}J_1(mc\sqrt{t^2 - r^2/c^2}).} \quad \text{(J.27)}$$

Note that result (J.27) is obtained from result (J.25) by $m \to -im$ according to a remark on p. 194 and using the relations

$$J_0(x) = I_0(-ix) \quad \text{and} \quad J_1(x) = iI_1(-ix)$$

between Bessel functions.

◊ **Chapter 4** ◊

☐ **Solution of exercise 4.1. Asymptotic behaviour of Bessel function J_0**

Apply formula (4.14) for stationary phase, p. 221 to find:

$$\theta_c = 0, \qquad \frac{\partial^{(2p)}\varphi}{\partial\theta^{(2p)}}(\theta_c;\lambda) = (-1)^{p+1}\lambda, \qquad \frac{\partial^{(2p+1)}\varphi}{\partial\theta^{(2p+1)}}(\theta_c;\lambda) = 0.$$

We then have

$$J_0(\lambda) \simeq \frac{1}{\pi}\operatorname{Re}\left[\sqrt{2\pi}e^{i\lambda}(i\lambda)^{-1/2}\right] \simeq \sqrt{\frac{2}{\pi\lambda}}\operatorname{Re}\left\{\exp\left[i(\lambda - \frac{\pi}{4})\right]\right\},$$

where we took the determination (4.24), p. 226 for the function $Z^{-1/2}$. The validity of this approximation can easily be established since

$$\frac{\partial^p\varphi}{\partial\theta^p}(\theta_c;\lambda)\left[\frac{\partial^2\varphi}{\partial\theta^2}(\theta_c;\lambda)\right]^{-p/2} = O(\lambda^{1-p/2}).$$

☐ **Solution of exercise 4.2. Binomial coefficients**

1. This relation is a direct consequence of the relation

$$(1+z)^n = \sum_{p=0}^{n} C_n^p z^p$$

and the residue theorem (here in $z = 0$).

2. We start by writing

$$C_n^{nx} = \frac{1}{2\pi i}\oint dz \frac{1}{z}\exp\left[-nf(z)\right] \qquad \text{with} \quad f(z) = -\ln(1+z) + x\ln z.$$

This integral is of the type (4.27), p. 227, with $\ell(z) = 1/z$ and a multiplicative control parameter, $\lambda = n$, so that it is legitimate to apply the saddle point method. A calculation shows that the derivative of f vanishes for

$$z_c = \frac{x}{1-x} \qquad \text{with} \quad f''(z_c) = -\frac{(1-x)^3}{x}.$$

Let us apply saddle point formula (4.28), noticing first that the integration contour $\mathcal{C}$ does not encounter the singularity in $z = 0$ as it is deformed to pass through saddle point z_c (see Figure J.10).

One then gets

$$C_n^{nx} \simeq \frac{1}{2\pi i}\frac{1}{z_c}\left(\frac{2\pi}{nf''(z_c)}\right)^{1/2}\exp\left[-nf(z_c)\right],$$

and eventually result (4.86), using determination (4.24), p. 226.

☐ **Solution of exercise 4.3. Asymptotic behaviour of Helmholtz Green's function**

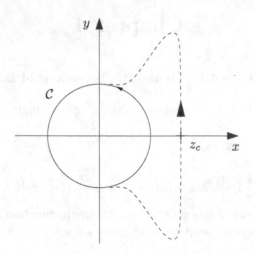

Fig. J.10   Original integration contour $\mathcal{C}$ is deformed to go through saddle point z_c. Note that angle θ, defined by equation (4.18), p. 224 is π here, so that the path of steepest descent is parallel to the y axis, see equation (4.21).

Let us apply the saddle point method to

$$f(t;r) = \frac{r^2}{4t} + m^2 t + \frac{d}{2}\ln t$$

with $r \to \infty$. We get

$$t_c(r) = \frac{r}{2m}\big(1 + O(1/r)\big) \qquad \text{and} \qquad \frac{\partial^2 f}{\partial t^2}(t_c(r);r) = \frac{4m^3}{r} + O(1/r^2) > 0,$$

$$f(t_c(r);r) = mr + \frac{d}{2}\ln\Big(\frac{r}{2m}\Big) + O(1/r).$$

The saddle point method then leads to:

$$G_\infty(\mathbf{r}) \simeq \frac{1}{(4\pi)^{d/2}} e^{-mr}\Big(\frac{r}{2m}\Big)^{-d/2}\sqrt{\frac{2\pi r}{4m^3}},$$

and so to result (4.87). It is also easy to check the validity of this approximation in the asymptotic regime. We note in conclusion that for $d = 3$ we recover the behaviour in $e^{-mr}/(4\pi r)$ seen on p. 74.

□ Solution of exercise 4.4. Isothermal-isobaric ensemble

For this exercise, one only has to make exactly the same analysis as in section 4.2.2, p. 236. To do so one should apply the saddle point method to function $\beta_0 G^*(E, V, N)$ in the limit $N \to \infty$, with

$$G^*(E, V, N) = P_0 V + E - T_0 S(E, V, N).$$

This function depends on two variables E and V. We start by determining the point (E_0, V_0) corresponding to an extremum. It is given by relations

$$\frac{P_0}{T_0} = \frac{\partial S}{\partial V}(E_0, V_0, N) \qquad \text{and} \qquad \frac{1}{T_0} = \frac{\partial S}{\partial E}(E_0, V_0, N).$$

Saddle point formula (4.34), p. 230, then gives

$$Q_N(P_0, T_0) \simeq \frac{2\pi}{\sqrt{\mathrm{Det}\left[A_0(N)\right]}} e^{-\beta_0(P_0 V_0 + E_0 - T_0 S(E_0, V_0, N))}, \qquad (\text{J.28})$$

where $A_0(N)$ is the matrix of second derivatives of $\beta_0 G^*(E, V, N)$ with respect to E and V, evaluated in (E_0, V_0). Assuming that the system is extensive, $\ln \mathrm{Det} A_0(N)$ is proportional to $\ln N$. In a similar manner to the discussion on p. 239 on the equivalence between microcanonical and canonical ensembles, the difference between Gibbs free energies by particle for microcanonical ensemble and for the isothermal-isobaric ensemble thus goes as $(\ln N/N)$. The two ensembles are therefore equivalent in the thermodynamic limit.

☐ **Solution of exercise 4.5. Evolution of a wave packet and group velocity**

Apply stationary phase method, or more specifically the complex saddle point method to an integral of the type (4.27), p. 227 with a multiplicative parameter. This is done by identifying:

$$\lambda \to t \quad ; \quad \ell(z) \to F(z) \quad ; \quad f(z; \lambda) \to -i(zv - \omega(z))t.$$

Since $(\partial f/\partial z) = -i(v - \omega'(z))t$, saddle point z_c corresponds to a wave vector k_0 supposed to be unique, so that group velocity $d\omega/dk(k_0)$ is equal to v. Saddle point formula (4.28) then gives:

$$P(x(t), t) \simeq \sqrt{2\pi} F(k_0) e^{it(k_0 v - \omega(k_0))} \left(it\omega''(k_0)\right)^{-1/2}$$

$$\simeq \sqrt{\frac{2\pi}{t}} F(k_0) e^{it(k_0 v - \omega(k_0)) + i\pi/4} \omega''(k_0)^{-1/2}.$$

Note that the amplitude decays as $1/\sqrt{t}$.

☐ **Solution of exercise 4.6. From Cattaneo to diffusion Green's function**

To recover the Green's function of the diffusion equation, take the limit $c \to \infty$ and then apply complex saddle point formula (4.25) to

$$f(z; c) = -mc\sqrt{t^2 - x^2/c^2} \sin z$$

in $z_c = \pi/2$. One finds $(\partial^2 f/\partial z^2)(z_c; c) = mc\sqrt{t^2 - x^2/c^2}$ and then, in the limit $c \to \infty$,

$$g_\infty^+(x; t) \simeq \frac{c}{4\pi} \sqrt{2\pi} e^{mc\sqrt{t^2 - x^2/c^2}} (mc\sqrt{t^2 - x^2/c^2})^{-1/2}.$$

It is enough to keep the dominant term of this expression, which gives back $G_\infty^+(x; t)$, the Green's function of diffusion equation.

□ Solution of exercise 4.7. Ising model with long-range interactions

1. The answer to the first question is simply obtained by evaluating the Gaussian integral

$$\left(\frac{N\beta}{2\pi}\right)^{1/2} \int_{-\infty}^{\infty} d\lambda \ \exp(-N\beta\lambda^2/2 + \beta\lambda \sum_i S_i) = \exp\left(\frac{\beta}{2N} \sum_{i,j} S_i S_j\right).$$

The sum over all configurations of spin variables $\{S_i\}$ with $S_i = \pm 1$ gives

$$Z = \left(\frac{N\beta}{2\pi}\right)^{1/2} \int_{-\infty}^{\infty} d\lambda \ \exp\left(-N\beta\lambda^2/2\right) \prod_i \left(\sum_{S_i=\pm 1} e^{\beta(\lambda+h)S_i}\right)$$

$$= \left(\frac{N\beta}{2\pi}\right)^{1/2} \int_{-\infty}^{\infty} d\lambda \ \exp\left(-N\beta\lambda^2/2\right) \left[2\,\mathrm{ch}(\beta(\lambda+h))\right]^N. \qquad (J.29)$$

One then identifies function $A(\lambda)$:

$$A(\lambda) = \frac{\lambda^2}{2} - \frac{\ln\left(2\ \mathrm{ch}(\beta(\lambda+h))\right)}{\beta}.$$

2. As we are interested in the limit where N is very large, we can apply the saddle point method (with multiplicative parameter) to the integral (J.29). Saddle point λ_0 corresponds to

$$A'(\lambda_0) = \lambda_0 - \mathrm{th}(\beta(\lambda_0+h)) = 0 \qquad (J.30)$$

and

$$A''(\lambda_0) = 1 - \beta/\mathrm{ch}^2(\beta(\lambda_0+h)) = 1 - \beta(1-\lambda_0^2) > 0. \qquad (J.31)$$

Formula (4.3) then gives:

$$Z = \left(\frac{1}{A''(\lambda_0)}\right)^{1/2} \exp(-N\beta A(\lambda_0))\ (1 + O(1/N)).$$

Free energy f is given by $A(\lambda_0)$ in the limit $N \to \infty$. Note that function $A(\lambda)$ may have a secondary minimum, but in the limit $N \to \infty$, the contribution around the absolute minimum dominates. This result then allows to interpret equations (J.30) and (J.31) as the conditions for the system to adopt a configuration corresponding to the free energy minimum. These equations coming from a purely mathematical condition thus take a concrete physical meaning.

3. The physical meaning of λ_0 only appears through the following expression for magnetisation m,

$$m = -\frac{\partial f}{\partial h} = \mathrm{th}(\beta(\lambda_0+h)) + O(1/N) = \lambda_0 + O(1/N).$$

Finally, equation (J.30) determining the saddle point becomes identical, for large N, to that obtained by the molecular mean field approximation. This mean field approximation then becomes exact when $N \to \infty$. One finds, in the absence of applied field ($H = 0$),

$$m = \text{th}(\beta m),$$

which has a unique solution for $\beta < 1$ in $m = 0$ (we easily verify that (J.31) is satisfied). However, for $\beta > 1$, two other non-zero solutions appear, and condition (J.31) indicates that these solutions are those to use for the saddle point method to be valid. We therefore have a non-zero value of magnetisation for $\beta > 1$. Point $\beta_c = 1$ is peculiar ($A''(\lambda_0)$ is zero at this point), it corresponds to the transition point where the partition function, estimated to first order by the saddle point method, is singular for $h = 0$. Note here that the order of limits is important. We first took $N \to \infty$ and then $h \to 0$.

□ Solution of exercise 4.8. Bernoulli random walk

1. The relation between $\widehat{P}_N(k)$ and $\widehat{P}_{N-1}(k)$ is simply given by

$$\widehat{P}_N(k) = \cos k \ \widehat{P}_{N-1}(k).$$

Moreover, as $P_1(m) = \frac{1}{2} [\delta_{m,-1} + \delta_{m,1}]$, its Fourier transform is $\widehat{P}_1(k) = \cos k$ and therefore $\widehat{P}_N(k) = \cos^N k$.

2. One gets an expression for $P_N(m)$ by inverse Fourier transform. This yields to the formula

$$I_N(m) = \int_{-\pi/2}^{\pi/2} dk e^{-ikm} \cos^N k.$$

Note that, by construction, for all times N even (resp. odd), the particle is on a site m even (resp. odd). It is therefore legitimate to find that the probability of $P_n(m)$ vanishes for odd $N + m$.

3. The integral can be evaluated for large N by the saddle point method:

$$P_N(m) \simeq \frac{1 + (-1)^{N+m}}{\sqrt{2\pi N \left(1 - \left(\frac{m}{N}\right)^2\right)}} \left(\frac{1 + \left(\frac{m}{N}\right)}{1 - \left(\frac{m}{N}\right)}\right)^{-\frac{m}{2}} \frac{1}{(1 - \frac{m^2}{N^2})^{\frac{N}{2}}}.$$

Note that the link with Green's function (3.41), p. 141 of one-dimensional diffusion equation appears in the continuous limit with $m = x/a$, $t = N\tau$ and x/a of order $\sqrt{N}$ where a is the lattice constant, also setting $D = a^2/(2\tau)$.

□ Solution of exercise 4.9. Harmonic oscillator and number theory

1. Since the harmonic oscillators are independent, the partition function $Z(\beta)$ is nothing but the product of each oscillator partition function:

$$Z(\beta) = \prod_{n=1}^{\infty} z_n \quad \text{with} \quad z_n = \frac{1}{1 - e^{-\beta n}}.$$

Another way to compute $Z(\beta)$ is to think in terms of energy. Thus, in expression (4.91) for $Z(\beta)$, the multiplicity of each integer m involved in $\Omega(n)$ (for example, the multiplicity of $m = 1$ written as $4 = 1 + 1 + 1 + 1$) corresponds to the level at which is excited the oscillator of frequency $\hbar\omega_m = m$.

2. One simply gets

$$\Omega(n) = \frac{1}{2\pi i} \oint \frac{d\xi}{\xi^{n+1}} Z(\beta = -\ln\xi),$$

where the integration contour is a circle around the origin.

3. For $n \to \infty$, we can apply the saddle point method in the complex plane to evaluate the integral. In addition, we use Euler-MacLaurin expression to expand the argument of the exponential in the integrand. The validity of this expansion is confirmed *a posteriori* by noticing that the saddle point equation gives β_{saddle} getting smaller when n increases. Keeping only the dominant term and the contribution of fluctuations, we eventually get the Hardy-Ramanujan formula:

$$\Omega(n) \simeq \frac{\exp\left[\pi\sqrt{\frac{2n}{3}}\right]}{4\sqrt{3}n}.$$

Appendix K

References

◊ **General references for mathematics** ◊

There are many mathematical books presenting useful concepts in physics. The reader is referred to the books [3], [4], [17], [67] and [86]. For specific topics, see the table below:

Topic	References
Complex variable	[13], [16], [29], [80]
Distributions	[42], [81]
Differential geometry	[12], [24], [68], [88]

◊ **References for Chapter 1** ◊

General references References on analyticity and its relation with causality are also numerous. A classic reference is [95]. A general reference containing many examples of response functions, including integral equations is [76]. In classical electrodynamics, where KK relations are of great importance to study wave propagation in dispersive media, the books [36], [64] or [89] should be mentioned. In particular, the reader will find discussions on dielectric properties. The article [57] present a detailed discussion about sum rules used in optics. One could find a comprehensive presentation on the implementation of KK relations in hydrodynamics and plasma physics in [40] and [41]. Finally, regarding linear response in field theory, two useful references are [23] and [32].

References concerning the examples

Section	Topic	References
1.2.3	Fluid mechanics	[28, 53]
	Bessel functions	[4, 26, 67]
1.2.4	Vlasov equation	[56]
	Plasmas	[34]
1.2.5	Liouville equation	[73]
	Kubo formula	[1], [47], [48], [62]
	Kubo formula in condensed matter physics	[11], [58]

References concerning the exercises The article [74] contains a derivation from the Kramers-Kronig of the lower limit of the volume occupied by the grains composing interstellar medium.

◊ References for Chapters 2 and 3 ◊

General references Reference books on Green's functions associated with differential operators, especially for Laplace equation, diffusion and propagation are [4], [17], [67], [72], [79], [86], [87], [93] and [96]. The reader will find many identities concerning Laplacian and d'Alembert operators in [76]. In [20], there is a classification of the various differential equations appearing in physics, and a detailed presentation of corresponding numerical methods. For further study of conformal transformations in the complex plane, we can refer to [13], [16], [29], [69] and [80]. One can find applications of Green's functions for the diffusion equation in [7] and [87]. For Brownian motion, the reader may refer to statistical mechanics books like [75], or those related to stochastic processes such as [43] or [78]. Green's functions in quantum mechanics are discussed in [19]. Moreover, quantum mechanics books like [14], [50], [55], [65], [66] and [82] present the use of Green's function for diffusion problems. Finally, [37] is entirely devoted to diffusion theory in quantum mechanics. One could find a presentation of the Green functions in electrodynamics, especially for Laplace and d'Alembert equations, in [36] and [84].

References concerning the examples

Section	Topic	References
2.2.2	Meissner effect	[18], [91]
3.2.2	Diffraction	[10], [71]
3.2.5	Cattaneo equation	[38], [67]
3.2.6	Hydrogen atom	[15]

References concerning the exercises

Ex.	Topic	References
2.5	Reciprocity relations	[44]
2.6	Sum rules	[90]
3.9	Robin BC for vibrating strings	[20]
3.10	Green's function with Robin BC	[9]
3.12	Klein-Gordon equation	[8], [35], [97]

◊ References for Chapter 4 ◊

General references The saddle point method is covered in books on mathematical methods for physics like [4] or [67]. For functional integral one can see [60], [98], [83] and [85].

References concerning the examples

Section	Topic	References
4.2.2	Thermodynamic ensembles	[33]
4.2.3	Harmonic crystal	[6], [45], [52], [59]
4.2.4	Ising model	[2], [54], [70], [97]
4.2.5	Semi-classical approximation	[22], [46], [77], [98]

References concerning the exercises   The article [92] gives a detailed discussion of the connection between the partition of an integer and the partition function of a harmonic oscillator.

Bibliography

[1] A. Akhiezer and S. Péletminski, *Les Méthodes de la Physique Statistique*, Mir (1980).

[2] D. J. Amit, *Field Theory, the Renormalization Group, and Critical Phenomena*, McGraw-Hill (1978).

[3] W. Appel, *Mathematics for Physics and Physicists*, Princeton University Press (2007).

[4] G. B. Arfken and H. J. Weber, *Mathematical Methods for Physicists 4th Ed.*, Academic Press (1995).

[5] V. I. Arnold, *Lectures on Partial Differential Equations*, Springer Verlag (2004).

[6] N. W Ashcroft and N. D. Mermin, *Solid State Physics*, Saunders (1976).

[7] J. V. Beck, *Heat Conduction Using Green's Function*, Taylor & Francis (1992).

[8] J. D. Björken and S. D. Drell, *Relativistic Quantum Fields*, McGraw-Hill (1965).

[9] J. D. Bondurant and S. A. Fulling, *The Dirichlet-to-Robin Transform*, J. Phys. A **38** p.1505 (2005).

[10] M. Born and E. Wolf, *Principles of Optics : Electromagnetic Theory of Propagation, Interference and Diffraction of Light*, Cambridge University Press (2002).

[11] J. Callaway, *Quantum Theory of the Solid State*, Academic Press (1991).

[12] Y. Choquet-Bruhat, Y. Dewitt-Morette and M. Dillard-Bleick, *Analysis, Manifolds and Physics, Volumes 1 and 2*, North-Holland (1989).

[13] R. V. Churchill and J. W. Brown, *Complex Variables and Applications*, McGraw-Hill, (1996).

[14] C. Cohen-Tannoudji, F. Laloe and B. Diu, *Quantum Mechanics*, 2 volumes, Wiley-VCH (1991).

[15] C. Cohen-Tannoudji, J. Dupont-Roc and G. Grynberg, *Photons and Atoms: Introduction to Quantum Electrodynamics*, Wiley-VCH (1997) and *Atom-Photon Interactions: Basic Processes and Applications*, Wiley-VCH (1998).

[16] J. B. Conway, *Functions of One Complex Variable I* (Graduate Texts in Mathematics 11), Springer (1978).

[17] R. Courant and D. Hilbert, *Methods of Mathematical Physics, Volumes 1 and 2*, Wiley (1991).

[18] P. G. de Gennes, *Superconductivity of Metals and alloys*, Addison-Wesley (1989).

[19] E. N. Economou, *Green's Functions in Quantum Physics*, Springer-Verlag (1979).

[20] S. J. Farlow, *Partial Differential Equations for Scientists and Engineers*, Dover (1993).

[21] A. L. Fetter and J. D. Walecka, *Quantum Theory of Many-Particle Systems*, International series in pure and applied physics, McGraw-Hill (1971).

[22] R. P. Feynman and A. R. Hibbs, *Quantum Mechanics and Path Integrals*, McGraw-Hill (1965).

[23] S. Gasiorowicz, *Elementary Particle Physics*, Wiley (1965).

[24] M. Göckeler and T. Schücker, *Differential Geometry, Gauge Theories, and Gravity* (Cambridge Monographs on Mathematical Physics) (1990).

[25] H. Goldstein, *Classical Mechanics*, 2nd edition, Addison-Wesley (1980).

[26] I. S. Gradshteyn and I. M. Ryzhik, *Table of Integrals, Series and Products*, Academic Press (1965).

[27] M.C. Gutzwiller, *Chaos in Classical and Quantum Mechanics*, Interdisciplinary Applied Mathematics, Volume 1, Springer (1990).

[28] E. Guyon, J.-P. Hulin and L. Petit, *Physical Hydrodynamics*, Oxford University Press (2015).

[29] H. J. Hamilton, *A Primer of Complex Variables, with an Introduction to Advanced Techniques*, Wadsworth Publishing Co. (1966).

[30] W. Heitler, *The quantum theory of radiation*, Dover (1984).

[31] J.-P. Hansen and I. R. McDonald, *Theory of Simple Liquids*, Academic Press (2006).

[32] J. Hilgevoord, *Dispersion Relations and Causal Description: An Introduction to Dispersion Relations in Field Theory*, North Holland (1960).

[33] K. Huang, *Introduction to Statistical Physics*, Taylor and Francis (2001).

[34] S. Ichimaru, *Basic Principles of Plasma Physics: A Statistical Approach*, Frontiers in Physics, W. A. Benjamin Advanced Book Program (1973).

[35] C. Itzykson and J.-B. Zuber, *Quantum Field Theory*, McGraw-Hill (1980).

[36] J. D. Jackson, *Classical Electrodynamics 3rd Ed.*, Wiley (1999).

[37] C. J. Joachain, *Quantum Collision Theory*, North-Holland (1987).

[38] D .D. Joseph and L. Preziosi, *Heat Waves*, Reviews of Modern Physics **61**, 41 (1989).

[39] J. V. José and E. J. Saletan, *Classical Dynamics: A Contemporary Approach*, Cambridge University Press (1998).

[40] L. P. Kadanoff and P. C. Martin, *Hydrodynamic Equations and Correlation Functions*, Annals of Physics **24**, p. 419 (1963).

[41] L. P. Kadanoff, *Statistical Physics: Statics, Dynamics and Renormalization*, World Scientific (2000).

[42] R. P. Kanwal, *Generalized Functions: Theory and Applications*, Birkhauser (2004).

[43] I. Karatzas and S. E. Shreve, *Brownian Motion and Stochastic Calculus*, 2nd edition, Springer (2006).

[44] K.-J. Kim and J. D. Jackson, *Proof that the Neumann Green's Function in Electrostatics can be Symmetrized*, Am. J. Phys. **61** (12), p. 1144 (1993).

[45] C. Kittel, *Introduction to Solid State Physics*, 8th edition, Willey (2004).

[46] H. Kleinert, *Path Integrals in Quantum Mechanics, Statistical Physics and Polymer Phyics*, World Scientific (1990).

[47] H. J. Kreuzer, *Non Equilibrium Thermodynamics and its Statistical Foundations*, Oxford University Press (1981).

[48] R. Kubo, M. Toda, and N. Hashitsume, *Statistical Physics II: Non Equilibrium Statistical Mechanics*, Springer-Verlag (1983).

[49] L. Landau and E. Lifshitz, *Course of Theoretical Physics II: The Classical Theory*

of Fields, Pergamon Press (1962).

[50] L. Landau and E. Lifshitz, *Course of Theoretical Physics III: The Classical Theory of Fields*, Pergamon Press (1958).

[51] L. Landau and E. Lifshitz, *Course of Theoretical Physics IV: Quantum electrodynamics*, 4th edition, Pergamon (1982).

[52] L. Landau and E. Lifshitz, *Course of Theoretical Physics V: Statistical Physics*, Pergamon Press (1969).

[53] L. Landau and E. Lifshitz, *Course of Theoretical Physics VI: Fluid Mechanics*, Pergamon Press (1959).

[54] M. Le Bellac, *Quantum and Statistical Field Theory*, Oxford University Press (1991).

[55] M. Le Bellac, *Quantum Physics*, Cambridge University Press (2006).

[56] R. L. Liboff, *Kinetic Theory: Classical, Quantum and relativistic Description*, Prentice Hall (1990).

[57] V. Lucarini, F. Bassani, K.-E. Peiponen and J. J. Saarinen, *Dispersion Theory and Sum Rules in Linear and Nonlinear Optics*, Rivista Del Nuovo Cimento **26**, 12 (2003).

[58] G. D. Mahan, *Many-Particle Physics*, 2nd edition, Plenum Press (1990).

[59] M. P. Marder, *Condensed Matter Physics*, John Wiley & Sons (2000).

[60] P. A. Martin, *L'intégrale fonctionnelle*, Presses Polytechniques Universitaires Romandes (1996).

[61] P. A. Martin and F. Rothen, *Many-Body Problems and Quantum Field Theory: An Introduction*, 2nd edition, Springer (2004).

[62] J. A. McLennan, *Introduction to Non Equilibrium Statistical Mechanics*, Prentice Hall (1989).

[63] M. L. Mehta, *Random Matrices*, 3rd edition, Academic Press (2004).

[64] D. B. Melrose and R. C. McPhedran, *Electromagnetic Processes in Dispersive Media*, Cambridge University Press (1991).

[65] E. Merzbacher, *Quantum Mechanics*, 3rd edition, John Wiley and Sons (1998).

[66] A. Messiah, *Quantum Mechanics*, Dover (2014).

[67] P. M. Morse and H. Feshbach, *Methods of Theoretical Physics, Part I and II*, McGraw-Hill (1953).

[68] M. Nakahara, *Geometry, Topology and Physics*, IOP (1992).

[69] Z. Nehari, *Conformal mapping*, McGraw-Hill, (1952).

[70] G. Parisi, *Statistical Field Theory*, Addison-Wesley (1988).

[71] J.-P. Pérez, *Optique: Fondements et Applications*, Dunod (2000).

[72] V. P. Pikulin and S. I. Pohoazev, *Equations in Mathematical Physics: A Practical Course*, Birkhäuser Verlag (2001).

[73] N. Pottier, *Nonequilibrium Statistical Physics: Linear Irreversible Processes*, Oxford University Press (2009).

[74] E. M. Purcell, *On the Absorption and Emission of Light by Interstellar Grains*, The Astrophysical Journal 158, p. 433 (1969).

[75] F. Reif, *Fundamentals of Statistical and Thermal Physics*, McGraw-Hill (1965).

[76] G. F. Roach, *Green's Functions*, Cambridge University Press (1982).

[77] G. Roepstorff, *Path Integral Approach to Quantum Physics*, Springer-Verlag (1994).

[78] L. C. G. Rogers and D. Williams, *Diffusions, Markov Processes, and Martingales*, 2nd edition, Cambridge University Press (2000).

[79] I. Rubinstein and L. Rubinstein, *Partial Differential Equations in Classical Mathematical Physics*, Cambridge University Press (1998).

[80] W. Rudin *Real and Complex Analysis*, (Higher Mathematics Series), McGraw-Hill (1986).

[81] L. Schwartz, *Mathematics for the Physical Sciences*, Addison-Wesley (1966).

[82] L. Schiff, *Quantum Mechanics*, McGraw-Hill (1955).

[83] L. Schulman and S. Lawrence, *Techniques and Applications of Path Integration*, Wiley (1981).

[84] J. Schwinger, *Classical Electrodynamics*, Perseus (1998).

[85] B. Simon, *Functional Integration and Quantum Physics*, Pure and applied mathematics (vol. 86), Academic Press (1979).

[86] R. Snieder, *A Guided Tour of Mathematical Methods for the Physical Sciences*, Cambridge University Press (2004).

[87] A. Sommerfeld, *Partial Differential Equations in Physics*, Academic Press (1949).

[88] M. Spivak, *Calculus on Manifolds*, Benjamin (1965).

[89] J. A. Stratton, *Electromagnetic Theory*, McGraw-Hill (1963).

[90] C. V. Sukumar, *Green's Functions and a Hierarchy of Sum Rules for the Eigenvalues of Confining Potentials*, Am. J. Phys. **58** (6), p. 561 (1990).

[91] M. Tinkham, *Introduction to Superconductivity*, Huntington (1975).

[92] M. N. Tran, M. V. N. Murthy and R. K. Bhaduri *On the Quantum Density of States and Partitioning an Integer*, Annals of Physics **311** (2004) 204.

[93] A. S. Vladimirov, *Equations of Mathematical Physics*, Mir (1986).

[94] S. Weinberg, *Gravitation and Cosmology: Principles and Applications of the General Theory of Relativity*, J. Wiley (1972).

[95] E. P. Wigner (Ed.), *Dispersion Relations and their Connection with Causality*, Academic Press (1960).

[96] H. W. Wyld, *Mathematical Methods for Physics*, Parseus Books, Advanced Books Classics (1999).

[97] J. Zinn-Justin, *Quantum Field Theory and Critical Phenomena*, Oxford Science Publications (1989).

[98] J. Zinn-Justin, *Path Integrals in Quantum Mechanics*, Oxford Graduate Texts, (2004).

Index

Printed in the United States
By Bookmasters